STRUCTURE OF CRYSTALLINE POLYMERS

STRUCTURE OF CRYSTALLINE POLYMERS

Edited by

I. H. HALL

Department of Pure and Applied Physics, The University of Manchester Institute of Science and Technology, Manchester, UK

ELSEVIER APPLIED SCIENCE PUBLISHERS
LONDON and NEW YORK

ELSEVIER APPLIED SCIENCE PUBLISHERS LTD
Ripple Road, Barking, Essex, England

Sole Distributor in the USA and Canada
ELSEVIER SCIENCE PUBLISHING CO., INC.
52 Vanderbilt Avenue, New York, NY 10017, USA

British Library Cataloguing in Publication Data
Structure of crystalline polymers.
1. Polymers and polymerization 2. Crystals
I. Hall, I. H.
547.7   QD380

ISBN 0-85334-236-9

WITH 13 TABLES AND 103 ILLUSTRATIONS

© ELSEVIER APPLIED SCIENCE PUBLISHERS LTD 1984

The selection and presentation of material and the opinions expressed in this publication are the sole responsibility of the authors concerned

All rights reserved. No part of this publication may be reproduced, stored in a retrieval system, or transmitted in any form or by any means, electronic, mechanical, photocopying, recording, or otherwise, without the prior written permission of the copyright owner, Elsevier Applied Science Publishers Ltd, Ripple Road, Barking, Essex, England

Filmset and printed in Northern Ireland at The Universities Press (Belfast) Ltd.

PREFACE

The theme of this book is the shape of the molecular chains in crystalline polymers. The approach is that of the experimental physicist who is concerned that his experimental techniques should discriminate between various contending theories, rather than seek confirmation of one particular model. Thus the emphasis is on the experimental techniques employed to study polymer structure and the uncertainties and ambiguities which remain after they have been critically applied, rather than on possible structural models. Sadly, this rigorous attitude has not been adopted in some published work in this area, and the book is intended as a stimulus to a more critical interpretation of experimental information. It starts by examining the conformation of the chains on the scale of a few ångströms, and proceeds up the morphological scale to that level of organisation which is visible in the optical microscope.

The first chapter is concerned with the use of the digital computer in the interpretation of wide-angle X-ray diffraction photographs. Using raster-scanning digital microdensitometers, the diffracted intensity can be measured at every point on a fine lattice superimposed on the photograph and the information stored in a computer. Methods of analysing this immense amount of data are discussed. At present structural models are developed to satisfy a few features of the diffraction pattern, these methods should enable them to be tested against its entirety.

The next chapter is concerned with the experimental study, using wide-angle X-ray diffraction, of the chain conformation in aromatic polyesters. These have been chosen because they have received systematic study over the past few years, and because they show the difficulties, uncertainty and dangers of misinterpretation which still exist in this type of work.

The next four chapters are concerned with experimental methods of studying the larger-scale organisation of crystalline material: they cover electron microscopy (Chapter 3), neutron scattering (Chapter 4), and X-ray scattering at small angles using both conventional wavelengths (Chapter 5) and long wavelengths (Chapter 6). Although Chapter 6 discusses work done with long-wavelength radiation from a conventional generator, the case it makes is equally valid if synchrotron radiation is used. Throughout, the emphasis is on the dangers of misinterpretation which may trap the unwary experimenter, on the limitations of the experimental techniques, and on their ability to discriminate between possible structural models. As well as providing a critique of the experimental method, these chapters will also assess and develop current views on the morphology of crystalline polymers.

Up to this point the discussion has been centred on experimental techniques. In the final chapter it is concentrated on a particular material. Low molecular weight fractions of poly(ethylene oxide) are well suited to the study of chain folding and have been the subject of extensive and rigorous investigation using optical microscopy, small-angle X-ray scattering, and differential scanning calorimetry. This has enabled certain critical issues concerning the nature of chain-folding to be clarified. But are the results of a study on such a model applicable to crystalline polymers in general? Not only is the importance of careful, systematic and critical experimentation demonstrated, but this more general issue is also discussed.

I. H. HALL

CONTENTS

Preface v

List of Contributors ix

1. Computer Analysis of X-ray Diffraction Patterns   1
   R. D. B. FRASER, E. SUZUKI and T. P. MACRAE

2. The Determination of the Structures of Aromatic Polyesters from their Wide-angle X-ray Diffraction Patterns 39
   I. H. HALL

3. Transmission Electron Microscopy of Polymers 79
   E. L. THOMAS

4. Neutron Scattering by Crystalline Polymers: Molecular Conformations and their Interpretation 125
   D. M. SADLER

5. The Ability of Small-angle X-ray Scattering to Distinguish between Morphological Models of Crystalline Polymers . . 181
   I. H. HALL and M. TOY

6. Long-wavelength X-ray Scattering to Study Crystal Morphology 229
   H. K. HERGLOTZ

7. Chain-folding in Polymer Crystals: Evidence from Microscopy and Calorimetry of Poly(ethylene oxide) 261
   C. P. BUCKLEY and A. J. KOVACS

Index 309

LIST OF CONTRIBUTORS

C. P. BUCKLEY
Department of Mechanical Engineering, University of Manchester Institute of Science and Technology, PO Box 88, Manchester M60 1QD, UK

R. D. B. FRASER
CSIRO Division of Protein Chemistry, 343 Royal Parade, Parkville, Victoria 3052, Australia

I. H. HALL
Department of Pure and Applied Physics, University of Manchester Institute of Science and Technology, PO Box 88, Manchester M60 1QD, UK

H. K. HERGLOTZ
Engineering Physics Laboratory, Engineering Research and Development Division, E. I. du Pont de Nemours and Company, Wilmington, Delaware 19898, USA

A. J. KOVACS
CNRS, Centre des Recherches sur les Macromolécules, 6 rue Boussingault, 67083 Strasbourg, France

T. P. MACRAE
CSIRO, Division of Protein Chemistry, 343 Royal Parade, Parkville, Victoria 3052, Australia

D. M. SADLER
H. H. Wills Physics Laboratory, University of Bristol, Royal Fort, Tyndall Avenue, Bristol BS8 1TZ, UK

E. SUZUKI
CSIRO Division of Protein Chemistry, 343 Royal Parade, Parkville, Victoria 3052, Australia

E. L. THOMAS
Polymer Science and Engineering Department, University of Massachusetts, Amherst, Massachusetts 01003, USA

M. TOY
Department of Physics, Mathematics and Computing, John Dalton Faculty of Technology, Manchester Polytechnic, Chester Street, Manchester M1 5GD, UK

Chapter 1

COMPUTER ANALYSIS OF X-RAY DIFFRACTION PATTERNS

R. D. B. Fraser, E. Suzuki and T. P. MacRae

CSIRO, Division of Protein Chemistry, Parkville, Victoria, Australia

1 INTRODUCTION

Many polymeric materials have a fibrous texture in which elongated particles with an ordered internal structure are preferentially aligned parallel to a particular direction termed the fibre axis. Frequently the particles are embedded in a matrix which may consist of amorphous polymer or may be a material of unrelated composition. X-ray diffraction patterns obtained from specimens with such a fibrous texture are potentially capable of providing structural information about the dimensions of the particles and their internal structure; the particle orientation density function; the nature of interparticle ordering; and the nature of the matrix.

The extraction of structural information from fibre diffraction patterns is complicated both by difficulties of data collection and by difficulties of interpretation. In recent years however the application of computer-assisted data collection and reduction has enabled quasi-continuous intensity data to be collected over a central section of the intensity transform of a specimen. Thus all the available intensity information can be assembled in the computer in a convenient form for further analysis. The principles involved in computer-assisted data collection are discussed in Section 3.

The extraction of structural information from the intensity data is complicated by several factors. These include the arcing of reflections due to imperfect alignment of the particle axes with the fibre axis; the natural breadths of reflections, associated with the small size of the particles; and the overlapping of reflections in the cylindrically averaged particle intensity transform. Two approaches are possible: we may attempt to extract such features as the structure factors for the particle by devising means of collecting the intensity belonging to particular reflections from the smeared-out version of the particle intensity transform that is available to us in the specimen intensity transform; or we may attempt to model the internal structure of the particle and the factors which tend to smear its intensity transform out and compare the results directly with the specimen intensity transform. In Section 2 the theory relevant to the understanding of fibre diffraction patterns is outlined and in Section 4 procedures for extracting information about orientation density functions and structure factors are reviewed. In the final section the simulation of fibre diffraction patterns is briefly discussed.

2 BACKGROUND THEORY

2.1 The Intensity Transform of a Crystalline Particle

2.1.1 The Intensity Transform of a Stationary Particle

In discussing the diffraction of X-rays by crystalline polymers it is convenient to use the reciprocal lattice concept[1] and to define position in reciprocal space by means of a vector **D** (Fig. 1). The relationship between **D** and the direction of the scattered wave is illustrated in Fig. 2, and it follows that the magnitude of **D** is given by:

$$D = \frac{2 \sin \theta}{\lambda} \quad (1)$$

where 2θ is the angle between the normals to the incident and the scattered waves and λ is the wavelength of the radiation. It is convenient to define a second vector **d** in the coordinate system $Sxyz$ of Fig. 2 to describe position in real space. The amplitude scattered by a crystalline particle in a direction defined by **D** can then be determined by evaluating the Fourier transform $T(\mathbf{D})$ of the distribution of electron density $\rho(\mathbf{d})$ giving:

$$T(\mathbf{D}) = \int \rho(\mathbf{d}) \exp(2\pi i \mathbf{D} \cdot \mathbf{d}) \, dv \quad (2)$$

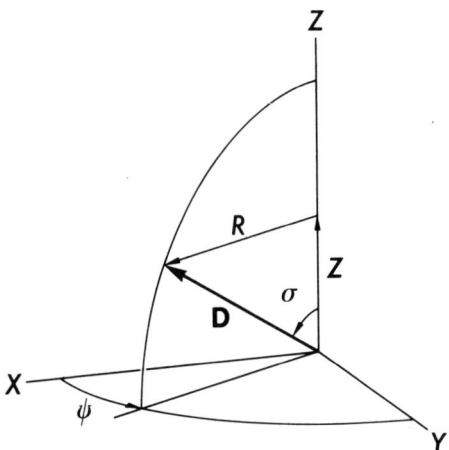

Fig. 1. Diagram illustrating the relationship between the reciprocal space position vector **D** and the Cartesian, cylindrical polar and spherical polar reciprocal space coordinate systems.[2]

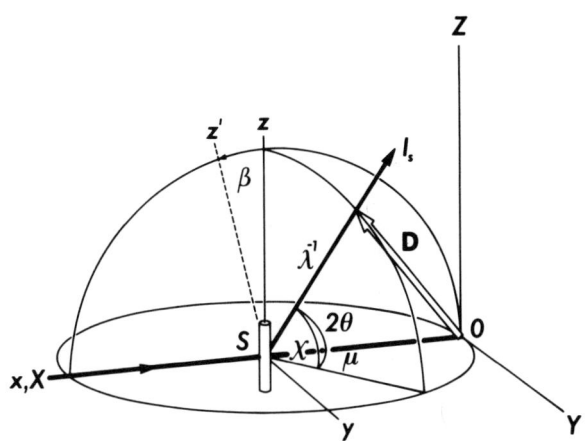

Fig. 2. Diagram illustrating the relationship between the angles μ and χ, which define the 'longitude' and 'latitude' of the intersection of a scattered ray with the sphere of reflection of radius λ^{-1}, and the reciprocal space position vector **D**.[2]

where dv is a volume element in the particle and the integration extends over the irradiated volume. $T(\mathbf{D})$ will be a complex number which describes the amplitude and phase of the scattered wave relative to that from a single electron at the origin. The intensity of the scattered radiation is proportional to the square of the magnitude of $T(\mathbf{D})$ and it is convenient to define an intensity transform:

$$I(\mathbf{D}) = T(\mathbf{D}) T^*(\mathbf{D}) \tag{3}$$

where the asterisk denotes a complex conjugate.

In the present treatment we consider the case of a particle with a three-dimensionally periodic structure in which the unit cell is orthorhombic with edges a, b and c parallel respectively to the x, y and z axes in Fig. 2. The extension to other types of cell is straightforward.[1] In a crystalline particle the density function $\rho(\mathbf{d})$ may be regarded as the convolution of the contents of the unit cell ρ and a point lattice ρ_L and if F and F_L are the corresponding Fourier transforms it follows[3] that:

$$T(\mathbf{D}) = F(\mathbf{D}) F_L(\mathbf{D}) \tag{4}$$

F is termed the structure factor and $I_L = F_L F_L^*$ the interference function. If the crystalline particle has the form of a right prism with edges parallel to Sx, Sy and Sz (Fig. 2) the interference function has the form:

$$I_L(X, Y, Z) = \frac{\sin^2(\pi p_x X) \sin^2(\pi p_y Y) \sin^2(\pi p_z Z)}{\sin^2(\pi a X) \sin^2(\pi b Y) \sin^2(\pi c Z)} \tag{5}$$

where p_x, p_y and p_z are the particle dimensions parallel to Sx, Sy and Sz respectively.

The interference function has maxima at the reciprocal lattice points, defined by $X = h/a$, $Y = k/b$ and $Z = l/c$, where h, k and l are the Miller indices. The value at these points is:

$$I_L(h/a, k/b, l/c) = p_x^2 p_y^2 p_z^2 / V_{cell}^2 = N^2 \tag{6}$$

where V_{cell} is the volume of the unit cell and N is the number of unit cells in the particle.

Except in the immediate vicinity of the reciprocal lattice points the interference function is small and the interference function is well approximated[4] by the expression:

$$I_L(X, Y, Z) = N^2 \exp\left\{-\pi\left[p_x^2\left(X - \frac{h}{a}\right)^2 + p_y^2\left(Y - \frac{k}{b}\right)^2 + p_z^2\left(Z - \frac{l}{c}\right)^2\right]\right\} \tag{7}$$

The structure factor will only vary very slowly within the ranges of X, Y and Z values for which the interference function has appreciable magnitude and so the intensity transform of the particle can be represented by:

$$I(X, Y, Z) = I(X_0, Y_0, Z_0) \exp\{-\pi[p_x^2(X-X_0)^2 + p_y^2(Y-Y_0)^2 + p_z^2(Z-Z_0)^2]\} \quad (8)$$

where $X_0 = h/a$, $Y_0 = k/b$ and $Z_0 = l/c$. The intensity transform for a stationary crystalline particle can thus be represented by the expression:

$$I(X, Y, Z) = N^2 FF^*(X_0, Y_0, Z_0) \exp\{-\pi[p_x^2(X-X_0)^2 + p_y^2(Y-Y_0)^2 + p_z^2(Z-Z_0)^2]\} \quad (9)$$

The integrated intensity associated with a particular reflection is proportional to the volume of the particle[1] and it is convenient to redefine $I(X, Y, Z)$ as the intensity per unit volume of the specimen where the unit of volume equals that of a unit cell. The right-hand side of Expressions (7) and (9) are thus reduced by a factor of N. If Expression (9) is integrated over the region around the reciprocal lattice point (X_0, Y_0, Z_0) for which the integrand has significant magnitude we obtain the important result:

$$FF^*(X_0, Y_0, Z_0) = V_{cell} \iiint_{reflection} I(X, Y, Z) \, dX \, dY \, dZ \quad (10)$$

relating the square of the modulus of the structure factor at the reciprocal lattice point to the integrated intensity per unit cell volume.

2.1.2 The Effects of Disorder

In the derivation of Expression (9) it was tacitly assumed that the particle was an ideal crystal, but in actual particles the atoms will always be displaced from their idealised positions, either due to thermal vibrations or to static disorder. The effects on the intensity transform have been treated elsewhere in detail,[5-7] in terms of two idealised types of positional disorder. In the first type it is assumed that the individual atoms are subject to random displacements about their positions in a perfect crystal, and that no correlation exists between the displacements of different atoms. In this case the intensities at the reciprocal lattice points decrease with increasing D but the integral breadths are not affected. If the displacements are uniform with respect to direction in the particle and all atoms have similar normal distributions of displacements the effects can be simulated by replacing the atomic scattering factors f by modified factors according to the expression:

$$f'(D) = f(D) \exp(-\tfrac{1}{4}BD^2) \tag{11}$$

The parameter B is related to the mean square of the atomic displacements δ_i parallel to $\mathbf{D}$ by the expression:

$$B = 8\pi^2 \langle \delta_i^2 \rangle \tag{12}$$

The value of B in crystalline polymers is generally in the range 5–15 Å, corresponding to values of $\langle \delta_i^2 \rangle^{1/2}$ of 0·25–0·44 Å. If the displacements are non-uniform with respect to direction a more complex term involving up to six parameters is required.[8] Although insufficient data are generally available from polymer diffraction patterns to introduce the required number of parameters it must be recognised that temperature factors for individual atoms in the cell may vary widely. Since the integral breadths of the reflections are not affected the integrated intensity will be reduced by the same factor as the peak intensity. The 'lost' intensity appears as a slowly varying background component.[1]

In the second idealised type of positional disorder the lattice vectors are supposed to fluctuate both in length and direction. The effects on atomic positions are cumulative and the displacement of an atom from its idealised position increases with distance from the origin. No completely satisfactory treatment of this problem exists at present[9] but the overall result is that the peak heights are progressively reduced and the integral breadths progressively increased with increasing D.[6,7,9–12] In addition the 'lost' intensity forms a background which progressively increases with D.

In treating the effects on the intensity transform of positional disorder of the second type it is convenient to specify a relative fluctuation parameter for interplanar distance defined, for example for the $[h00]$ reflections, by:

$$g_{100} = \frac{(\langle d_{100}^2 \rangle - \langle d_{100} \rangle^2)^{1/2}}{\langle d_{100} \rangle} \tag{13}$$

where $\langle d_{100} \rangle$ is the mean interplanar distance and $\langle d_{100}^2 \rangle$ is the mean square of the interplanar distance. Provided that the quantity $2\pi^2 g_{100}^2 h^2 \ll 1$, the treatment given by Hosemann and Wilke[13] leads to the prediction that the integral breadth, which has a value of l/p_x in the absence of disorder, will increase with h according to the expression:

$$\frac{1}{(p'_x)^2} = \frac{1}{(p_x)^2} + \frac{(\pi g_{100} h)^4}{\langle d_{100} \rangle^2} \tag{14}$$

In the derivation of Expression (14) it was assumed that both the original profile and the distribution of d_{100} values were Gaussian. If instead these distributions are assumed to be Lorentzian the expression becomes:

$$\frac{1}{p'_x} = \frac{1}{p_x} + \frac{\pi^2 g_{100}^2 h^2}{\langle d_{100} \rangle} \tag{15}$$

Measurements of the change in integral breadth in a series of reflections with related Miller indices thus provides a means of estimating the values of both crystallite size in a direction perpendicular to the set of planes and the relative fluctuation in interplanar distance.[11-14] The reduction in peak height with increase in D, as measured above the continuous background, is not readily incorporated in a specific way into Expression (9), but for small values of g the effect is absorbed by an increase in the value of B in Expression (11).

In addition to positional disorder, crystalline particles of polymer molecules are subject to other types of imperfection involving coordinated movements of all the atoms in a molecule[15,16] and irregularities in chain direction.[17,18] The effects on the intensity transform are discussed in detail in the references cited above and by Arnott.[19] The effects on the intensity transform of other types of packing defects and of distortion are discussed by Wilson,[5] Caspar and Holmes,[20] Makowski and Caspar,[21] Squire,[22] Holmes et al.[23] and Luther and Squire.[24]

2.1.3 The Cylindrically Averaged Intensity Transform

Crystalline particles of polymer molecules are generally elongated parallel to one of the crystallographic axes and for the purposes of the present discussion this will be taken as the c-axis. In typical fibre specimens the particles with c-axes parallel to any particular direction in space will exhibit a uniform distribution of azimuthal orientations about that direction and their contribution to the scattering from the specimen will be given by the sum of the individually cylindrically averaged intensity transforms.[25] If we consider, as before, the case of a particle having the form of a right prism with edges parallel to Sx, Sy

and Sz (Fig. 2) and an orthorhombic unit cell, the required cylindrical average $I_p(R, Z)$ of the intensity transform is given by:

$$I_p(R, Z) = \frac{1}{2\pi} \int I(X_0, Y_0, Z_0) \exp\{-\pi[p_x^2(X - X_0)^2 + p_y^2(Y - Y_0)^2 + p_z^2(Z - Z_0)^2]\} \, d\psi \quad (16)$$

From Fig. 1 it follows that $X = R \cos \psi$, $Y = R \sin \psi$ and provided that h and k are not both zero, and the distance over which the integrand has significant value is small compared with R, we may write:

$$X - X_0 = (R - R_0) \cos \psi_0 - R_0 \sin \psi_0 (\psi - \psi_0) \quad (17)$$

$$Y - Y_0 = (R - R_0) \sin \psi_0 + R_0 \cos \psi_0 (\psi - \psi_0) \quad (18)$$

where:

$$\tan \psi_0 = kb^*/ha^* \quad (19)$$

and a^* and b^* are reciprocal cell edge lengths. With these approximations the integral may be evaluated and yields the result:

$$I_p(R, Z) = \frac{I(X_0, Y_0, Z_0) \exp\{-\pi[p_r^2(R - R_0)^2 + p_z^2(Z - Z_0)^2]\}}{2\pi R_0 (p_x^2 \sin^2 \psi_0 + p_y^2 \cos^2 \psi_0)^{1/2}} \quad (20)$$

where:

$$p_r^2 = \frac{p_x^2 p_y^2}{p_x^2 \sin^2 \psi_0 + p_y^2 \cos^2 \psi_0} \quad (21)$$

For reflections centred on the axis of rotation both h and k are zero and an exact solution of the integral in Expression (16) is possible. The result is:

$$I_p(R, Z) = I(0, 0, Z) \exp\{-\pi[\tfrac{1}{2}(p_x^2 + p_y^2)R^2 + p_z^2(Z - Z_0)^2]\} \times I_0[\tfrac{1}{2}\pi R^2(p_x^2 - p_y^2)] \quad (22)$$

where I_0 is a modified Bessel function of the second kind of order zero. Expressions (20)–(22) describe the profiles of the reflections in the cylindrically averaged intensity transform of a particle with an orthogonal unit cell. Expressions for other types of cell are more complex but can be derived in a similar manner.

2.2 The Intensity Transform of a Fibrous Assembly of Crystalline Particles

In the commonest type of fibrous specimen elongated microscopic particles are preferentially aligned parallel to a particular direction termed the fibre axis. In addition the azimuthal orientations of the particles around their elongated direction are generally random. The intensity transform of the specimen will therefore be cylindrically symmetrical about the fibre axis and will be denoted by $I_s(R, Z)$. Again it is convenient to define I_s as the intensity per unit volume of the fibrous assembly of particles and to choose the unit volume as the volume of the unit cell. It has been shown[25] that I_s can be calculated by summing the contributions of the individual particles, and where necessary we shall distinguish between coordinates in particle reciprocal space and specimen reciprocal space by using the suffixes p and s respectively.

If the alignment of the particle axes with the fibre axis was perfect and all the particles were of similar dimensions I_s would be described by the expressions for I_p given in Expressions (20)–(22). In practice however there is always a distribution of particle axis directions which is conveniently described by means of an orientation density function $G(\alpha)$ which is normalised in such a way that the fraction of particle axis directions contained in the elementary solid angle $d\omega$ (Fig. 3) is

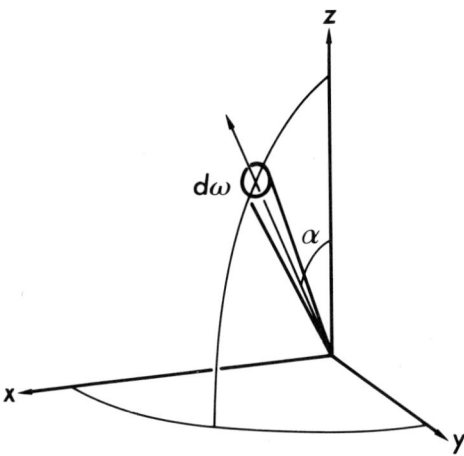

Fig. 3. The orientation density function $G(\alpha)$ is defined in such a way that the fraction of particle axes contained in the elementary solid angle $d\omega$ is $G(\alpha) \cdot d\omega$.[2]

$G(\alpha)\,d\omega$. Thus:

$$2\pi \int G(\alpha) \sin \alpha \, d\alpha = 1 \qquad (23)$$

Various expressions for $G(\alpha)$ have been derived which are applicable to idealised situations. Kratky[26] showed that the density function applicable to orientation produced by stretching an array of initially randomly oriented rodlets embedded in an amorphous matrix is:

$$G(\alpha) = \frac{s^3}{4\pi[1+(s^3-1)\sin^2\alpha]^{3/2}} \qquad (24)$$

where s is the ratio of the stretched to the unstretched length. A second expression, applicable to an oriented sol in which the particles are subject to Brownian motion has the form:

$$G(\alpha) = \frac{\exp(-\alpha^2/2\alpha_0^2)}{2\pi\alpha_0^2} \qquad (25)$$

where α_0 is a parameter defining the degree of departure from parallelism. This type of orientation density function has been found to be an adequate model for many synthetic and naturally occurring fibrous specimens. Holmes and Barrington Leigh[27] showed that for a distribution of this type the intensity transform of the specimen could be related, for small values of α_0, to the intensity transform of the particle by the expression:

$$I_s(D, \sigma_s) = \frac{1}{\alpha_0^2} \int I_p(D, \sigma_p) \exp\left[\frac{-(\sigma_p-\sigma_s)^2}{2\alpha_0^2}\right] i_0\left[\frac{\sin\sigma_s \sin\sigma_p}{\alpha_0^2}\right] \sin\sigma_p \, d\sigma_p \qquad (26)$$

where $i_0(x) = \exp(-x) I_0(x)$, $I_0(x)$ as before is a modified Bessel function of the second kind of order zero, and D and σ are spherical polar coordinates (Fig. 1). The integral in Expression (26) can only be evaluated if assumptions are made about the form of I_p. Over the small ranges of R and Z for which Expression (20) has appreciable value we may write:

$$R - R_0 = (D - D_0)\sin\sigma_0 + D_0 \cos\sigma_0 (\sigma - \sigma_0) \qquad (27)$$

$$Z - Z_0 = (D - D_0)\cos\sigma_0 - D_0 \sin\sigma_0 (\sigma - \sigma_0) \qquad (28)$$

where $(D_0, \sigma_0) \equiv (R_0, Z_0)$. The expression then becomes:

$$I_p(D, \sigma) = I_p(D_0, \sigma_0) \exp\{-\pi[p_r^2 p_z^2 (D-D_0)^2/(p_r^2 \cos^2\sigma_0 + p_z^2 \sin^2\sigma_0) \\ + D_0^2(p_r^2 \cos^2\sigma_0 + p_z^2 \sin^2\sigma_0)(\sigma-\sigma_d)^2]\} \qquad (29)$$

where:

$$\sigma_d = \sigma_0 - \frac{(D-D_0)(p_r^2-p_z^2)\sin\sigma_0\cos\sigma_0}{D_0(p_r^2\cos^2\sigma_0+p_z^2\sin^2\sigma_0)} \quad (30)$$

$$I_p(D_0,\sigma_0) = \frac{NFF^*(X_0,Y_0,Z_0)}{2\pi R_0(p_x^2\sin^2\psi_0+p_y^2\cos^2\psi_0)^{1/2}} \quad (31)$$

for an intensity transform normalised to the unit cell volume of the specimen.

Two approaches to the evaluation of the integral obtained by substituting the expression given for $I_p(D,\sigma)$ in Expression (29) into Expression (26) are possible. The integral may be evaluated numerically[28] or alternatively approximate solutions may be obtained applicable to certain ranges of D and σ_s. An approximation, applicable for values of σ_s remote from zero has been obtained[29] and has the following form:

$$I_s(D,\sigma_s) = I_p(D_0,\sigma_0)\exp\left\{-\pi\left[\frac{p_r^2 p_z^2(D-D_0)^2}{p_r^2\cos^2\sigma_0+p_z^2\sin^2\sigma_0}\right.\right.$$

$$\left.\left.+\frac{(\sigma_s-\sigma_d)^2}{2\pi\alpha_0^2+1/(D_0^2 p_r^2\cos^2\sigma_0+D_0^2 p_z^2\sin^2\sigma_0)}\right]\right\} \quad (32)$$

$$\times\left(\frac{\pi^{1/2}}{2\gamma\alpha_0^2}\right)\{\text{erf}\,[\gamma(\pi-\sigma_m)]$$

$$+\text{erf}\,[\gamma\sigma_m]\}i_0\left(\frac{\sin\sigma_s\sin\sigma_m}{\alpha_0^2}\right)\sin\sigma_m$$

where erf (x) denotes the error function and:

$$\gamma = [1/(2\alpha_0^2)+\pi D_0^2(p_r^2\cos^2\sigma_0+p_z^2\sin^2\sigma_0)]^{1/2} \quad (33)$$

$$\sigma_m = \frac{\sigma_s/(2\pi\alpha_0^2)+D_0^2(p_r^2\cos^2\sigma_0+p_z^2\sin^2\sigma_0)\sigma_d}{1/(2\pi\alpha_0^2)+D_0^2(p_r^2\cos^2\sigma_0+p_z^2\sin^2\sigma_0)} \quad (34)$$

Where disorientation leads to overlap between reflections, additional terms appropriate to other values of D_0 and σ_0 must be introduced.

Inspection of Expression (32) reveals that the principal factor determining reflection profile in the specimen intensity transform is the exponential term and this offers the potential for obtaining estimates of α_0, p_x, p_y, and p_z by profile analysis, and thus defining reflection boundaries for the purpose of background correction and for estimating values of FF^* either by numerical integration or by profile fitting.

Computational procedures for carrying out these operations are discussed in detail in Section 4.1 and these are based on the observation that provided the exponential term is the only rapidly varying term in Expression (32) it can be cast in the form:

$$Q(U, V) = \frac{I_s(D, \sigma_s)}{I_s(D_0, \sigma_0)} = \exp[-\pi(U^2 + V^2)] \qquad (35)$$

where:

$$U^2 = \frac{p_r^2 p_z^2 (D - D_0)^2}{p_r^2 \cos^2 \sigma_0 + p_z^2 \sin^2 \sigma_0} \qquad (36)$$

$$V^2 = \frac{(\sigma - \sigma_d)^2}{2\pi\alpha_0^2 + 1/[D_0^2(p_r^2 \cos^2 \sigma_0 + p_z^2 \sin^2 \sigma_0)]} \qquad (37)$$

The significance of Expression (35) is that all reflections should have the same profile when plotted in 'Q'-space. From Expression (35) it follows that the scaling factor by which the background-corrected values of $I_s(D, \sigma)$ associated with a particular reflection must be multiplied to give the best fit to the function $Q(U, V)$ yields an estimate of $[I_s(D_0, \sigma_0)]^{-1}$ and thus of $|F(X_0, Y_0, Z_0)|$ via Expressions (31) and (32).

2.3 Helical Particles

In the case of a particle with helical symmetry the structure is only periodic in one dimension and if the symmetry axis is taken as the c-axis the intensity transform I_p will be negligible except in the vicinity of a series of planes in reciprocal space which satisfy the condition:

$$Z = l/c \qquad (38)$$

where l is a layer line index and c is the period along the particle axis. The intensity distribution over a particular layer line will be a continuous function of X and Y. If, as before, I_p is defined in terms of the scattered intensity per structural repeating unit of the particle the value of I_p for $Z = l/c$ is related to the cylindrically averaged square of the modulus of the structure factor:

$$\langle FF^*(R, l/c) \rangle_\psi = \frac{1}{2\pi} \int_0^{2\pi} F(R, \psi, l/c) F^*(R, \psi, l/c) \, d\psi \qquad (39)$$

by the expression:

$$I_p(R, l/c) = N \langle FF^*(R, l/c) \rangle_\psi \qquad (40)$$

where F is the structure factor for all the helically related units in the repeat distance c, and N is the number of repeats.

The variation of intensity with Z can again be closely approximated[4] by writing:

$$I_p(R, Z) = I_p(R, l/c) \exp\left[-\pi p_z^2 (Z - l/c)^2\right] \quad (41)$$

where p_z is the length of the particle. If intensity is integrated with respect to Z for a particular layer line we obtain, from Expressions (40) and (41)

$$FF^*\langle(R, l/c)\rangle_\psi = c \int_{\text{layer line}} I_p(R, Z) \, dZ \quad (42)$$

This may be compared with Expression (10) which is applicable to a stationary crystalline particle. For a cylindrically averaged transform Expression (10) becomes

$$\sum_{i=1}^{m} F_i F_i^*(h/a, k/b, l/c) = 2\pi V_{\text{cell}} \iint_{\text{reflection}} I_p(R, Z) R \, dR \, dZ \quad (43)$$

where m is a multiplicity factor. Provided that h and k are not both zero and that the value of p_r is not small this will be well approximated by

$$\sum_{i=1}^{m} F_i F_i^*(h/a, k/b, l/c) = 2\pi R_0 V_{\text{cell}} \iint_{\text{reflection}} I_p(R, Z) \, dR \, dZ \quad (44)$$

In certain types of disorder both discrete and continuous scattering are present on the layer lines and it follows from Expressions (42) and (44) that an integration of the intensity contributed by a discrete reflection to a central section of the particle intensity transform may be used to estimate an isolated value of $\langle FF^*(R, l/c)\rangle_\psi$ by multiplying by a factor:[30–32]

$$2\pi R_0 V_{\text{cell}}/cm \quad (45)$$

Up to the present time virtually all integrations of intensity have been carried out in the specimen intensity transform and, when reflections are spread over significant ranges of R, Expression (44) will be a poor approximation to (43). With the advent of the computer mapping procedures described in the next section the calculation of integrated intensities is not restricted to summations of $I_s(R, Z)$, and the R-weighted integral in Expression (43) can be evaluated numerically from the digitised data and a value of $(1/m)\sum_{i=1}^{m} F_i F_i^*$ obtained which may be used directly as an estimate of $\langle FF^*(R, l/c)\rangle_\psi$. Note however that the average is only based on m samples and may differ from a more finely sampled average.

The cylindrically averaged intensity transform of a fibrous assembly of helical particles can be predicted using Expression (26) provided that the form of the particle intensity transform is known.[27] If Expression (41) is used to represent the Z-distribution of intensity about the layer lines and it is assumed that $I_p(R, l/c)$ varies only slowly in the range for which the exponential term has a significant value we obtain[4]

$$I_s(D, \sigma_s) = \frac{I_p(D, \sigma_0)}{\alpha_0^2} \int_0^\pi \exp\left[-\pi p_z^2 D^2(\cos \sigma_p - \cos \sigma_0)^2 - \frac{(\sigma_p - \sigma_s)^2}{2\alpha_0^2}\right]$$

$$\times i_0\left[\frac{\sin \sigma_s \sin \sigma_p}{\alpha_0^2}\right] \sin \sigma_p \, d\sigma_p \qquad (46)$$

As in the case for assemblies of crystalline particles two approaches are possible: approximate formulae applicable in special conditions can be derived[4,33,34] or a numerical integration can be carried out using a computer program.[35] For values of σ_s remote from zero, Expression (46) is well approximated† by:

$$I_s(D, \sigma_s) = I_p(D, \sigma_0)\left(\frac{\pi^{1/2}}{2\gamma\alpha_0^2}\right) \exp\left[\frac{-(\sigma_s - \sigma_0)^2}{2\alpha_0^2 + 1/(\pi p_z^2 R_0^2)}\right]$$

$$\times i_0\left(\frac{\sin \sigma_s \sin \sigma_m}{\alpha_0^2}\right)\{\operatorname{erf}[\gamma(\pi - \sigma_m)] + \operatorname{erf}[\gamma \sigma_m]\} \sin \sigma_m \qquad (47)$$

where γ is defined in Expression (33) and σ_m in Expression (34) for $p_r = 0$ and $\sigma_d = \sigma_0$. In cases where the exponential term varies rapidly compared with the remaining terms Expression (47) may be written:

$$\frac{I_s(D, \sigma_s)}{I_s(D, \sigma_0)} = \exp\left[\frac{-(\sigma_s - \sigma_0)^2}{2\alpha_0^2 + 1/(\pi p_z^2 R_0^2)}\right] \qquad (48)$$

This expression provides a basis for the deconvolution of overlapping layer lines in the specimen intensity transform, as discussed in Section 4.2.

† There is a typographical error in eqn. (68) of Reference 2. This is corrected in the present Expression (47).

3 COLLECTION OF INTENSITY DATA

The great majority of X-ray diffraction studies of fibrous specimens are carried out using an experimental arrangement in which the specimen and the film are both stationary[2,32] and the present description of computer-aided data collection is restricted to patterns recorded in this manner. The use of other types of cameras and of diffractometers are discussed in detail by Alexander[11] and by Kakudo and Kasai.[12]

3.1 Mapping in Film Space

The instrument most commonly used for mapping optical density in film space is the rotating drum microdensitometer.[36-38] Factors which must be taken into consideration in choosing spot size include the graininess of the film and averaging errors encountered when the optical density changes shows significant variation over the area of the measuring spot.[39-42] A spot size of 100 μm is commonly used, and this is also chosen as the raster interval; with fine-grain film a spot size and raster interval of 50 μm is preferred.

The optical density at any raster point is used as an indication of the total energy incident on the film over the area of the measuring spot. Corrections need to be applied to compensate for (a) non-linearity between optical density and digital output of the microdensitometer, (b) non-linearity between optical density and energy per unit area of the film for normal beam incidence and (c) the enhancement of optical density for oblique passage through the emulsion. Corrections for (a) and (b) are most simply applied by exposing a piece of film to the X-ray beam for varying times and plotting energy versus optical density. The resulting relationship can then be either used in the form of a look-up table or alternatively fitted to a suitable function and individual corrections calculated for each optical density reading. Matthews et al.[42] used a relationship of the form:

$$E = a_1 D + a_2 D^2 \tag{49}$$

to represent the relationship between energy E and optical density D measured relative to unexposed film, where a_1 and a_2 are adjustable parameters. An alternative procedure is to correct the measured optical densities for non-linearity using a fitted expression of the form:

$$D' = D[1 + c_1 \exp(c_2 D)] \tag{50}$$

where c_1 and c_2 are adjustable parameters. At some later stage a local background D_B is determined and then the energy is calculated from the expression

$$E = D' - D'_B \qquad (51)$$

The increased photographic action due to the additional path in the emulsion for oblique rays[43] may be corrected for[44] by multiplying the value of E obtained from Expressions (49) or (51) by the factor $C(0)/C(\tau)$, where τ is the angle of incidence on the film. In the case of a double-sided film:

$$C(\tau) = 1 - \exp(-k_e \sec \tau) + \exp[-(k_e + k_b) \sec \tau]$$
$$- \exp[-(2k_e + k_b) \sec \tau] \qquad (52)$$

where k is the product of the mass absorption coefficient and mass per unit area, and the suffixes e and b refer to film emulsion and base respectively. Expressions for the angle τ for various experimental arrangements are given by Fraser et al.[2] In some instances additional corrections are warranted to take account of absorption suffered by the X-ray beam in the specimen[45] and in air, where this is present in the camera.

It is convenient to define a set of Cartesian coordinates Ouv on the flattened-out film such that the intersection of the direct beam with the film has coordinates $(0, 0)$ and the v-axis corresponds to the intersection of the plane containing the fibre axis and the main beam with the film. The raster coordinates of the main beam can be determined by reference to a calibration ring[46] or from a recording made during the exposure. If neither of these are available the raster coordinates of the main beam must be determined from the observed pattern.[2,23] In general the v-axis will not be accurately aligned with a raster direction and the angle of rotation between these two directions must be determined. This can be done by means of a computer program which minimises the variance between selected mirror-related features in the pattern as a function of the angle of rotation. If the raster coordinates of the main beam are unknown it is necessary to co-refine the angle of rotation with the lateral raster coordinate for an assumed longitudinal raster coordinate.

The longitudinal raster coordinate cannot in general be determined from symmetry in the pattern since this is destroyed by any tilt of the fibre axis with respect to the normal to the X-ray beam. A procedure for determining tilt in patterns containing sharp reflections has been described by Franklin and Gosling[47] but this can only be applied if the raster coordinates of the main beam are known. A generally applicable procedure has been devised[2] in which the specimen tilt and longitudinal raster coordinate of the direct beam are co-refined. It is based on the fact that values of $I_s(R, Z)$ and $I_s(R, -Z)$ derived from the pattern should be identical, and values of the specimen tilt and the longitudinal raster coordinate of the direct beam are chosen which minimise the root mean square difference between $I_s(R, Z)$ and $I_s(R, -Z)$ for selected areas of the pattern. If the value of the longitudinal raster coordinate of the main beam so obtained differs substantially from that used to determine the lateral coordinate and pattern rotation a further cycle of refinement of both processes may be required.

3.2 Mapping in Reciprocal Space

The specimen intensity transform $I_s(\mathbf{D})$ provides a convenient means of summarising the intensity data collected from a specimen. The transform is cylindrically symmetrical and as discussed earlier a convenient measure of its value, for a particular value of $\mathbf{D}$, is the scattered intensity per structural repeating unit, corrected for specimen absorption, relative to the scattered intensity from a single electron at S (Fig. 2). In the conventional arrangement for collecting fibre diffraction data the specimen is irradiated by a collimated beam of monochromatic radiation passing along xS (Fig. 2) and the scattered intensity is measured for a range of values of the angles μ and χ. The values of I_s so obtained are distributed over a spherical surface (the sphere of reflection) centred on S with radius λ^{-1}, where λ is the wavelength of the radiation. In order to explore the region around the Z axis in the transform the specimen is rotated about the y axis and this is accompanied by a corresponding rotation of the transform about the Y axis. The angle β through which the specimen is rotated (Fig. 2) is chosen so that the region of interest intersects the sphere of reflection. This procedure must be repeated for each layer line for which data is required in the vicinity of $R = 0$.

The angles μ and χ are respectively the 'longitude' and 'latitude' of the intersection of the diffracted beam with the sphere of reflection and are related to the reciprocal space coordinates as follows:†

$$\sin \chi = [\lambda Z - \tfrac{1}{2} \sin \beta \lambda^2 (R^2 + Z^2)]/\cos \beta \qquad (53)$$

$$\cos \mu = [1 - \tfrac{1}{2} \lambda^2 (R^2 + Z^2)]/\cos \chi \qquad (54)$$

† There is a typographical error in eqn. (25) of Reference 2, which is corrected in the present Expression (53).

The reciprocal space coordinates may be expressed in terms of μ and χ as follows:

$$D = [2(1 - \cos \mu \cos \chi)]^{1/2}/\lambda \qquad (55)$$

$$Z = [\sin \beta (1 - \cos \mu \cos \chi) + \cos \beta \sin \chi]/\lambda \qquad (56)$$

$$R = (D^2 - Z^2)^{1/2} \qquad (57)$$

$$\sin \psi = \sin \mu \cos \chi / \lambda R \qquad (58)$$

and formulae relating μ and χ to the film coordinates u and v for various types of camera have been given by Fraser et al.[2]

Since the specimen intensity transform as discussed earlier is cylindrically symmetrical, it follows that all the information contained in it can be summarised by computing the distribution of intensity in a central section, and this is conveniently done at equispaced intervals of R and Z. In order to carry out this mapping it is necessary to establish a relationship between $I_s(R, Z)$ and the energy per unit area incident on the film.

By definition the energy scattered during the exposure into the elementary solid angle $d\omega$ included in the limits μ and $\mu + d\mu$, χ and $\chi + d\chi$ will be proportional to $API_s \, d\omega$, where A is a term which takes account of absorption within the specimen and P is the polarisation factor. If $E(u, v)$ is a number calculated from the measured optical density that is proportional to the energy per unit area of the flattened out film we have:

$$KE(u, v) \, da = API_s \, d\omega \qquad (59)$$

where da is the area of film illuminated by the solid angle $d\omega$ and K is a constant of proportionality.

Since:

$$d\omega = \cos \chi \, d\chi \, d\mu \qquad (60)$$

and:

$$da = \left(\frac{\partial u}{\partial \mu} \frac{\partial v}{\partial \chi} - \frac{\partial u}{\partial \chi} \frac{\partial v}{\partial \mu} \right) d\chi \, d\mu \qquad (61)$$

(see for example Massey and Kestelman[48]) we obtain:

$$I_s = KE(u, v) \left[\frac{\partial u}{\partial \mu} \frac{\partial v}{\partial \chi} - \frac{\partial u}{\partial \chi} \frac{\partial v}{\partial \mu} \right] / (AP \cos \chi) \qquad (62)$$

This formula provides a completely general transformation from film space to transform space. Evaluation of the Jacobian in Expression (62) for the case of a flat film camera gives

$$I_s = \frac{Kr^2 E(u, v)}{AP \cos^3 \mu \cos^3 \chi} \quad (63)$$

where r is the specimen–film distance measured along the main beam. Formulae applicable to other types of camera have been given by Fraser et al.[2]

In the foregoing treatment it has been assumed that point-focused beams have been used in recording the diffraction pattern. In cases where the dimensions of the focus are such that detail present in the specimen intensity transform is degraded in the recorded pattern a correction for instrumental broadening must be applied either when mapping intensities or when extracting information from the map.[4,11,12,49] A typical application of the procedure to a fibre diffraction pattern is illustrated in Fig. 4.

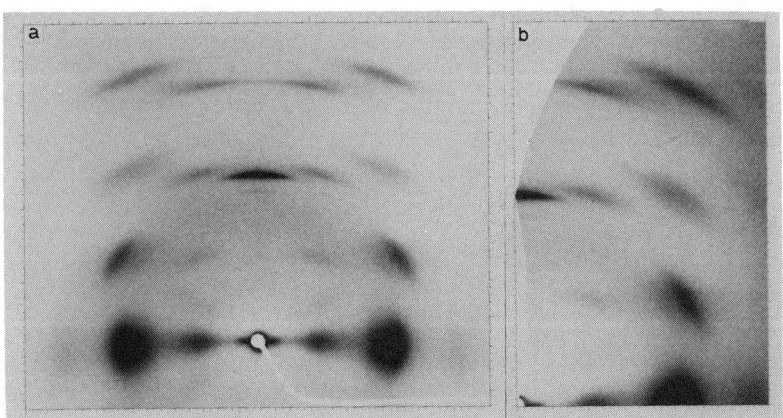

Fig. 4. (a) Analog representation, obtained using a Photowrite instrument, of the contents of a disc file containing digitised optical density values obtained by scanning an X-ray diffraction pattern of Bombyx mori silk fibroin with a $50 \times 50 \, \mu\text{m}$ raster. The fibre axis was inclined at an angle of $12 \cdot 5°$ to the normal to the X-ray beam. Each scale division corresponds to 1 mm. (b) Analog representation of the contents of a disc file containing intensity values computed for a central section of the specimen intensity transform obtained from the optical density data in (a). No information on the blank areas is available from a specimen tilted at $12 \cdot 5°$ but this information can be obtained by varying the specimen tilt. Each scale division corresponds to $0 \cdot 01 \, \text{Å}^{-1}$.

4 EXTRACTION OF INFORMATION FROM INTENSITY MAPS

4.1 Crystalline Particles

4.1.1 Orientation Density Function
The use of mappings of the specimen intensity transform for determining the orientation density functions for individual reciprocal lattice points is well documented.[11,12,50] Single-crystal diffractometry is often used in these studies and the results presented in the form of so-called pole figures which are stereographic projections of the orientation density function for particular reciprocal lattice points. In the present treatment we are restricting attention to fibre-type orientation and in this case a central section of the cylindrically symmetrical specimen intensity transform $I_s(R, Z)$ contains all the information required to construct a pole figure for any reciprocal lattice point. A correction for the effects of particle size may be necessary.

The background theory in Section 2 was developed on the assumption that the orientation density function for particle directions could be adequately represented by a Gaussian function. If the effects of particle-size broadening can be neglected Expression (26) can be approximated by

$$I_s(D_0, \sigma) = I_s(D_0, \sigma_0) \exp\left[-\frac{(\sigma - \sigma_0)^2}{2\alpha_0^2}\right] \tag{64}$$

provided σ is not small[27,51] and so α_0 can be determined from the σ-variation of intensity.[35,52,53]

4.1.2 Structure Factor Data—Current Practice
Three methods of estimating structure factors are commonly used. In the first, peak values of intensity above background are estimated for each reflection and corrections applied both for diffraction geometry and for the spread of intensity in reciprocal space associated with imperfect alignment of the particle axes with the fibre axis.[47,54] In the second, radial traces of optical density in film space (corresponding to D-traces in a central section of the specimen intensity transform) are corrected for background on an *ad hoc* basis and the integrated optical densities multiplied by factors designed to correct for both diffraction geometry and arc length.[55-57] In the third method both radial and circumferential traces of optical density are recorded for each reflec-

tion and after background correction the integrated areas are multiplied together, divided by the peak height and geometrical corrections appropriate to the peak position applied to give an estimate of the integrated intensity.[30,51] Meridional reflections require special treatment in all these methods.[32,56,58,59]

With the advent of high-speed digital microdensitometers the distribution of optical density in film space could be precisely mapped in two dimensions and the methods based on integrated optical density in film space refined.[60–65] As explained in Section 2 the problem with integrations carried out in film space or in $I_s(R, Z)$ is that the geometrical correction factors used to convert this integral to an integrated intensity are only valid for the centre of the reflection and appreciable errors are introduced when this method is applied to near-meridional reflections, particularly if α_0 is not small. This source of error can be completely eliminated by calculating an R-weighted integral of $I_s(R, Z)$ according to Expression (43)[2,46] or by applying individual Lorentz corrections in film space.[45] A particularly important advantage of these latter procedures is that they may be applied directly to meridional reflections. These reflections can not be treated in the same manner as other reflections when average Lorentz factors are used and must be scaled separately[56] thus introducing an additional parameter into any refinement process. This is avoided by applying individual corrections to each element of the integral.

The R-weighted numerical integration of $I_s(R, Z)$ is theoretically capable of yielding integrated intensities which approach, in accuracy, those obtained from single-crystal studies. However, problems encountered in the estimation of background and in correcting for the effects of overlap between reflections prejudice the chance of achieving such accuracy. Background correction is usually made on an *ad hoc* basis by attempting to identify regions close to reflections where the contribution to the observed optical density from the discrete reflections is assumed to be negligible and interpolating background values over the range of integration.[23,45,46,55,61,64] In some instances there is a radially symmetric component of background, which may be instrumental or due to diffuse rings associated with amorphous material, and local background correction is simplified if this is first removed.[45,66] The use of two-dimensional cubic spline fitting for interpolating local background has been advocated by Meader *et al.*[46]

Two procedures can be used when reflections partially overlap. In the first the intensity is integrated over the group of reflections and used to estimate the sum of the individual integrated intensities. In the second the integration is carried out over parts of the reflections free from overlap and an estimate made of the missing part of the integral.[61] The first method is generally applicable; the second can not be used when the overlap is extensive. More sophisticated procedures for dealing with overlap are discussed in Section 4.1.3.

4.1.3 Structure Factor Data—Profile Fitting

The advantages of profile fitting in the extraction of structure factors from single-crystal intensity data are well established.[67,68] In this method it is assumed that each reflection on the film has the same normalised intensity distribution, or profile, over a suitably chosen two-dimensional grid of points about its centre. The integrated intensity is derived from the scale factor which minimises the discrepancies between the observed intensities and the scaled profile function. In addition to a substantial improvement in accuracy this procedure provides objective criteria both for background correction and for processing partial or overlapped reflections.[69]

Makowski[52,53] has shown that profile fitting can be used for angular deconvolution of overlapping layer lines and this is discussed in Section 4.2. In order to apply profile fitting, as practised with single-crystal data, to patterns obtained from assemblies of crystalline particles it is necessary to define a two-dimensional grid of points over which the profiles of all the reflections are the same. Provided that σ is not small this can be done by means of Expressions (35)–(37). In the implementation of this procedure[70] values of R and Z are first predicted at equispaced intervals around a circle in Q-space, for example, where $\exp\{-\pi(U^2+V^2)\} = 0.01$, corresponding to 1% of the peak value. Intensity values extracted at these (R, Z) values are then used to derive a local background correction by fitting a plane:

$$I_B(R, Z) = b_0 + b_1 R + b_2 Z \tag{65}$$

The accuracy with which Expression (65) will represent the true background can be improved by first removing any radially symmetric component according to the procedure described by Fraser *et al.*[66] A least squares procedure is then used to optimise the fit between the background-corrected intensities chosen over grids with a range $-0.85 < U, V < 0.85$ for each reflection and the profiles generated by Expression (35), multiplied by a set of adjustable scale factors. The parameters which are concurrently refined are:

(a) the unit cell dimensions
(b) the crystallite dimensions
(c) the set of intensity scale factors
(d) the orientation parameter α_0.

An example of such a refinement is illustrated in Fig. 5.

Overlap between reflections is treated by storing the set of (R, Z) values for each reflection which define the $Q = 0\cdot 01$ contour and testing for overlap. An expression for I_B is determined by combining the non-overlapped portions of the $Q = 0\cdot 01$ contours for the two bands and the background-corrected intensity is apportioned between reflections on the basis of the values of the product of the Q-value and current scale factor for each band at that point in the transform (Fig. 6).

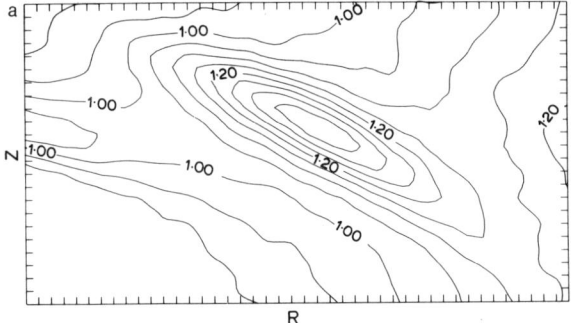

Fig. 5. An example of the procedure outlined in the text for automatic background fitting and profile fitting in patterns obtained from fibres containing crystalline particles. (a) Contour map of one of a set of observed reflections in the intensity transform of a specimen of Bombyx mori silk fibroin. (b)–(d) Contour maps of the observed intensity distribution plotted in terms of the variables U and V, defined in Expressions (36) and (37), for one member of the set showing the progressive refinement of parameters. In (b) trial values of the reciprocal cell parameters and crystallite dimensions were assumed and the scaling factors refined; in (c) the scaling factors together with α_0, a^* and b^* were co-refined and in (d) scaling factors, α_0, a^*, b^*, c^*, p_x and p_y were refined. The broadening due to p_z was too small to be incorporated. In (e) the idealised form of the scaled Q-function, defined in Expression (35), for this particular reflection is shown. The maps are plotted at equispaced intervals of U and V in the ranges $-0\cdot 85$ to $0\cdot 85$, so that the value of $Q(U, V)$ at the corners of the square is $0\cdot 01$.

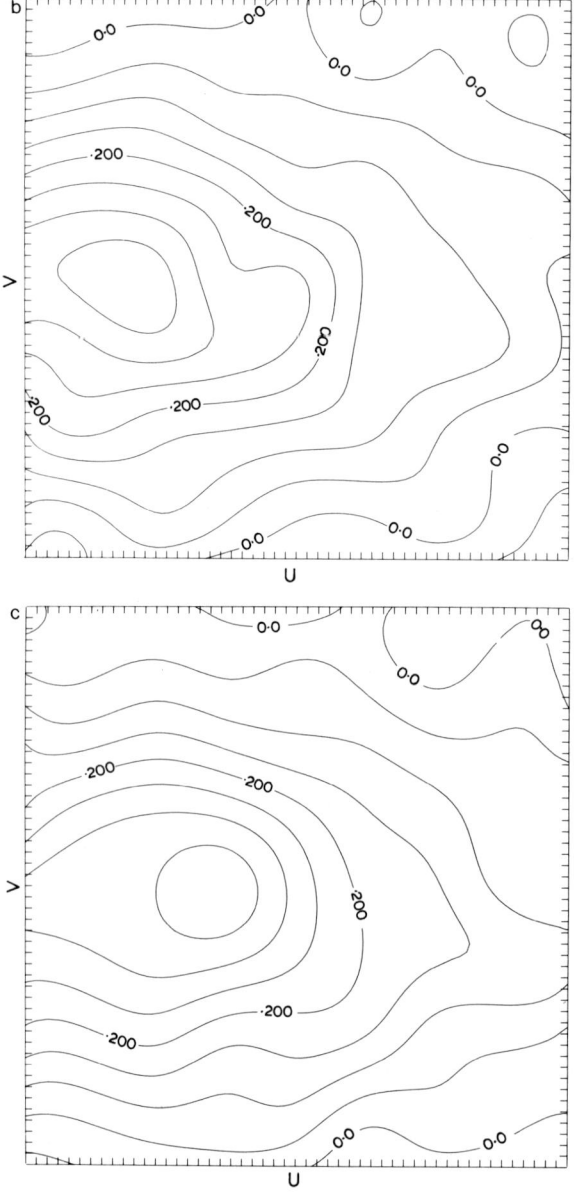

Fig. 5.—*contd.*

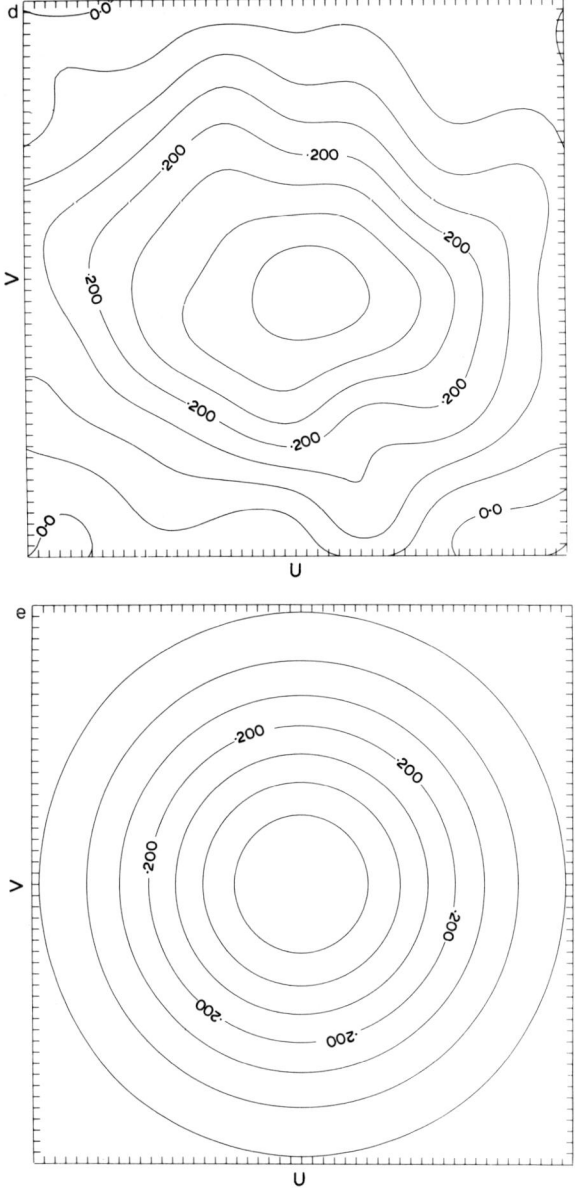

Fig. 5.—*contd.*

It will be recalled that the expression used to evaluate Q involves various approximations but these can be avoided by using the full numerical integral in Expression (32) if the increased computation time can be tolerated. Thus the $Q = 0.01$ profile can be determined by a search procedure and precise values of Q can be generated over the same grid of (U, V) values.

4.2 Helical Particles

With assemblies of helical particles the distribution of intensity along the layer line is continuous. The cylindrically averaged modulus of the structure factor $\langle FF^*(R, l/c)\rangle_\psi$ is related to the Z-integral of intensity across the layer line in the particle intensity transform $I_p(R, Z)$ by Expression (42), and this provides a means of obtaining structure factor data. If the layer lines in the particle intensity transform are sufficiently sharp Holmes and Barrington Leigh[27] showed that the intensity measured along the centre of the layer line in the specimen intensity transform, $I_s(R, l/c)$ could be used to estimate the Z-integrated value in the particle intensity transform by:

$$\int_{\text{layer line}} I_p(R, Z) \, dZ = I_s(R, l/c)(2\pi)^{1/2}\alpha_0 R \qquad (66)$$

provided that R was not small. This treatment was extended by Stubbs[4] to give approximations which take into account the effects of finite layer line width and finite beam size. Other approximations have been given by Fraser and Suzuki.[33] The range of applicability of both these treatments is subject to restrictions which can be troublesome and the more generally applicable treatment given by Provencher and Glöckner[34] is more useful since it gives high accuracy over a wide range of parameter values. However when α_0 is large and p_z is small it may be necessary to use numerical integration to evaluate the integral in Expression (46). Approximate methods of extracting integrated intensities directly have been discussed by Fraser and MacRae,[32] Fraser et al.[2] and by Holmes et al.[23]

As mentioned earlier Makowski[52,53] has described a method of profile fitting in film space which enables integrated intensities to be determined and also permits the resolution of contributions from overlapping layer lines. This latter feature is most important since the resolution of the X-ray data is limited in the R-coordinate direction by layer line overlap. This method has been successfully applied to several aggregates of macromolecules.[52,71]

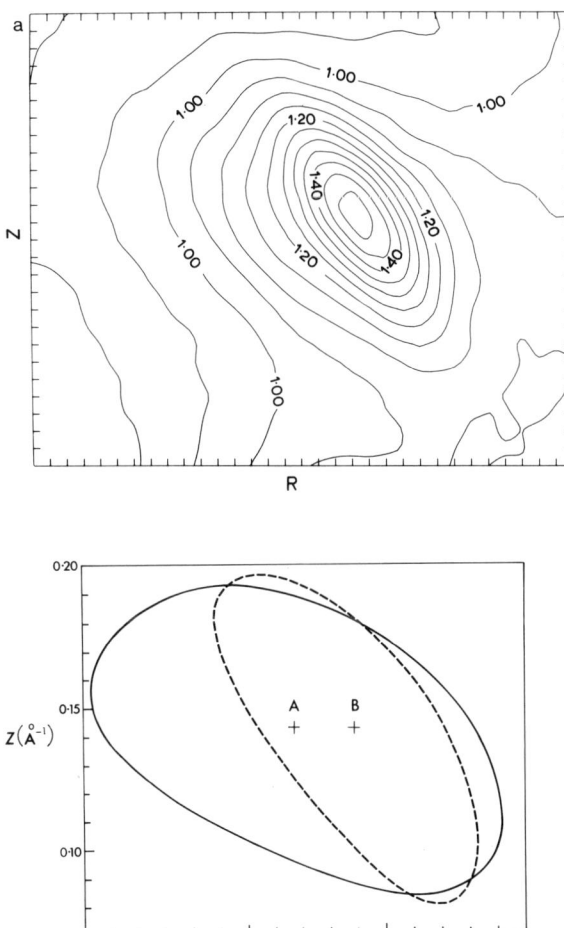

Fig. 6. An illustration of the procedure described in the text for resolving the individual contributions of overlapping reflections. (a) Contour map of an overlapped pair of reflections in the intensity transform of a specimen of *Bombyx mori* silk fibroin. (b) Estimated 1% of peak height contours for component A (full line) and component B (broken line). Automatic background correction is achieved by fitting a plane to the portions of the contours outside the zone of overlap. (c) and (d) Maps of the intensity distribution plotted in terms of the variables U and V, after parameter refinement, for components A and B respectively. (e) and (f) Maps of the corresponding idealised scaled Q-functions.

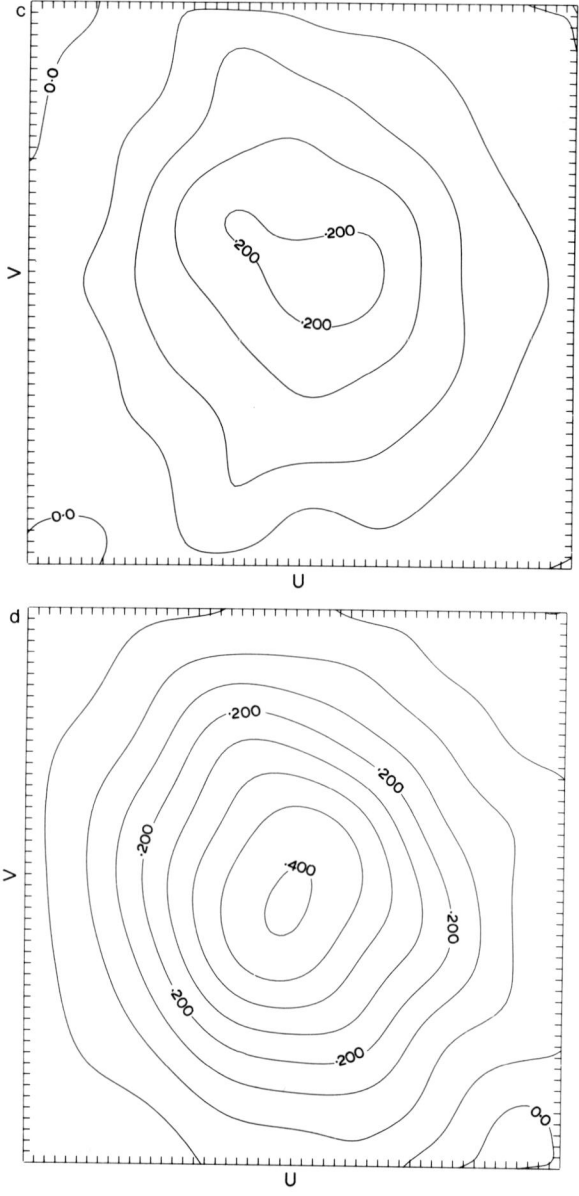

Fig. 6.—*contd.*

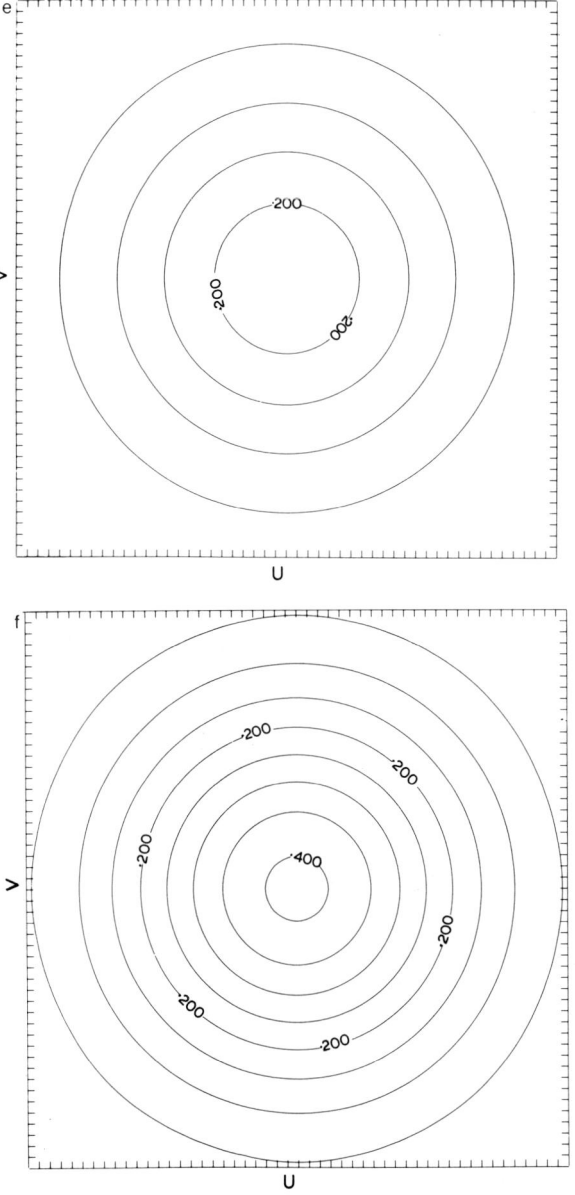

Fig. 6.—contd.

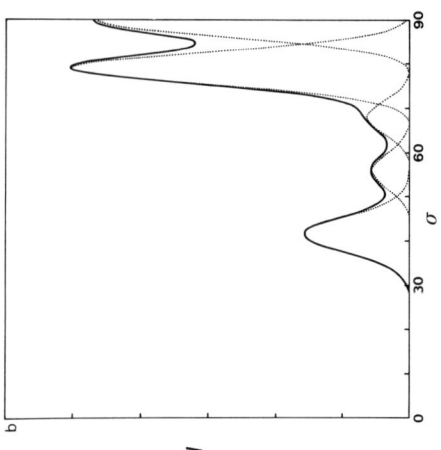

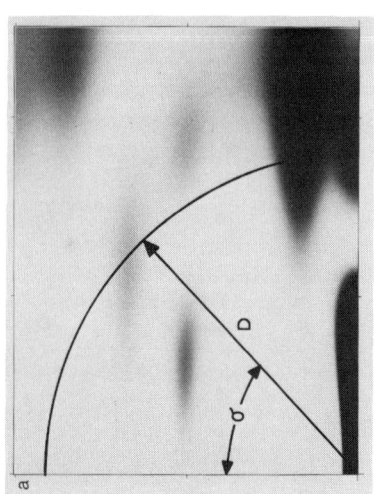

Fig. 7. (a) Part of the simulated diffraction pattern calculated for a collagen model.[35,73] (b) Intensity data extracted along a circular arc with $D = 0.18$ Å^{-1} (full line) and contributions from individual layer lines (dotted lines) determined by a deconvolution method similar to that described by Makowski;[52,53] except that a more complete representation of the intensity distribution was used, as in Expression (46), and also the unit height, unit twist and particle length were treated as independent variables in addition to α_0.

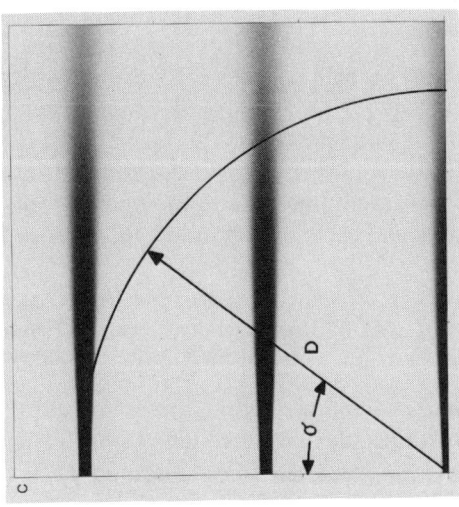

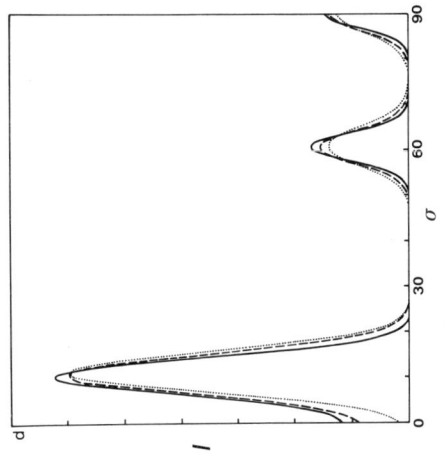

Fig. 7.—contd. (c) Simulated diffraction pattern calculated for an assembly of particles where there is a uniform distribution of intensity along layer lines spaced at intervals of 0·125 Å^{-1} in the particle transform, $\alpha_0 = 3°$ and $p_z = 200$ Å. (d) Comparison of deconvolution procedures using the full numerical integral given in Expression (46) (full line), the approximate function given in Expression (48) which makes allowance for both α_0 and finite particle length (broken line), and the approximate function given in Expression (64) in which no allowance is made for the spread of intensity due to finite particle length (dotted line). The first method accurately reproduces the test data, yielding values of $\alpha_0 = 3°$ and $p_z = 200$ Å. Both approximate methods give a poor representation of layer line profiles at small σ. In the present example the effects have been exaggerated by choosing an isolated D value which has a high concentration of intensity at small σ. The result in this instance is that the fitted value of α_0 is increased to 3·56° for the approximation which takes account of particle length, and 4·25° for the approximation which ignores this factor. In actual cases, where the distribution of intensity along the layer lines in the particle intensity transform is non-uniform, all three methods are subject to errors if used to represent the specimen intensity transform for small values of σ.

The treatment given by Makowski[52] takes no account of finite layer line thickness but this can readily be incorporated, in an approximate fashion by using Expression (48), or more exactly by using the full integral in Expression (46). Both these methods have been used on maps of the specimen intensity transform[72] and the refinement can readily be extended to include the unit height h and unit twist t of the helix as parameters (Fig. 7).

5 SIMULATION OF FIBRE DIFFRACTION PATTERNS

An alternative to attempting to extract information about the particle intensity transform from the smeared-out data in the specimen intensity transform is to produce a simulated specimen intensity transform based on a model, using the methods outlined in Section 2.[73] A quantitative comparison of the simulated and observed transforms can then be used as a basis for automated refinement of such features as the internal structure and dimensions of the particle, and various types of distortion and disorder. This approach is essentially an extension of the Rietveld, or whole-pattern fitting procedure that has been successfully used with powder patterns.[74-77]

The procedure used by Suzuki et al.[73] to calculate specimen intensity transforms involves three distinct stages:

1. The intensity transform for an infinite particle is calculated with appropriate allowances for random fluctuations in atomic positions, packing defects, and solvent scattering.
2. A mapping of the particle intensity transform is constructed by modifying the transform of an infinite particle to incorporate the effects of finite particle dimensions and of intraparticle lattice disorder, if present.
3. A mapping of the specimen intensity transform is then constructed by integrating over arcs of appropriate length in the particle intensity transform.

With crystalline particles, values of the structure factors for each reflection are computed and if the molecules are surrounded by solvent the effect on the structure factors is approximated by replacing $f(D)$ in Expression (11) by:

$$f''(D) = f(D) - v\rho \exp[-\pi(v^{1/3}D)^2] \tag{67}$$

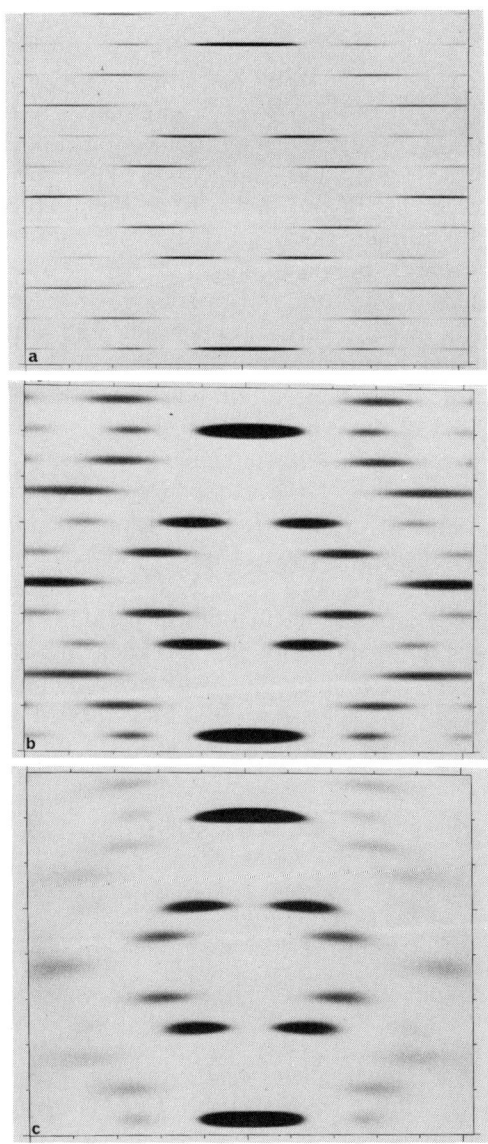

Fig. 8. An illustration of the procedure used to simulate fibre diffraction patterns. Intensity transforms are illustrated for a structure consisting of points uniformly distributed on a helix with radius $r = 4 \cdot 5$ Å, unit height $h = 3$ Å, unit twist $t = 108°$. (a) Particle intensity transform for $p_z = 500$ Å; (b) $p_z = 100$ Å; (c) Specimen intensity transform for $p_z = 100$ Å and $\alpha_0 = 3°$. (Reference 73, p. 64).

where v is the volume of solvent displaced by the atom, and ρ is the mean electron density of the solvent.[78] Alternative methods of approximating the effects of solvent scattering have been described by Ninio et al.,[79] Fedorov et al.[80] and Marvin and Wachtel.[81] The effects of random displacements of the atoms from their lattice positions are incorporated by further modifying the atomic scattering factors as outlined in Section 2. In constructing the map of a central section of the particle intensity transform the broadening of reflections due to the combined effects of finite particle dimensions and lattice disorder are incorporated using Expressions (29)–(31).

A similar procedure is used for particles with helical symmetry except that values of the cylindrically averaged intensity transform for the centre of the layer line are computed at intervals of $R = 1/(2r_{max})$, where r_{max} is the radius of the excribed cylinder around the helical particle. The intensity for other R values is obtained by interpolation. In constructing a central section of the cylindrically averaged particle intensity transform, allowance is made for Z-broadening associated with finite helix length and fluctuations in period.

In both cases the mapping of the specimen intensity transform is carried out by numerical integration over arcs of appropriate length in the map of I_p, using Expression (26). Examples of results obtained using this procedure are illustrated in Fig. 8.

REFERENCES

1. James, R. W. (1954). *The Optical Principles of the Diffraction of X-Rays*, Bell & Sons, London.
2. Fraser, R. D. B., MacRae, T. P., Miller, A. and Rowlands, R. J. (1976). *J. Appl. Cryst.*, **9**, 81–94.
3. Jennison, R. C. (1961). *Fourier Transforms and Convolutions*, Pergamon, Oxford.
4. Stubbs, G. J. (1974). *Acta Cryst.*, **A30**, 639–45.
5. Wilson, A. J. C. (1962). *X-Ray Optics*, Methuen, London.
6. Hosemann, R. and Bagchi, S. N. (1962). *Direct Analysis of Diffraction by Matter*, North-Holland, Amsterdam.
7. Vainshtein, B. K. (1966). *Diffraction of X-Rays by Chain Molecules*, Elsevier, London.
8. Ibers, J. A. and Hamilton, W. C. (1974). *International Tables for X-Ray Crystallography*, Vol. IV, Kynoch Press, Birmingham.
9. Welberry, T. R., Miller, G. H. and Caroll, C. E. (1980). *Acta Cryst.*, **A36**, 921–9.

10. Zernike, F. and Prins, J. A. (1927). Z. Phys., **41**, 184–94.
11. Alexander, L. E. (1969). *X-Ray Diffraction Methods in Polymer Science*, Wiley-Interscience, New York.
12. Kakudo, M. and Kasai, N. (1972). *X-Ray Diffraction by Polymers*, Elsevier, Amsterdam.
13. Hosemann, R. and Wilke, W. (1968). *Makromol. Chem.*, **118**, 230–49.
14. Hindeleh, A. M., Johnson, D. J., and Montague, P. E. (1980). In *Fiber Diffraction Methods* (A. D. French and K. H. Gardner (Eds)) American Chemical Society, Washington, pp. 149–82.
15. Clark, E. S. and Muus, L. T. (1962). *Z. Kristallogr.*, **17**, 108–18.
16. Tanaka, S. and Naya, S. (1969). *J. Phys. Soc. Japan*, **26**, 982–93.
17. Arnott, S., Dover, S. D. and Elliott, A. (1967). *J. Mol. Biol.*, **30**, 201–208.
18. Fraser, R. D. B., MacRae, T. P., Parry, D. A. D. and Suzuki, E. (1971). *Polymer*, **12**, 35–56.
19. Arnott, S. (1980). In *Fiber Diffraction Methods* (A. D. French and K. H. Gardner (Eds)) American Chemical Society, Washington, pp. 3–30.
20. Caspar, D. L. D. and Holmes, K. C. (1969). *J. Mol. Biol.*, **46**, 99–133.
21. Makowski, L. and Caspar, D. L. D. (1978). In *The Single-Stranded DNA Phages* (D. T. Denhardt, D. Dressler and D. S. Ray (Eds)) Cold Spring Habor Laboratory, pp. 627–43.
22. Squire, J. M. (1979). In *Fibrous Proteins: Scientific, Industrial and Medical Aspects* (D. A. D. Parry and L. K. Creamer (Eds)) Vol. 1, Academic Press, London, pp. 27–70.
23. Holmes, K. C., Tregear, R. T. and Barrington Leigh, J. (1980). *Proc. Roy. Soc. (London)*, **B207**, 13–33.
24. Luther, K. P. and Squire, J. M. (1980). *J. Mol. Biol.*, **141**, 409–39.
25. Deas, H. D. (1952). *Acta Cryst.*, **5**, 542–6.
26. Kratky, O. (1933). *Kolloid Z.*, **64**, 213–22.
27. Holmes, K. C. and Barrington Leigh, J. (1974). *Acta Cryst.*, **A30**, 635–8.
28. Suzuki, E. (1978). Unpublished.
29. Fraser, R. D. B. and Suzuki, E. Unpublished results.
30. Marvin, D. A., Spencer, M., Wilkins, M. H. F. and Hamilton, L. D. (1961). *J. Mol. Biol.*, **3**, 547–65.
31. Yonath, A. and Traub, W. (1969). *J. Mol. Biol.*, **43**, 461–77.
32. Fraser, R. D. B. and MacRae, T. P. (1973). *Conformation in Fibrous Proteins*, Academic Press, New York.
33. Fraser, R. D. B. and Suzuki, E. (1976). *J. Appl. Cryst.*, **9**, 91–4.
34. Provencher, S. W. and Glöckner, J. (1981). *EMBL Technical Report DA01*, pp. 1–14; (1982) *J. Appl. Cryst.*, **15**, 132–5.
35. Fraser, R. D. B., MacRae, T. P. and Suzuki, E. (1979). *J. Mol. Biol.*, **129**, 463–81.
36. Abrahamsson, S. (1966). *J. Sci. Instrum.*, **43**, 931–3.
37. Xuong, N. (1969). *J. Phys. E.*, **2**, 485–9.
38. Nockolds, C. E. and Kretsinger, R. H. (1970). *J. Phys. E.*, **3**, 842–6.
39. Jones, R. N. (1952). *J. Am. Chem. Soc.*, **74**, 2681–3.
40. Wooster, W. A. (1964). *Acta Cryst.*, **17**, 878–82.
41. Goldstein, D. J. (1971). *J. Microsc.*, **93**, 15–42.

42. Matthews, B. W., Klopfenstein, C. E. and Colman, P. M. (1972). *J. Phys. E.*, **5**, 353–9.
43. Cox, E. G. and Shaw, W. F. B. (1930). *Proc. Roy. Soc. (London)*, **A127**, 71–88.
44. Hellner, E. (1954). *Z. Kristallogr.*, **106**, 122–45.
45. Miller, D. P. and Brannon, R. C. (1980). In *Fiber Diffraction Methods* (A. D. French and K. H. Gardner (Eds)) American Chemical Society, Washington, pp. 93–112.
46. Meader, D., Atkins, E. D. T., Elder, M., Machin, P. A. and Pickering, M. (1980). In *Fiber Diffraction Methods* (A. D. French and K. H. Gardner (Eds)) American Chemical Society, Washington, pp. 113–38.
47. Franklin, R. E. and Gosling, R. G. (1953). *Acta Cryst.*, **6**, 678–85.
48. Massey, H. S. W. and Kestelman, H. (1959). *Ancillary Mathematics*, Pitman, London.
49. Stokes, A. R. (1948). *Proc. Phys. Soc. (London)*, **61**, 382–91.
50. Aspden, R. M., and Hukins, D. W. L. (1979). *J. Appl. Cryst.* **12**, 306–311.
51. de Wolff, P. M. (1962). *J. Pol. Sci.*, **60**, S34–36.
52. Makowski, L. (1978). *J. Appl. Cryst.*, **11**, 273–83.
53. Makowski, L. (1980). In *Fiber Diffraction Methods* (A. D. French and K. H. Gardner (Eds)) American Chemical Society, Washington, pp. 139–48.
54. Cella, R. J., Lee, B. and Hughes, R. E. (1970). *Acta Cryst.*, **A26**, 118–24.
55. Langridge, R., Wilson, H. R., Hooper, C. W., Wilkins, M. H. F. and Hamilton, L. D. (1960), *J. Mol. Biol.*, **2**, 19–37.
56. Arnott, S. (1965). *Polymer*, **6**, 478–82.
57. Guss, J. M., Hukins, D. W. L., Smith, P. J. C., Winter, W. T., Arnott, S., Moorhouse, R. and Rees, D. A. (1975). *J. Mol. Biol.*, **95**, 359–84.
58. Crist, B. and Worthington, C. R. (1980). *J. Appl. Cryst.*, **13**, 585–590.
59. Worthington, C. R., Worthington, A. R. and Wang, S. K. (1980). *J. Appl. Cryst.*, **13**, 273–9.
60. Wachtel, E. J., Wiseman, R. L., Pigram, W. J. and Marvin, D. A. (1974). *J. Mol. Biol.*, **88**, 601–18.
61. Hall, I. H. and Pass, M. G. (1975). *J. Appl. Cryst.*, **8**, 60–4.
62. Kolpak, F. J. and Blackwell, J. (1976). *Macromolecules*, **9**, 273–8.
63. Folkhard, W., Leonard, K. R., Malsey, S., Marvin, D. A., Dubochet, J., Engel, A., Achtman, M. and Helmuth, R. (1979). *J. Mol. Biol.*, **130**, 145–60.
64. Blackwell, J., Gardner, K. H., Kolpak, F. J., Minke, R. and Claffey, W. B. (1980). In *Fiber Diffraction Methods* (A. D. French and K. H. Gardner (Eds)) American Chemical Society, Washington, pp. 315–34.
65. Eikenberry, E. F. and Brodsky, B. (1980). *J. Mol. Biol.*, **144**, 397–404.
66. Fraser, R. D. B., MacRae, T. P., Suzuki, E. and Tulloch, P. A. (1977). *J. Appl. Cryst.*, **10**, 64–6.
67. Diamond, R. (1969). *Acta Cryst.*, **A25**, 43–55.
68. Ford, G. C. (1974). *J. Appl. Cryst.*, **7**, 555–64.
69. Rossman, M. G. (1979). *J. Appl. Cryst.*, **12**, 225–38.
70. Fraser, R. D. B., Suzuki, E. and MacRae, T. P. In preparation.
71. Makowski, L., Caspar, D. L. D. and Marvin, D. A. (1980). *J. Mol. Biol.*, **140**, 149–81.

72. Suzuki, E., Fraser, R. D. B. and MacRae, T. P. (1981). Unpublished data.
73. Suzuki, E., Fraser, R. D. B. and MacRae, T. P. (1980). In *Fiber Diffraction Methods* (A. D. French and K. H. Gardner (Eds)) American Chemical Society, Washington, pp. 61–67.
74. Rietveld, H. M. (1967). *Acta Cryst.*, **22,** 151–2.
75. Rietveld, H. M. (1969). *J. Appl. Cryst.*, **2,** 65–71.
76. Cheetham, A. K. and Taylor, J. C. (1977). *J. Solid State Chem.*, **21,** 253–75.
77. Young, R. A., Lundberg, J. L. and Immirzi, A. (1980). In *Fiber Diffraction Methods* (A. D. French and K. H. Gardner (Eds)) American Chemical Society, Washington, pp. 69–91.
78. Fraser, R. D. B., MacRae, T. P. and Suzuki, E. (1978). *J. Appl. Cryst.*, **11,** 693–4.
79. Ninio, J., Luzzati, V. and Yaniv. M. (1972). *J. Mol. Biol.*, **71,** 217–29.
80. Fedorov, B. A., Ptitsyn, O. B. and Voronin. L. A. (1974). *J. Appl. Cryst.*, **7,** 181–6.
81. Marvin, D. A. and Wachtel, E. J. (1976). *Phil. Trans. Roy. Soc. (London)*, **B276,** 81–98.

Chapter 2

THE DETERMINATION OF THE STRUCTURES OF AROMATIC POLYESTERS FROM THEIR WIDE-ANGLE X-RAY DIFFRACTION PATTERNS

I. H. HALL

Department of Pure and Applied Physics, UMIST, Manchester, UK

1 INTRODUCTION

In this review we shall be concerned with the crystalline structure, as determined by wide-angle X-ray diffraction, of materials belonging to the family of compounds having the structural formula:

$$\left[-\!\!\left\langle\!\!\bigcirc\!\!\right\rangle\!\!-COO-(CH_2)_m-COO- \right]_n$$

and abbreviated mGT. Poly(ethylene terephthalate) (2GT) is the best known member of this series but others have recently received considerable attention. Three independent studies have been published giving the unit cell and chain conformation of 4GT;[1-3] in two of these it is shown that under stress there is a phase change which is reversible on relaxation, and the structure of the new phase is also given.[2,3] There are two independent reports of the structure of 3GT,[4,5] and one of 5GT,[6] which also changes under stress to another phase, though this is only partially reversible on relaxation.[7] No phase transition has been observed with 3GT. There are several structural reports on 6GT though the chain conformation has not been determined. The material is polymorphic in the unstressed state;[8-10] three polymorphs have been identified and their unit cells determined, although more than one phase was always present in the test samples. Only very limited structural information is available on other members of this series[11,12] and they will not be considered in this review.

Intercomparison of the independent studies enables the techniques of structure determination to be critically evaluated. This will be our first concern. We shall then look at the structures themselves and, where they are available, use the independent determinations to assess the accuracy with which the structural parameters are known. Finally, by comparison of the structures of different members of the series, we shall infer which features of conformation and packing are common to all.

2 METHOD OF STRUCTURE DETERMINATION

2.1 Determination of the Unit Cell

The first stage in any structure determination is to obtain the unit cell, and to do this the location of as many reflections as possible must be measured. With a crystalline polymer, the most highly ordered test-piece available is usually an oriented fibre, and if the X-ray beam is

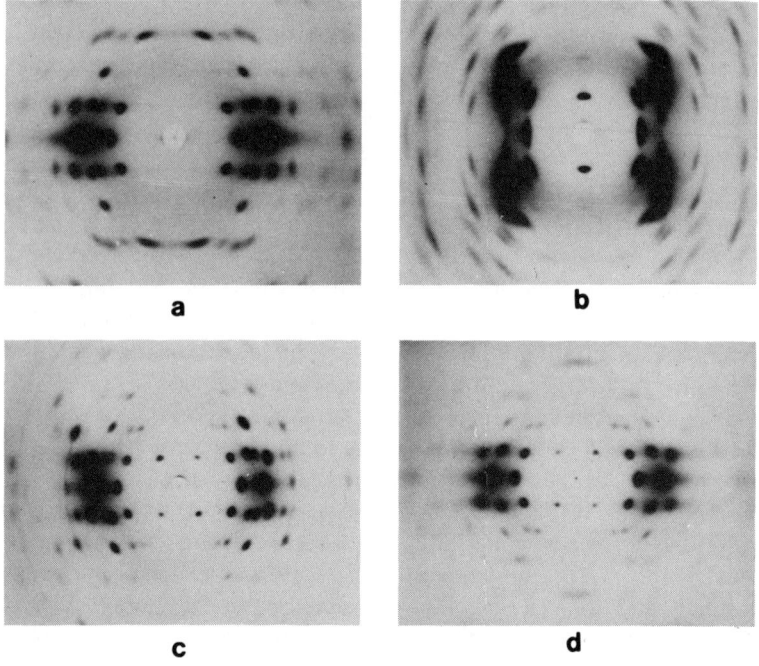

Fig. 1. X-ray diffraction patterns of mGT polymers (a) 2GT (b) 3GT (c) α-4GT (d) β-4GT (e) α-5GT (f) β-5GT (g) α-6GT (h) β-6GT (i) γ-6GT.

normal to the fibre axis (the arrangement commonly used), the diffraction spots lie on layer lines as shown in Fig. 1. These photographs were taken using a semi-cylindrical film with its axis coincident with the fibre axis; the other common arrangement is to use a flat film normal to the X-ray beam, in which case the layer lines are curved and the outer reflections are more diffuse and less intense. We will not be concerned

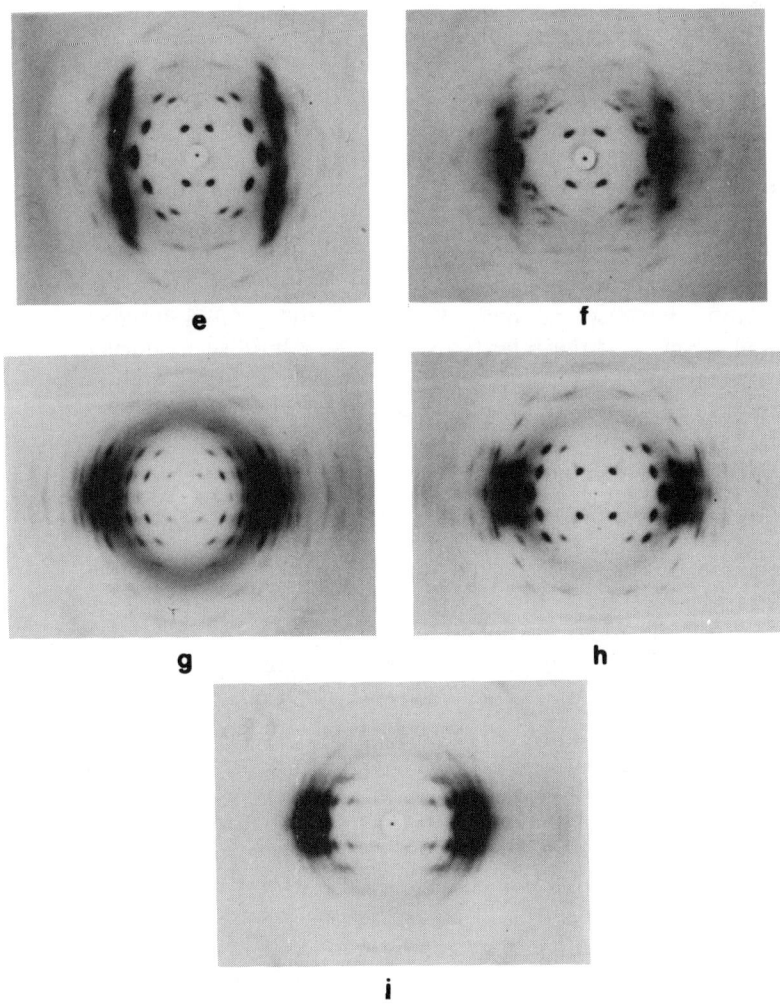

Fig. 1.—*contd.*

with the details of the experimental procedure for obtaining the diffraction patterns. It is sufficient to note that, if these are to yield reliable structural parameters, they must comprise a large number of small, well-resolved diffraction spots with a minimum of background intensity. Preparation of test-piece, type and adjustment of camera, and conditions of photographic processing are all important in achieving this.

Reciprocal space coordinates (ξ, ζ) are calculated from the film coordinates, and the equatorial reciprocal lattice plane is constructed from the equatorial reflections. To do this the three smallest values of ξ are chosen, and a triangle constructed with edge lengths equal to these values. The two shortest edges are taken as a^* and b^* and the angle between them as γ^*. (The usual convention, which we shall follow, is to take the c-axis as the fibre axis). The distances from the origin of points on this reciprocal lattice should correspond with the other measured values of ξ for the equatorial spots, which may thus be assigned their Miller indices, although since several lattice points will sometimes have closely similar values of ξ this can not always be done unambiguously. If the reflections lie on row lines (lines normal to the layer lines), c^* will be perpendicular to the a^*b^* plane, and the lattice planes of all layer lines will be identical with that of the equator (Fig. 2a). Thus the values of ξ for all layer lines should correspond with distances of lattice points from the origin.

If the reflections do not lie on row lines, then c^* is not perpendicular to the a^*b^* plane, although the fibre axis still is (Fig. 2b). In this case, the lattice net is the same for all layer lines, but the location of the origin changes from line to line. Taking the first layer line, the innermost reflection is assumed to be (001) and so ξ is the distance of this lattice point from the origin, which is thus located on a circle centred on (001), radius ξ. The next innermost reflection is now assumed to be one of the lattice points surrounding (001) and another circle drawn, as above. The origin must now be at one of the points of intersection of these two circles. One or two more reflections should now be tested to see if they can be located on lattice points for one of these possible origins; if not, the second reflection is assigned to another of the lattice points surrounding (001). The trial and error procedure is continued until all reflections on this layer line can be indexed. Once the origin has been located for the first layer line, it can be located for any other by simple geometry, and the lattice can be confirmed by the fact that all reflections on these can be satisfactorily

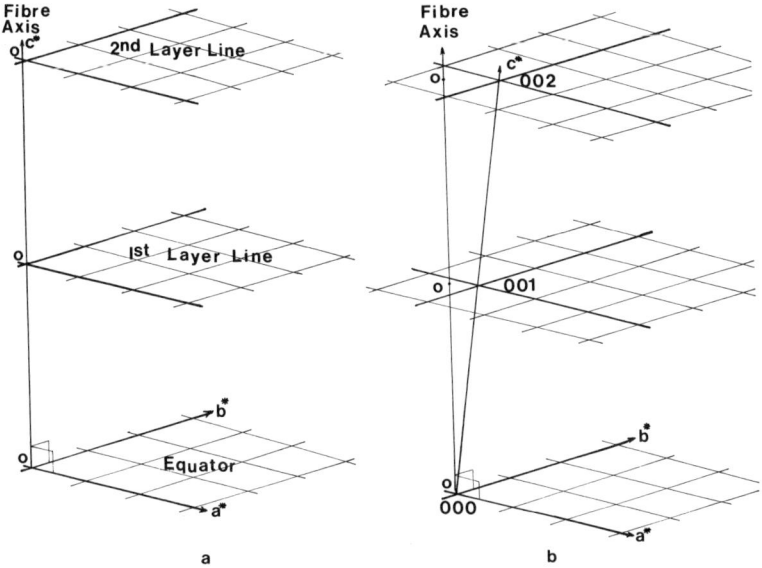

Fig. 2. Reciprocal lattice nets and the fibre axis (a) c^* perpendicular to the a^*b^* plane (b) c^* inclined to the a^*b^* plane.

indexed. At some stage during the procedure it may be necessary to halve a^* or b^* to index some reflections.

From this brief description, it is clear that in order to fix the lattice three or four reflections are necessary on the equator and first layer line, and that to confirm it many more should be measurable on these and higher layer lines. Providing these conditions are met, then the procedure can often be followed straightforwardly to the determination of a lattice which is essentially correct. This was the case with 2GT (Fig. 1a) and the α-phase of 4GT (Fig. 1c). (The α-phase is that which exists when the fibre is relaxed; when it is extended a reversible transition to another, β-phase, occurs.) When they are not met, the determination becomes more uncertain, and gross errors become possible. If a unit cell is to be determined with confidence, it is essential to have a highly oriented, well-crystallised specimen so that reflections may be located on layer lines without ambiguity, closely neighbouring spots may be resolved and the locations of a large number of reflections (more than about thirty) may be measured. Unless the time and trouble is taken to develop such a specimen, the uncertainty in the determined structure will be great—though this is seldom admitted in published work.

Examples of the difficulties and errors likely in cell determination may be found in published reports on the mGT series. The diffraction pattern of the β-phase of 4GT is shown in Fig. 1d. It will be seen that the reflections are more blurred than those for the α-phase shown in Fig. 1c; in particular, whereas the second and third reflections on the equator are resolved in Fig. 1c, they are not in Fig. 1d. Even on the original film they are not resolved; the inner reflection is only apparent from the asymmetry of the diffraction spot. Hall and Pass[3] use this as evidence to assign them different values of ξ, Yokouchi et al.[2] treat them as an unresolvable, overlapping pair and arrive at a different unit cell. They give a table of lattice plane spacings calculated from the observed positions of diffraction spots (d_o), and from the unit cell parameters (d_c); Hall and Pass do not, but if one is compiled from their unpublished data, the overall agreement between d_o and d_c is not significantly different. The choice between the cells rests solely on this pair of reflections, and it is only when their calculated positions on the film are compared with the location and extent of the observed spots[13] that it becomes clear that Yokouchi et al. must be wrong.

In the diffraction pattern of 3GT (Fig. 1b), the spots appear to be on row lines, but there are no well-separated layer lines. This suggests a unit cell in which c^* is perpendicular to the a^*b^* plane, and in which the c-axis is long. Extensive attempts to determine such a cell were unsuccessful, and it eventually transpired that the cell was triclinic, but instead of the chain axes (the unit cell c-axes) being parallel to the fibre axis they were systematically misoriented from it by a few degrees and in such a direction that c^* was almost parallel to it. This tilted crystal orientation will be discussed in more detail shortly, but one of its effects is to displace reflections up or down from their layer line positions. This material could not be obtained in a very highly oriented state and the lengths of its diffraction arcs were comparable with the layer line spacings (compare the lengths of the diffraction arcs in Figs 1b and 1c); thus the layer line locations were unclear. It transpired that the strong meridional reflection lies on the second layer line.

With 5GT the principal difficulty was that, unless great care was taken, reflections from two different crystalline phases of the same material were present in the same diffraction pattern. Figure 1e is the pattern for the fibre annealed free to relax, and is that of a pure phase called the α-phase. Figure 1f was obtained with a fibre which had been held at a constant extension of 14% whilst annealed and had its diffraction pattern recorded without relaxing this strain. It is apparent that some strong reflections in Fig. 1e are also present in Fig. 1f, and so a significant proportion of α-phase remains in this fibre. However, many of the reflections are new, and these belong to a different phase, called the β-phase. Unless care is taken in specimen preparation, a significant proportion of β-phase is present in a relaxed fibre and confuses cell determination. This behaviour was only discovered and elucidated because experiments were performed in which specimens were annealed under different conditions in an attempt to improve the clarity of the diffraction pattern. It was noted that annealing under different conditions caused the shape and relative intensity of reflections to change. Joly et al.[14] observed none of this, but used a specimen from which only 14 reflections were visible, whereas Hall and Rammo[6] observed 40 reflections with the α-phase, and 39 with the β. An additional difficulty with this material was that there were too few equatorial reflections visible to determine a^*, b^* and γ^*, and so a large amount of trial and error was necessary before arriving at satisfactory values. Although the cells which were eventually obtained satisfactorily passed the tests which were applied to them, for the above reasons they must be regarded as less certain than the others for this series.

The difficulty of identifying different phases present in the same fibre is even more acute with 6GT.[8,9] There are three allomorphs (Figs 1g–1i), and although conditions of preparation may be chosen which favour a particular phase, a significant proportion of the other two is always present. Since all have the same layer line spacing, it is not immediately apparent to which phase a particular reflection belongs and correct assignment was only achieved by carefully observing changes in relative intensities with different methods of specimen preparation and testing the resulting cells by using the checking procedure to be discussed shortly. Two investigations in which this was not done[11,14] obtained incorrect cells. Figures 1g–1i are the diffraction patterns of the purest forms obtained.

From the foregoing discussion it is clear that even when the procedure for cell determination has been satisfactorily followed, and the observed positions of all reflections are reasonably close to lattice points, further checks on the correctness of the cell are desirable.

The density may be calculated from the cell dimensions, the chemical composition, and the number of monomers in the cell (which is usually clear from the dimensions). This should be near to the measured value, and is usually slightly greater since there is disordered material present in the test-piece.

All members of this series crystallise with the preferred direction of the chain axes tilted from the fibre axis by a few degrees, this occurring by rotation about a given line with respect to the unit cell. As will be seen from Fig. 3, a reflection will be displaced from the layer line by an amount and in a direction which depends on the position of its reciprocal lattice point with respect to the tilt axis. Indices will have been assigned to all reflections from the unit cell determination; the observed direction of their displacement can be marked on a reciprocal lattice net such as that in Fig. 3. If the cell is correct it should be possible to draw a line, the tilt axis, dividing those displaced upwards from those displaced downwards. The magnitude of the displacement of one reflection can be used to determine the angle of tilt, and it is then possible to determine the magnitude and direction of displacement of all reflections and compare this with that which is observed.[15] Agreement provides powerful confirmation of the correctness of the cell.

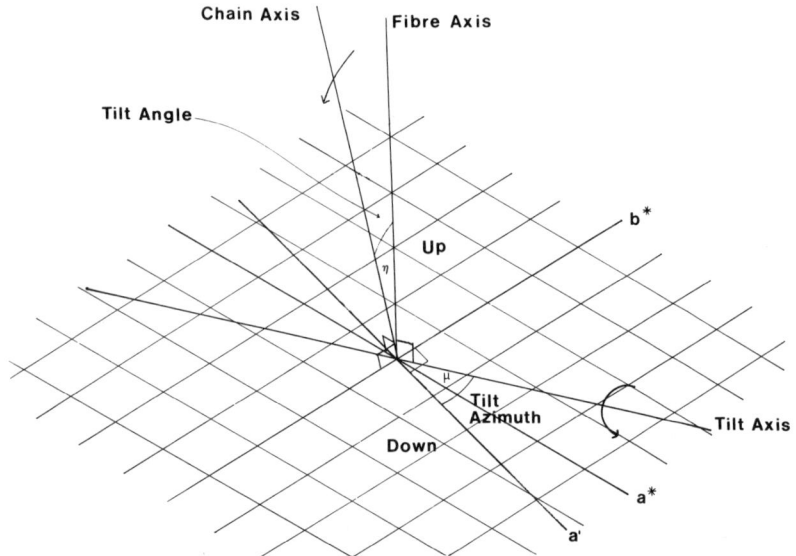

Fig. 3. Displacement of lattice points by tilting.

A major reason for the ambiguities in unit cell determination using an oriented fibre is that all directions normal to the fibre axis are equally likely for the a^*- and b^*-axes. (If tilting occurs a^* and b^* are not normal to the fibre axis, but are still uniformly distributed around it). Thus in reciprocal space, the lattice points lie on concentric circles, from whose radii the lattice must be deduced. If a preferred orientation can be introduced normal to the fibre axis (biaxial orientation) these circles will separate into arcs, and diffraction photographs taken using a Weissenberg camera should directly give the two-dimensional lattice. Biaxial orientation can often be induced by passing a specimen between rollers to reduce its thickness. In practice, it is seldom possible to induce sufficient orientation this way to determine a unit cell, but an existing cell can be tested.

A final method of confirmation that has sometimes been used[4,9,10] is to grow single crystals from solution, and to use these to obtain the electron diffractogram of the a^*b^* plane.

The published unit cells of the mGT series all have densities greater than the measured value (by between 1 and 6%), and all satisfactorily describe the observed displacement of reflections from layer lines. Electron diffraction was used to confirm the cell of 3GT, biaxial orientation was used to confirm the α-phase unit cell of 4GT, and all methods were used extensively to identify the different phases of 6GT.

In the method which has been described, the lattice parameters are fixed using the innermost reflections on the equatorial and first layer line, and the remaining reflections merely confirm that it is satisfactory. Because reflection centres cannot be located with complete precision, all will not be exactly at lattice points and it is advisable to utilise some refinement procedure. One which has been commonly used adjusts lattice parameters to minimise squares of discrepancies between the observed and calculated lattice plane spacings. There are two objections to this. First it only locates the reflection on a sphere in reciprocal space and so does not utilise the information available about layer line spacings, and second because there is a non-linear relationship between d and the film coordinates of the reflection, the uncertainty in d is not proportional to the uncertainty in these coordinates. Reflections are therefore incorrectly weighted in the refinement procedure. A method which minimises the squares of the discrepancies between the observed and calculated film coordinates is preferable and has been used by Hall and co-workers (see Ibrahim[9]).

In the original determination of the unit cell of 2GT by Daubeny *et al.*[15] no refinement procedure was used (indeed this determination was made before digital computers, which have made such procedures possible, were generally available) and their parameters have been adjusted by Fakirov *et al.*,[17] using an undefined procedure, to improve the agreement between d_o and d_c for a selection of reflections. The present author has refined the parameters of Daubeny *et al.* to minimise squares of the discrepancies between observed and calculated film coordinates for a larger selection of reflections (though still for an incomplete set).

In a least-squares refinement procedure, each residual $((p_c - p_o)^2$ where $p_c - p_o$ is the discrepancy being minimised, p_c the calculated value of the parameter and p_o its observed value) should be given a weight inversely proportional to the standard deviation of p_o. If this is done, the standard deviations of the cell parameters may be calculated. As far as I am aware, this procedure has never been followed with polymeric fibres; indeed it would be very difficult to do so because high-order reflections are more diffuse than low-order and the accuracy with which they are located is not easily estimated. Until it is done, it is futile to discuss the significance of small changes in cell parameters caused by different treatments of a material, and unwise to place undue reliance on quantities such as crystalline density which depend critically upon them.

2.2 Measurement of Intensity of X-ray Reflections

Whilst the locations of the various X-ray reflections are controlled by the unit cell dimensions, their intensities are controlled by the atomic arrangement within the cell. Thus the accuracy with which this arrangement is determined will depend upon the accuracy of intensity measurements.

In early work, such as the determination of the structure of 2GT, the intensities were estimated from the diffraction photograph by eye. This was superseded by the use of single-scan optical microdensitometers, which plot the optical density along a line (usually a layer line, or line of constant azimuth across the photograph). Provided exposure is made within the linear range of the photographic emulsion the optical density is proportional to the X-ray intensity, and so the intensity profile is obtained along a line through the reflection. To obtain the integrated intensity the area under this profile is determined and some correction factor applied to allow for the azimuthal spread, which will depend on the film–specimen geometry being used, and upon the disorientation of the specimen. 'Arc correction factors' have been determined[18,19] which make allowance for disorientation assuming that it is of equal magnitude in all ($hk0$) planes.

These corrections can only be avoided if the intensity is measured over the whole area of the reflection, and the advent of high-speed raster scanning microdensitometers has made this possible. Their application has been discussed in the first chapter of this book.

With a polymeric fibre the reflections are superimposed upon a background intensity which varies continuously from point to point, and however it is measured the intensity must be apportioned between these two components. This is frequently a subjective judgement on the part of the experimenter but attempts have recently been made to quantify the procedure. For example Neisser,[20] as part of a procedure which uses computer analysis of data produced by raster scanning, assumes that the background intensity varies linearly over the region of a reflection. This is probably an oversimplification and Atkins and Meader[21] use a more elaborate 'cubic-spline' method of curve fitting and extrapolation. As yet there is insufficient experience in the application of these methods to assess the degree of complication necessary, but hopefully the quantification of procedures will enable this question to be investigated.

Because the scatterer possesses only fibre symmetry, reflections from different lattice planes may overlap and form only one spot on the diffraction photograph. Whenever possible its intensity should be apportioned between the contributing reflections. Where this has been attempted, it is again, usually, the subjective judgement of the experimenter, but Hall and Neisser[20] have developed a quantitative procedure which rests on the assumption that the intensity profile along any line through the centre of a spot is Gaussian. This method can be used provided the peak intensities of the overlapping members are resolved.

From this brief discussion it will be clear that the accuracy of the determined structure is very dependent upon the accuracy with which intensity is measured, but that present techniques of intensity measurement are very unsatisfactory. Errors are likely to be large and unknown; indeed Hall[33] has shown that the differences between 3 independently measured intensity sets for α-4GT are equivalent to an R-factor of about 14%. The development of methods of computer analysis of data produced by raster scanning should lead to a reduction and quantification of error, and until this is done it must always be remembered that the parameters describing the atomic arrangement in a crystalline polymer are of very uncertain accuracy. In particular, choice can only confidently be made between competing structures when mistakes can be demonstrated in the determination of one of them.

Many of the methods mentioned above have been used in the determination of the structures of aromatic polyesters. Visual estimation was used by Daubeny et al.[15] and by Poulin-Dandurand et al.[4] Tadokoro and his co-workers have used single-scan microdensitometry, whilst investigations by Hall have used raster scans. One set of intensities have been determined from these using the computer-based procedures already mentioned, the others by integrating manually around iso-density contours.[22]

2.3 Assignment of Bond Lengths and Bond Angles

Methods of structure determination involve a refinement stage, in which parameters of a model are adjusted to improve the agreement between the observed and calculated reflection intensities. To do this there must be several times more data than parameters to be refined; a criterion which is rarely met if a polymer is regarded as a collection of independent atoms. Also, since intensities are usually only measurable at fairly small values of scattering angle, the resolution available is rarely adequate for this approach.

The usual practice is therefore to assign values to the bond lengths and bond angles, and to assume the complete structure of rigid units such as benzene rings, thus reducing the number of degrees of freedom possessed by the structure, which becomes a sequence of linked atoms defined by the conformation angles (Fig. 4). Since small changes in these angles can have quite a large effect on the overall shape, they can be determined more precisely than the low resolution of the data might suggest. An additional source of error is, however, introduced because the determined structure will depend upon the values chosen for the bond parameters, which are best estimated from the structure of a low molecular weight compound containing chemical groupings closely resembling those in the polymer. Brisse and co-workers[23-29] have determined the crystalline structures of several oligomethylene glycol derivatives enabling good choices to be made for aromatic polyesters, but since these did not become available until much of the work described here was completed comparison with values actually used provides an indication of the discrepancies which might exist. All values are listed in Table 1 (the parameters may be identified by

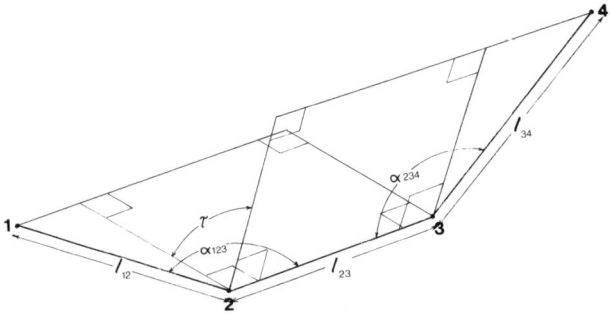

Fig. 4. Conformation angle of sequence of linked atoms. If the lengths l of the bonds joining the four atoms 1, 2, 3, 4, and the angles α between them are fixed, then the structure is defined by the conformation angle τ. This is positive if 3—4 is rotated clockwise from the plane 1, 2, 3 when the structure is held so that 1—2 is nearer the observer than 3—4.

reference to Fig. 5) and whilst similar values, lying within the range expected from the single-crystal measurements, are chosen for most parameters, there are some bad discrepancies. Of particular interest are those in Column J for which the bond lengths were taken from a standard reference table.[30] Several values lie outside the range to be expected from the single-crystal work, showing the limited usefulness of such tables and the need to seek determinations on compounds containing groups of bonds similar to those in the polymers.

The results of single-crystal studies can also be used to decide whether it is justified to regard particular groups of bonds as being rigid. For example, in Columns A to E of Table 1, the sums of α_3, α_4 and α_5 and of α_6, α_7 and α_8 are both very close to 360°. Thus each of these groups can be treated as rigid planar entities. However, since the plane of $O_1C_4O_2$ is always inclined by a few degrees to that of $C_2C_3C_1$, rotation must be allowed about the bond C_3C_4. Authors are not always explicit about their assumptions in these respects, and whilst fractional atomic coordinates give a full description of a structure they do not convey information on these items in a direct manner.

2.4 Generation of Trial Models

With a fibrous polymer it is not possible to proceed directly from the X-ray data to the atomic arrangement. A trial and error procedure must be followed in which a model of the structure is proposed, tested against such criteria as are readily available, and, if it is proved

Table 1. Values of bond parameters used in structural studies on aromatic polyesters.

Bond	Length (Å)												
	Single-crystal (measured)					Polymer (assumed)							
	A	B	C	D	E	F*	G	H	I	J	K	L	M
C_0C_1	1·378	1·379	1·387	1·373	1·376	1·36	1·378	1·59	1·40	1·395	1·38	1·39	1·375
C_1C_3	1·384	1·392	1·363	1·382	1·384	1·41	1·383	1·39	1·40	1·395	1·38	1·39	1·375
C_2C_3	1·389	1·392	1·383	1·391	1·380	1·34	1·390	1·39	1·40	1·395	1·39	1·39	1·38
C_4C_3	1·477	1·482	1·511	1·486	1·479	1·48	1·486	1·48	1·49	1·49	1·48	1·48	1·49
O_1C_4	1·339	1·344	1·327	1·329	1·331	1·35	1·330	1·34	1·26	1·36	1·34	1·34	1·33
O_2C_4	1·209	1·204	1·200	1·205	1·200	1·26	1·200	1·21	1·23	1·23	1·21	1·21	1·20
C_5O_1	1·442	1·452	1·452	1·450	1·452	1·45	1·450	1·44	1·41	1·43	1·44	1·44	1·45
C_6C_5	1·499	1·493	1·480	1·504	1·485	1·47	1·500	1·50	1·53	1·54	1·50	1·50	1·49
C_7C_6					1·523				1·53	1·54	1·50	1·54	1·52
C_8C_7					1·496								1·50

Angle (degrees)

	A	B	C	D	E	F	G	H	I	J	K	L	M
α_1	119·8	120·5	120·5	121·5	120·3	120	120·4				120	120	119·5
α_2	120·0	120·7	118·0	120·2	120·1	123	120·0				120	120	120·5
α_3	119·9	119·2	122·2	118·9	119·1	117	119·7	120	120	120	120	120	120
α_4	118·6	118·0	117·4	119·8	117·8	117	118·0	119	120	119	120	120	118
α_5	121·8	122·8	120·8	121·3	123·1	126	122·3	119	120	121	120	120	122
α_6	124·4	124·5	123·8	124·8	123·9	127	124·3	121	125	123	125	125	124
α_7	113·0	112·5	112·3	112·4	112·7	110	112·3	119	113	114	113	113	113
α_8	122·5	123·0	123·9	123·6	123·4	122	123·4	120	122	123	122	122	123
α_9	115·7	115·8	116·3	117·0	115·8	114	116·2	119	116	111	119	116	116
α_{10}	105·0	107·3	108·2	106·8	108·3	105	107·5	104	106	110	105	108	108
α_{11}					111·7		115·3	105	113	110	113	112	112
α_{12}					113·4								113

A, Ethylene glycol dibenzoate;[25] B, Ethylene glycol di-*para*-chlorobenzoate;[24] C, Trimethylene glycol dibenzoate;[29] D, Trimethylene glycol di-*para*-chlorobenzoate;[26] E, Hexamethylene glycol dibenzoate;[28] F, Poly(ethylene terephthalate);[15] G, Poly(trimethylene terephthalate);[4] H, Poly(trimethylene terephthalate);[5] I, Poly(tetramethylene terephthalate);[1] J, Poly(tetramethylene terephthalate);[2] K, Poly(tetramethylene terephthalate);[3] L, Poly(pentamethylene terephthalate);[6] M, Poly(hexamethylene terephthalate).[9]

* Because an incomplete set of bond parameters were included in the paper these were calculated from the fractional atomic coordinates and differ slightly from the bond lengths which were given.

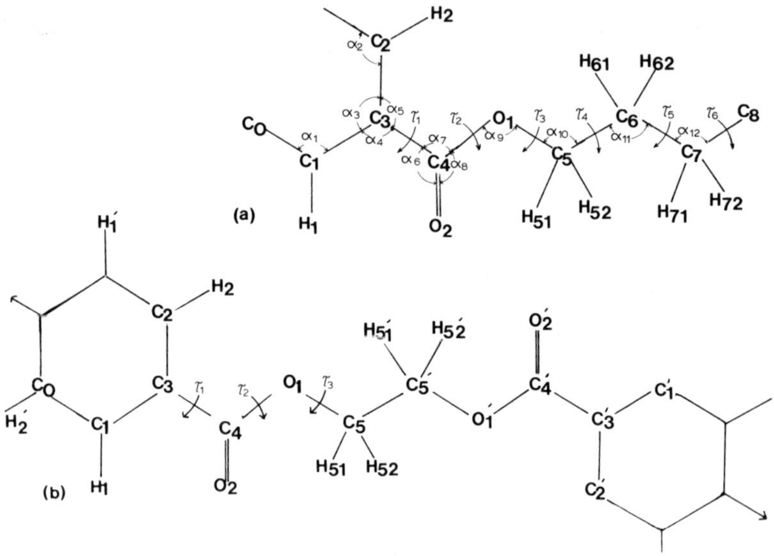

Fig. 5. Labelling scheme used to define atoms and bond parameters. (a) The scheme of numbering used for all m-GT polymers. The atomic sequence is symmetric about its mid-point and is terminated at the appropriate atom for the value of m. (b) Application of numbering scheme to 2GT.

satisfactory, then refined to minimise the discrepancies between its calculated X-ray structure factors and those observed.

Three criteria must be satisfied by a suitable model. Conformations and symmetries which occur commonly, either in low molecular weight compounds or in other polymers, should be retained; there should be no unacceptably short separations between non-bonded atoms; and the crystallographic repeat should agree with the observed value. This ensures that models chosen for X-ray refinement are those with low conformation and packing energy and which satisfy the known stereochemical constraints. The linked-atom refinement procedure[46,47] provides a convenient way of generating and subsequently refining such models, and this program was utilised in the author's own computation.

If these principles are applied to mGT polymers, then from studies on oligomers the atoms between C_0 and C_5 are always nearly coplanar and may be treated as a rigid plane in preliminary models. Methylene bonds (such as C_5—C_6 in Fig. 5a) are usually close to the *trans*- or one of the *gauche*-conformations and so only the angle about O_1C_5 is ill defined. Oligomers with m even[24,25,28] are centrosymmetric about the central methylene bond (e.g. the centre of bond C_5—C_5' in Fig. 5b) and this symmetry may be assumed in model building. When m is odd the symmetry of the monomer is not so clear; in the author's studies none was assumed, and the testing of possible symmetries will be discussed in Section 3.3. Also the crystallographic repeat distance comprises two monomers which are likely to be symmetrically related, but this could be either by a centre of symmetry at the benzene ring, or by a two-fold screw axis along the chain axis. The number of possible conformations is thus greater than for m even. Some of the conformations which satisfy the above requirements and the crystallographic repeat may be rejected because they lead to unacceptably small separations of non-bonded atoms.

If there is only one chain in the unit cell, it must be located with its axis coincident with the c-axis, and the angle of rotation of some reference plane in the molecule about this axis is all that remains to be determined to complete the generation of the model. For some conformations all orientations will give unacceptable packing and these may be rejected; for others, those giving the fewest unacceptably short atomic separations are retained for further investigation. Where there are two or more chains in the cell, space group consideration must be used to determine possible symmetry relations between them, and relative placings and orientations become variables in the model. Possible ranges can be identified by investigating atomic separation; the procedure is the same in principle as that already described, but is complicated by the greater number of variables. At the completion of this stage, several competing models will probably exist, but it will be possible to reject some of these because their calculated structure factors are in poor agreement with those determined experimentally.

More exact methods of conformational energy calculation are frequently used in model building. Thus with 4GT, Yokouchi *et al.*[2] determined the energy as a function of the dihedral angles τ_1, τ_2 and τ_3 whilst maintaining the chain length at the crystallographic repeat length. Six conformations giving energy minima were found; only one of these gave reasonable agreement with X-ray data even though in each case the molecule was rotated about the chain axis to find the orientation giving best agreement.

Poulin-Dandurand *et al.*[4] followed a very similar procedure with 3GT. The main difference being that they fixed τ_1 at 5°, taken from the structure of dimethyl terephthalate,[27] and that they used energy considerations to determine the chain packing in the unit cell,[31,32] rejecting one of the two possible conformations because it gave much higher packing energy than the other.

2.5 Refinement of Structure

Having determined a trial model which appears to be a likely approximation to the structure, it is refined (i.e. its parameters are adjusted) to minimise the discrepancy between observed and calculated structure factors. This is conveniently done by the linked–atom method already mentioned. The computer program LALS[47] which embodies this principle has many facilities and its strategy of use depends on the particular structure being studied; here we shall only comment on experience with refinement of the structure of aromatic polyester fibres.

Structure refinement on 4GT has been performed by Mencik,[1] Yokouchi et al.,[2] and Hall and Pass[3] and on other aromatic polyesters by Hall and co-workers.[5,6,9] Criticisms and comparisons of the 4GT structures[13,33,34] have also involved refinement. Two major points of difference are apparent in this work, the treatment of unobserved reflections, and the weighting of reflections.

Unobserved reflections were left out of the refinement procedure in all of the original studies of α-phase 4GT, but were utilised by Hall and Pass[3] with the β-phase and in some of the comparative studies.[13,33] They have also been utilised in refinement of the other members of the series.

A reflection is unobserved because its intensity is below the threshold of observation. This is an important experimental observation which should be utilised in the refinement. It is conveniently done by estimating the threshold at the point on the film where the reflection would be observed if it were sufficiently intense and including it in the refinement procedure at this intensity if it is calculated stronger. Otherwise it is omitted. Even if such reflections are not included in the refinement scheme, their intensities should be calculated and checked against threshold values. It has been my experience that where they are omitted completely, refinement can take a structure to unlikely stereochemical conformations, but calculation of the intensities of unobserved reflections then shows several sufficiently strong to be clearly seen enabling these conformations to be rejected.

Where refinement minimises the sum over all reflections of the squares of the discrepancies $((F_c - F_o)^2$, where F_c is the calculated and F_o the observed structure factor of a reflection), errors are minimised if each discrepancy is weighted by $1/\Delta F_o$ where ΔF_o is the standard deviation of the observed structure factor. Since, with the methods of measuring intensity which are currently available for polymeric fibres, ΔF_o cannot be determined, this cannot be done, and either all have to be weighted uniformly, or the Cruickshank scheme[35] has to be used.

To weight all uniformly is to assume that the standard deviation of all structure factors is the same, whereas the Cruickshank scheme is equivalent to the assumption that the relative standard deviations of the structure factors are the same, since on average it makes spots of all intensities contribute equally to the sum of the squares of the discrepancies. The correct one to use will depend upon the experimental method that has been followed. If only one photograph was taken (i.e. all reflections were recorded at the same exposure), the standard deviation of all spot intensities would be expected to be similar, and uniform weights should be used. However, the range of intensities of the spots would exceed the linear range of the emulsion and this would introduce serious error.

It is preferable to take a series of photographs of differing exposure and choose from them so that all measurements are made on spots of similar density. Not only are they then all within the linear range of the emulsion, but also their standard deviations would be expected to be proportional to the exposure-corrected intensities requiring Cruickshank weighting to be used, which is preferable because all data contribute equally to the refinement. However, even in this situation, the intensities of weak spots are likely to be less accurately measured than those of strong because both background subtraction and the assignment of a reflection boundary are more difficult, hence weak reflections will be overweighted. Hopefully, as experience is gained with the methods described in the first chapter, it will be possible to determine the values of the standard deviations of all reflections.

3 THE STRUCTURES OF AROMATIC POLYESTERS

3.1 The Isolation of Allomorphs

Allomorphs have been reported for three different glycol terephthalates.[2,3,6,9] With 4GT one phase (the α-phase) is always present in an unstressed oriented fibre, and transforms to a different one (the

Table 2. Published unit cells

									Reference		
	15	17	36	38	39	5	4	37	1	2	3
			2GT				3GT				4GT α-phase
a(Å)	4·56	4·48	4·52	4·44	4·52	4·59	4·64	4·58	4·83	4·83	4·89
b(Å)	5·94	5·85	5·98	5·91	5·92	6·21	6·23	6·22	5·96	5·94	5·95
c(Å)	10·75	10·75	10·77	10·67	10·70	18·31	18·64	18·12	11·62	11·59	11·67
α(degrees)	98·5	99·5	101	100·1	99·8	98·0	98·4	96·9	99·9	99·7	98·9
β(degrees)	118	118·4	118	117·0	117·5	90·0	93·1	89·4	115·2	115·2	116·6
γ(degrees)	112	111·2	111	111·8	111·4	111·7	111·2	111·0	113·8	110·8	110·9
No. of chains	1	1	1	1	1	1	1	1	1	1	1
No. of monomers	1	1	1	1	1	2	2	2	1	1	1
Volume (Å³)	219·0	210·6	216·0	210·6	215·5	479·4	493·4	478·1	260·0	260·4	262·8
Calculated density (gcm⁻³)	1·455	1·513	1·476	1·515	1·479	1·427	1·386	1·430	1·405	1·402	1·398
Measured density (gcm⁻³)	1·41					1·35	1·35		1·31	1·33	1·34

β-phase) when the material is stressed, reverting to the original form when the stress is released. 5GT undergoes a similar transition, but the details of the behaviour are different. In order to obtain an oriented fibre in pure α-phase it must be annealed free of any restraint. Reversible transformation to a β-phase is then observed when stress is applied. When stress is present during annealing a mixture of both phases is formed which remains when the stress is released. Stressing the annealed fibre increases the proportion of β-phase present.

The three different phases of 6GT which have been found are different in type from those just described in that transformations between them cannot be produced solely by the application of stress and none have been produced in a pure form. The α-phase is favoured by crystallisation from the melt under conditions of high stress. Thus oriented fibres which have been melt spun and collected with a high thread-line tension will have a high proportion of this form, as will those which have been cold-drawn to a high draw-ratio. Annealing these fibres unrestrained at high temperatures will enhance the proportion of a second β-phase but with the spin-oriented material will not reduce the total amount of α-phase material present. Materials crystallised from the melt under low tension will have a high proportion of

of aromatic polyesters.

13	14	38	3	13	37	6	37	6	9	9	37	9
			4GT β-phase			5GT α-phase		5GT β-phase	6GT α-phase	6GT β-phase		6GT γ-phase
4·87	4·87	4·82	4·69	4·73	4·73	4·96	4·68	4·96	9·06	4·75	4·91	5·3
5·99	5·76	5·93	5·80	5·83	5·75	5·79	5·77	5·82	17·24	5·73	5·68	13·9
11·67	11·71	11·74	13·00	12·90	13·11	24·66	24·42	28·16	15·51	15·68	15·44	15·5
99·8	100·1	100·0	101·9	101·9	104·2	111·5	113·5	125·8	127·3	104·4	103·4	123·6
116·2	116·6	115·5	120·5	119·4	120·8	94·0	87·2	73·6	90·0	116·0	117·8	129·6
110·9	110·3	111·0	105·0	105·1	100·9	105·4	105·9	119·7	90·0	107·8	108·8	88·0
1	1	1	1	1	1	1	1	1	6	1	1	2
1	1	1	1	1	1	2	2	2	1	1	1	1
262·6	261·9	260·7	268·9	274·6	274·7	590·2	580·6	572·7	1928	326·4	320·9	629
1·391	1·394	1·401	1·358	1·330	1·329	1·316	1·338	1·356	1·284	1·262	1·284	1·308
					1·283				1·242	1·251		1·256

β-phase and the amount of α-phase present will be reduced by unrestrained annealing. The third, γ-phase, is favoured by crystallisation from solution.

3.2 Unit Cells

The published unit cells of these materials are given in Table 2. As far as I am aware all published cells have been included with the exception of those which literature reports have already shown to be wrong. Best estimates of the cell of each material are given in Table 3, and the reasoning used in arriving at these estimates is given in the following paragraphs.

The Daubeny, Bunn and Brown (DBB) cell[15] for 2GT has been criticised because the calculated value of the d-spacing of ($\bar{1}05$) reflection does not agree with that observed. Heffelfinger and Schmidt[16] (using Cu Kα radiation) report that this corresponds with a diffraction angle 42·8°, and claim that the DBB cell corresponds to 46°. Their calculation must be in error; it corresponds to 42·7° and, for all cells given, this angle varies between 42·7° and 43·1°, so it is not possible to discriminate between them on that basis.

In Table 4 the calculated d-spacings are listed for the reflection

Table 3. Best estimates of cell parameters.

	2GT	3GT	α-4GT	β-4GT	α-5GT	β-5GT	α-6GT	β-6GT	γ-6GT
a(Å)	4.48±0.04	4.60±0.03	4.86±0.03	4.72±0.02	(4.69)	4.69	4.96	9.06	4.83±0.08	5.3
b(Å)	5.89±0.04	6.22±0.01	5.96±0.03	5.79±0.03	(6.41)	5.79	5.82	17.24	5.71±0.03	13.9
c(Å)	10.71±0.04	18.36±0.25	11.65±0.06	13.00±0.10	(24.66)	24.66	28.16	15.51	15.56±0.12	15.5
α(degrees)	99.8±0.3	97.8±0.6	99.7±0.6	102.7±1.0	(112.5)	111.5	125.8	127.3	103.9±0.5	123.6
β(degrees)	117.6±0.7	90.8±1.9	116.0±0.7	120.2±0.7	(86.0)	94.0	73.6	90.0	116.9±0.9	129.6
γ(degrees)	111.5±0.3	111.3±0.4	110.8±0.5	103.7±2.1	(119.5)	105.4	119.7	90.0	108.3±0.5	88.0
Volume (Å^3)	212.3±3.5	483.8±7.6	261.5±4.0	272.9±7.4		590.2	572.7	1928	324.2±7.9	629
Unit cell density (gcm^{-3})	1.501	1.414	1.397	1.338	1.316	1.356	1.281	1.270	1.309
Measured density (gcm^{-3})	1.41	1.35	1.34		1.28		1.24	1.25	1.26
No. of chains	1	1	1	1	1	1	6	1	2
No. of monomers	1	2	1	1	2	2	1	1	1
a'(Å)	3.97	4.60	4.37	4.07	(4.68)	4.68	4.76	9.06	4.31	4.08
b'(Å)	5.80	6.16	5.87	5.65	(5.96)	5.39	4.72	13.71	5.54	11.55
γ'(degrees)	120.7	111.6	119.0	114.3	(120.4)	108.3	115.1	90.0	119.2	119.6
t_a(Å)	2.1	0.1	2.1	2.4	(−0.3)	0.3	−1.4	0	2.2	3.4
t_b(Å)	1.0	0.8	1.0	1.3	(2.5)	2.1	3.4	10.4	1.4	7.7

STRUCTURES OF AROMATIC POLYESTERS

Table 4. Calculated and observed d-spacings for poly(ethylene terephthalate).

		Reference									
17	39	38		15		36		17		39	
d_o(Å)	d_o(Å)	hkl	d_c(Å)	hkl	d_c(Å)	hkl	d_c(Å)	hkl	d_c(Å)	hkl	d_c(Å)
4.99	5.05	0 1 0	4.99	0 1 0	5.06	0 1 0	5.03	0 1 0	4.97	0 1 0	5.03
3.87	3.93	1 $\bar{1}$ 0	3.88	1 $\bar{1}$ 0	3.94	1 $\bar{1}$ 0	3.92	1 $\bar{1}$ 0	3.86	1 $\bar{1}$ 0	3.92
3.41	3.45	1 0 0	3.39	1 0 0	3.47	1 0 0	3.42	1 0 0	3.40	1 0 0	3.45
1.94	1.95	$\bar{2}$ 2 0	1.94	$\bar{2}$ 2 0	1.97	$\bar{2}$ 2 0	1.96	$\bar{2}$ 2 0	1.93	$\bar{2}$ 2 0	1.96
		2 2 0		2 2 0	1.98	2 1 0	1.95	2 1 0	1.93	1 3 0	1.94
1.67	1.72	2 0 0	1.70	2 0 0	1.73	2 0 0	1.71	2 0 0	1.70	2 3 0	1.71
5.37	5.42	0 $\bar{1}$ 1	5.40	0 $\bar{1}$ 1	5.40	0 $\bar{1}$ 1	5.51	0 $\bar{1}$ 1	5.37	0 $\bar{1}$ 1	5.42
4.09	4.14	$\bar{1}$ 1 1	4.08	$\bar{1}$ 1 1	4.17	$\bar{1}$ 1 1	4.12	$\bar{1}$ 1 1	4.08	$\bar{1}$ 1 1	4.12
						$\bar{1}$ 0 1	4.11			$\bar{1}$ 0 1	4.11
3.72	3.72	0 1 1	3.71	0 1 1	3.78	0 1 1	3.69	0 1 1	3.70	0 1 1	3.73
3.14	3.19	1 $\bar{1}$ $\bar{1}$	3.18	1 $\bar{1}$ $\bar{1}$	3.20	1 $\bar{1}$ $\bar{1}$	3.18	1 $\bar{1}$ $\bar{1}$	3.14	1 $\bar{1}$ $\bar{1}$	3.20
2.71		0 $\bar{2}$ 1	2.72	0 2 1	2.74	1 $\bar{2}$ 1	2.70	0 $\bar{2}$ 1	2.70	0 $\bar{2}$ 1	2.73
	2.72	1 0 1		1 0 1	2.73	1 2 1					
				1 2 1	2.72						
2.68		$\bar{1}$ 2 1	2.68	$\bar{1}$ $\bar{1}$ 1	2.70	$\bar{1}$ 1 1	2.68	1 0 1	2.67	$\bar{1}$ 1 1	2.69
		1 2 1				$\bar{1}$ 0 1		1 2 1		1 2 1	
3.51	3.55	$\bar{1}$ 1 2	3.52	$\bar{1}$ 1 2	3.60	$\bar{1}$ 1 2	3.54	$\bar{1}$ 1 2	3.54	$\bar{1}$ 1 2	3.55
3.35	3.38	$\bar{1}$ 0 3	3.31	$\bar{1}$ 0 3	3.37	$\bar{1}$ 0 3	3.37	$\bar{1}$ 0 3	3.36	$\bar{1}$ 0 3	3.35
3.09	3.12	0 $\bar{1}$ 3	3.12	0 $\bar{1}$ 3	3.11	0 $\bar{1}$ 3	3.14	0 $\bar{1}$ 3	3.10	0 $\bar{1}$ 3	3.11
2.85	2.88	0 0 3	2.88	0 0 3	2.91	0 0 3	2.85	0 0 3	2.87	0 0 3	2.89
				1 1 3	2.85						
2.78	2.80	$\bar{1}$ 1 3	2.79	$\bar{1}$ $\bar{1}$ 3	2.76	$\bar{1}$ $\bar{1}$ 3	2.80	$\bar{1}$ 1 3	2.81	$\bar{1}$ 1 3	2.76
						1 1 3	2.79	$\bar{1}$ $\bar{1}$ 3	2.76		

planes for which Fakirov, Fischer and Schmidt (FFS) give observed d-spacings.[17] Observed values are also given from Hall's measurements on a diffraction photograph of 2GT.[39] It will be seen that these are systematically larger than FFS's values by about 1%, which could be because the different samples have genuinely different unit cells. However, it is more likely that small errors in measurements of film–specimen distance and differences in the choice of the point on the reflection used for measurement accounts for this. Almost all the calculated d-spacings of the DBB cell lie outside the range of the two sets of experimental values. Although most of those calculated from Tomashpol'skii's cell[36] are in good agreement, two lie badly outside the experimental range. These two cells can therefore be rejected. Of the remainder, all calculated d-spacings are in good agreement, but it should be noted that the 3rd layer line reflection indexed ($\bar{1}13$) for the cell of Bornschlegl and Bonart[38] is indexed ($\bar{1}\bar{1}3$) for the other cells. Because of the systematic misorientation of the chain axes a ($\bar{1}\bar{1}3$) reflection would be displaced downwards from the mean layer line position, which agrees with observation, whereas ($\bar{1}13$) would be displaced upwards. Two equatorial reflections in Hall's cell are indexed differently from the others, but it is not possible to test these by displacement, and it would be necessary to test them in a structure determination. The three cells giving satisfactory agreement between observed and calculated d-spacings are considered equally valid on arriving at the best estimate.

The three cells of 3GT are in good agreement; however, detailed comparison of the calculated d-spacings with published measured values[4,5] reveals more bad disagreements for the cells from Reference 4 and 37 than from 5. The mean parameters of all three are given in Table 3, and since these are fairly close to those of Reference 5, if there are errors in 4 and 37, they do not affect the mean values significantly.

A detailed comparison of the cells of α-4GT has already been published[13] and it was concluded that for all of them the calculated locations of the diffraction spots did not differ significantly from their observed positions and so all were equally valid. Since then one additional cell determination has been published,[38] but since its parameters are very close to the others, it does not affect the conclusion already drawn. The possible error in volume given in Table 3 is larger than that originally quoted, which did not take proper account of the uncertainty in the contributing dimensions.

The unit cell of β-4GT determined by Yokouchi et al.[2] shows bad disagreement with the values of α and γ from the other two determinations. Thus agreement with published observed d-spacings is also worse, but the reflections are rather blurred in this diffraction pattern and all three must be considered equally valid.

The parameters for α-5GT in Reference 37 (axes have been transformed to bring values into agreement with those of Reference 6) yield some bad disagreements with the only published set of observed d-spacings. It would be desirable to compare it with the set used in its determination, and until these become available only the cell from Reference 6 can be regarded as satisfactory. As already stated (Section 2.1) even these are less certain than those of other materials.

The two cells of β-6GT are in reasonably close agreement, and since both account satisfactorily for the observed reflections, must be considered equally valid.

For each of the remaining materials, only one cell determination has been published and these are therefore the best estimates currently available. Error estimates are not possible in these circumstances.

It is apparent from Table 3 that typical uncertainties are about 0·05 Å in cell lengths, 0·7° in angle, and 2% in cell volume. It does not seem to be generally appreciated that the uncertainty in the latter value can lead to very large errors in estimates of crystallinity by density. The differences between the calculated and measured density, and between measured and amorphous density enter into the calculation of this quantity, and a 2% error in the calculated density can cause an error as large as 20% in the crystallinity. (This assumes that the amorphous and observed densities are known exactly. The answer is also sensitive to small errors in these.)

Also given in Table 3 are the values (a', b', γ') of the lattice projection onto the plane normal to the chain axis. These show the lateral chain packing. For the purposes of the ensuing discussion, 2GT, β-4GT, and β-6GT, which have an even number of methylene units and a crystallographic repeat (c) close to the length of the most extended conformation possible, will be called Group 1. 3GT and α-5GT, which have an odd number of methylene units and a contracted conformation, will be called Group 2. All members of Group 1 have similar values of a', b', and γ'. Although α and γ-6GT might be expected to belong to this group, their projected lattices appear to be different. However, if with γ-6GT b' is halved, the values become close to those of 2GT. Since there are two chains in the cell of γ-6GT, this suggests that their lateral arrangement is similar to those of 2GT. In Fig. 6 it is shown that the projected lattice of α-6GT may be transformed to one similar to that of the β- and γ-forms, and so the 6 chains in the unit cell are also likely to have a similar lateral arrangement.

For the members of Group 2, the values of a', b', and γ' appear to differ, but the cell axes of α-5GT may be transformed to give the values in parentheses in Table 3. These are now close to those of 3GT and of Group 1. In fact it is only β-5GT which is exceptional, the value of b' being smaller than for the other materials. In general, both a' and b' tend to be about 0·5 Å larger for those materials having a contracted conformation, but the fluctuation in γ' cannot be correlated with either chain extention or the number of methylenes.

Two other parameters, t_a and t_b are listed in Table 3. These are the relative translation in the c-direction of neighbouring chains along the a-axis and b-axis respectively. For all members of Group 1 their values are about 2 and 1 Å respectively. Both members of Group 2

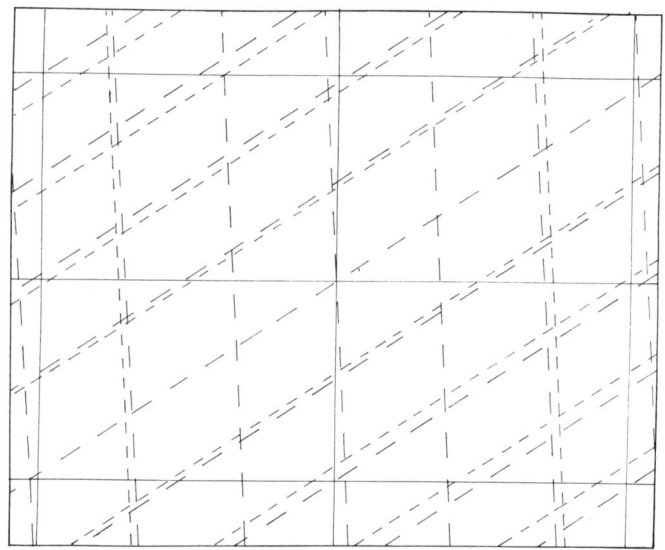

Fig. 6. Projection of 6GT lattices onto plane normal to c-axis. α-phase ——, β-phase ———, γ-phase -----. (Reproduced from Hall, I. H. and Ibrahim. B. A. (1982). *Polymer,* **23,** 805–16, by permission of the publishers, IPC Business Press Ltd.)

STRUCTURES OF AROMATIC POLYESTERS 65

have t_a near zero, but differing values of t_b. α-4GT has similar values to Group 1. For α-6GT, $t_a = 0$, corresponding with one of the observed values for Group 2. The average value of t_b per chain for α- and γ-6GT is about 3·5 Å, about the same as t_a for γ-6GT.

Thus to summarise, the three materials comprising Group 1 all have very similar packing. So does α-4GT even though its conformation is contracted. This suggests that this packing represents the most stable arrangement of the terephthaloyl parts of the molecules. The other phases of 6GT are still extended conformations, and have a similar lateral arrangement of chains, but the relative translation of neighbours is different. The lateral arrangement for the members of Group 2 is the same as for Group 1 but the relative c-translations of neighbours are different, and for β-5GT, the only phase having an odd number of methylenes in an extended conformation, both lateral arrangement and relative translation are different from those observed in other materials and phases.

It therefore appears that the stable packing of the terephthaloyl parts is disrupted, either by increasing the length of the methylene sequence or by an odd number of units in that sequence. We shall return to the question of packing of these units in Section 3.6.

3.3 Symmetry of Chain Conformation

In all the structure determinations of 2GT, 4GT and 6GT which have been published it has been assumed that centres of symmetry exist at the centres of the benzene rings and at the middle of the central methylene bond (e.g. C_5—C_5' in 2GT, Fig. 5b), and these assumptions have been confirmed by good agreement that has been obtained between observed and calculated structure factors.

With 3GT and 5GT the situation is less clear. The stereochemistry allows the centre of symmetry at the benzene ring to be retained, but it is no longer possible for there to be one anywhere else. From the unit cell dimensions, there are two monomers per crystallographic repeat, and it is unlikely that there will be no symmetry relationship between these. In the structural determinations performed by Hall and his co-workers[6,13] two possible relationships were tested, a 2/1 screw axis and a centre of symmetry on the benzene ring. The former was clearly rejected by both the packing and the structure factor requirements, but satisfactory models were obtained with the latter.

In their work with 3GT, Poulin-Dandurand et al.[4] further assume a two-fold rotation axis through the central carbon atom of the methylene sequence; this axis being in the plane of the two C–C bonds at this atom and bisecting the angle between them. The conformation angles of equivalent bond sequences either side of the symmetry point are then equal in both magnitude and sign. (In the centrosymmetric case they are of equal magnitude but opposite sign.) With this assumption, satisfactory agreement between observed and calculated structure factors was obtained. In both 3GT and 5GT it was assumed that the two parts of the chain were unrelated, and so it is worthwhile investigating the effect upon the R-factor of imposing this symmetry constraint. This has been done with 3GT and with both α- and β-5GT, and at the same time, the opportunity was taken to revise the bond parameters so that they were close to the means of Columns A to F of Table 1.

The results are shown in Table 5. In each case, the published structure was used as a starting model, the parameters were changed to those given in Table 5 and a refinement was performed without the symmetry constraint. The constraint was then applied and the refinement repeated to obtain the conformation angle and R-factors given in Column (b) of Table 5. Finally the constraint was removed and the refinement repeated to give the values in Column (a). In all cases unobserved reflections were included as recommended in Section 2.5, but the R-factors were calculated using only observed reflections.

Hamilton's test[40] was applied to the results and the comments in Table 5 indicate the conclusions. Quite clearly, whereas there is no significant improvement in R-factor as a result of relaxing the symmetry requirement for 3GT the opposite conclusion is drawn with α-5GT. Thus, for these two materials, the conformation angles will be taken to be those in Columns (b) and (a) respectively. With β-5GT the situation is less clear. The difference in R-factor would be significant in 97·5% of all cases. However, since with both phases of 5GT there was considerable disorder and there were few independent equatorial reflections, both intensity measurement and unit-cell determination were less accurate than normal. We shall therefore assume that the asymmetric structure of the β-phase has not been shown to give significantly better agreement between observed and calculated structure factors than the symmetric, and use the conformation angles given in Column (b).

Interatomic separations were also checked to ensure that there were no atoms unacceptably close. With 3GT, this was true of both the

STRUCTURES OF AROMATIC POLYESTERS 67

Table 5. **Investigation of symmetry within monomers of 3GT and 5GT.**

Bond lengths (Å)		Bond angles (degrees)		Conformation angles (degrees)					
				3GT		α-5GT		β-5GT	
				(a)	(b)	(a)	(b)	(a)	(b)
C_0—C_1	1·379	α_1	120·5	τ_1	−173·9	176·7	170·1	171·4	164·0	169·8
C_1—C_3	1·381	α_2	119·7	τ_1'	170·8		−175·0		−171·4	
C_1—C_3	1·387	α_3	119·8	τ_2	−179·1	177·7	170·2	164·0	155·9	178·3
C_2—C_3	1·487	α_4	118·3	τ_2'	−177·2		−149·8		−153·0	
C_4—C_3	1·334	α_5	121·9	τ_3	173·2	−169·1	156·4	169·2	126·6	159·9
O_1—C_4	1·204	α_6	124·2	τ_3'	−156·3		−131·8		166·0	
O_2—C_4	1·450	α_7	112·6	τ_4	−62·0	−60·0	−103·9	−89·4	171·1	176·9
C_5—O_1	1·492	α_8	123·2	τ_4'	−65·7		−122·6		163·8	
C_6—C_5	1·510	α_9	116·1	τ_5			−168·1	−108·0	−177·2	−91·7
		α_{10}	107·1	τ_5'			−105·7		148·3	
		α_{11}	114·1	R%	25·9	27·1	20·0	41·5	18·8	23·8
		α_{12}	114·1		Improvement will occur by chance in more than 50% of cases		Improvement will occur by chance in fewer than 0·5% of cases		Improvement will occur by chance in 2·5% of cases or fewer	

symmetric and asymmetric conformation. One bad intermolecular contact in the symmetric model was relieved in the asymmetric for α-5GT. With β-5GT the intermolecular contacts in the symmetric conformation were much to be preferred. Thus the choices made above on the grounds of structure factor are confirmed by consideration of contacts.

In all cases the symmetric conformation might be intuitively expected, and so the surprising result is that for α-5GT. The difference in R-factors is quite decisive and that for the asymmetric model is good by polymer standards. However, as has already been stated, the unit cell of this material was particularly difficult to determine, and because of the diffuse reflections the intensities are likely to be measured less accurately than is usual. For these reasons some doubt must remain concerning the result.

In the remainder of this review the structural models preferred from Table 5 will replace those published by the author and his co-workers.

3.4 Selection of Values of Parameters Describing Chain Conformation

Three independent structure determinations have been published for the α-phase of 4GT[1,2,3] and from these Hall[33] has assessed the contribution that has been made to the uncertainty in the structural parameters by differences in the values of the quantities used. There were some questionable choices of bond parameters and one structure factor set gave R-factors which were consistently significantly lower than those obtained using other sets, otherwise there was no justification for regarding the values used in one determination preferable to those used in any other. In this manner an average set of conformation angles and their uncertainties was determined. In the discussion of the structure of this material these will be used in preference to other published values.

This is the only case in which different published structures of the same material can be considered to be of equal value. Daubeny et al.[15] and Tomashpol'skii and Markova[36] have published the structure of 2GT, but the latter determination was made using electron diffraction and the stereochemistry of parts of the final model is very dubious so it will not be considered here. Poulin-Dandurand et al.[4] and Desborough et al.[5] have both published a structure for 3GT which are in close agreement, but the former group did not refine structural parameters, so their work will not be included in this discussion.

For the other materials in this series the only full structural determinations published in readily available reference sources have been by the author and co-workers. Other determinations have been made[37] but, with the details available, possible disagreements cannot be assessed and so these will not be considered further. This is particularly unfortunate for 5GT where the uncertainties already mentioned would justify a further study of the structure.

The values of the structural parameters of β-6GT are from a preliminary determination. Since it was not possible to obtain pure phases of this material, reflections of the α- and γ-phases were always present overlapping some of those of the β-phase, and so only those structure factors for which it was certain that the β-phase was the dominant contributor were included in the refinement. Less certain and unobserved ones, were omitted. For this reason the structure is likely to be less accurate than others in this series.

3.5 Chain Conformations of the Various Materials

If τ_1 (defined by the atoms $C_1C_3C_4O_1$ in Fig. 5a) is 180°, the carbonyl unit will be coplanar with the benzene ring. It is seen from Table 6 that for all the materials this angle differs from 180° by less than about 10°. However, the difference is likely to be significant since investigation of the accuracy with which conformation angles are determined[33] has indicated that the probable error in this one is less than 4°. τ_2 is always close to 180°, and for the most precisely determined structures is within 2° of this value. Since an error of up to 4° is again possible, these deviations might not be significant. Thus the atoms $C_3C_4O_2O_1C_5$ are almost exactly coplanar, a conformation which also occurs in oligomethylene glycol derivatives,[24–26,28,29] though again the small deviations are probably significant. It therefore appears that in both polymers and oligomers these angles are always close to, but not necessarily equal to, 180°, τ_1 generally being within 10° and τ_2 within 5° of this value. The magnitudes of the deviations do not appear to be related to other structural features.

With this conformation short separations will occur between non-bonded atoms, as illustrated in Fig. 7. These only vary slightly between compounds and indicate strong affinities between O_2 and C_1, O_1 and C_2, and O_2 and C_5.

All values of τ_3 are close to ±160°, except for α-4GT for which it is −88°; in the low molecular weight compounds they were all close to 180°. Sundararajan et al.[41] have calculated conformations of minimum energy for ethylene glycol diacetate, obtaining values of 180°, ±92°

Table 6. Structural parameters of mGT polymers.

	2GT	3GT		α-4GT	β-4GT	α-5GT		β-5GT		β-6GT	Terephthalic acid
τ_1 (deg)	−169.3	176.7		177.2	172.3	170.1	−175.0	169.8		167.6	
τ_2 (deg)	178.2	177.7		−177.5	−177.7	170.2	−149.8	178.3		166.1	
τ_3 (deg)	−158.7	−169.1		−88.0	−155.2	156.4	−131.8	159.9		176.0	
τ_4 (deg)		−60.9		−75.0	154.7	−103.9	−122.6	176.9		174.0	
τ_5 (deg)						−168.1	−105.7	−91.7		168.8	
						(−13.0)	(−160.6)				
θ (deg)	13.3	−178.4	−10.2	3.6	13.9	167.0	19.4	168.8	57.3	141.5	160.3
ϕ (deg)	24.1	43.8		25.8	26.6	50.8	42.6	40.2		28.1	0.2
ψ (deg)	82.2	52.7		73.3	85.7	46.0	65.1	66.3		87.5	89.3
q (Å)	3.55	3.70	3.56	3.56	3.77	4.00	3.40	3.71	2.92	3.46	3.11
r	0.45	0.46	0.72	0.58	0.26	0.90	0.26	0.60	0.98	0.16	0.73
$\alpha - \phi$ (deg)	75.7	54.0		74.1	76.1	69.9	61.7	85.6		76.3	126.0
μ (deg)	−4	−66		42	71	120		117		77	
η (deg)	−1	−3		−2	1	−4		−1		−1	
σ (deg)	45	87		44	41	99		114		48	
ν (deg)	36	9		34	38	24		36		37	
$\mu - \theta$ (deg)	−17	−56		38	57	133	−79		60	−65	
k/h	−0.82	−3.6		0.41	3.2	−2.5		−2.2		4.4	
$\mu - \sigma$ (deg)	−49	−153		−2	30	21		3		29	

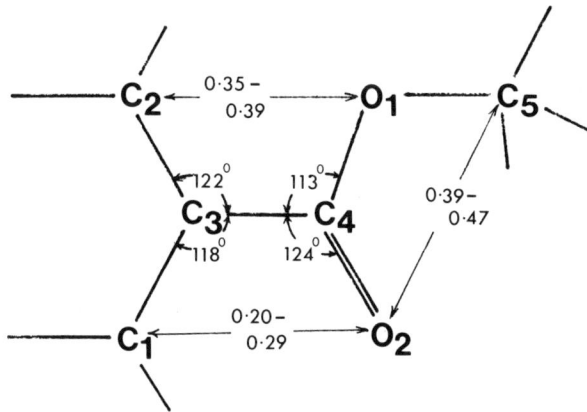

Fig. 7. Short atomic separations around carbonyl group. The numbers are the differences between the sum of the Van der Waals' radii and the observed separations of the pairs of atoms.

and $\pm 78°$ for the angle equivalent to τ_3. Thus the values around $\pm 160°$ for the polymer appear to differ significantly both from those for the oligomer and from Sundararajan's predictions. The value of $-88°$ for α-4GT lies close to a predicted value, and Turner-Jones and Bunn[42] obtained $-114°$ for the equivalent angle in poly(ethylene adipate).

τ_3 will affect the separation of the atoms H_{51}, H_{52} and C_6 from those of the carbonyl and aromatic units. For all the polymers, the separation between H_{51} and C_4 is about 0.21 Å shorter than the sum of the Van der Waals' radii of these atoms. For α-4GT, the different value of τ_3 causes an additional short separation between C_4 and C_6 (0.26 Å short).

With the exception of 5GT, τ_4 and τ_5 are close to either $-60°$ or $180°$, the gauche- and trans-values, and so the conformations of these materials comprise a section between C_0 and C_5 which is essentially planar, and a methylene sequence having a combination of gauche- and trans-conformations. The two parts are joined by the bond O_1C_5, the rotation about this being generally about $20°$ out of plane, though for α-4GT it is about $90°$. The conformation is stabilised by strong affinities between C_1 and O_2, C_5 and O_2 and C_4 and H_{51}, the separations of these pairs of atoms being appreciably shorter than the sums of their Van der Waals' radii.

3.6 Orientation of Molecules Within the Unit Cell

The orientation of the molecule within the unit cell is specified by the three angles θ, ϕ, and ψ. If the molecule is placed in the cell so that its chain axis coincides with the c-axis and the normal to the benzene ring inclined towards the positive c-direction lies in the (010) plane in the positive a-direction, θ is the rotation required to bring the molecule to its correct orientation and it is positive if this rotation is counterclockwise looking in the negative c-direction. This angle, together with the conformation angles, is sufficient to define the structure unambiguously. However ϕ and ψ are given to help visualise the molecular arrangement; ϕ is the angle between C_3C_4 and the c-axis, and ψ is the angle the normal to the benzene ring makes with the c-axis. All three are given in Table 6 (the values in parentheses for α-5GT apply to the transformed cell—see Table 3—and will be used in discussion since they conform better to the overall pattern of behaviour).

Those materials with two monomers per crystallographic repeat (3GT and 5GT) require a different value of θ for each ring. For 3 and α-5GT these are separated by about 180°, but for β-5GT the separation is closer to 90°. It will be seen from Table 6 that, with the exception of β-5GT and 6GT, values of θ are close to zero or 180° and hence the normals to the benzene rings lie within a few degrees of the ac plane. (With 3GT a shift of the unit cell origin by $c/2$ along the c-axis would make the smaller value of θ the first value and so make it more consistent with the other members of the series. Such a shift would change the signs of the τ values.)

The value of θ for 6GT is surprising. Both the chain comformation, and the c-axis projection of the unit cell are similar to those for 2GT and β-4GT, and so a similar arrangement of the benzene rings within the unit cell might be expected. However, the inaccuracies in the structure factor of 6GT have already been mentioned and these quantities alone determine θ, whereas for such a highly extended conformation geometric factors will play a large role in the determination of the conformation angles. Hence θ might be expected to be less accurately determined than the other quantities.

The values of ϕ and ψ can be grouped according to the chain conformation. For an extended chain with even numbers of methylene units (2GT, β-4GT, and β-6GT) the normal to the plane of the benzene ring is almost normal to the c-axis and the ring is rotated in its own plane by about 25° from the orientation in which C_3C_4 would lie parallel to the c-axis. In α-4GT this angle is little altered, but that between the normal and the c-axis decreases by about 15°. For 3GT and 5GT (with odd numbers of methylene units) the plane of the benzene ring is even more sharply tilted and its normal is now about 55° to the c-axis. A further rotation in the plane of the ring, of about 20°, also occurs.

Since the benzene ring is the most bulky part of the structure, it might be expected to control the packing, in which case the perpendicular separation of the planes of adjacent rings might be expected to be the same for all compounds and equal to the benzene ring thickness. These separations (q) have been calculated, and are listed in Table 6. In most cases they do not differ significantly from 3·5 Å, the exceptions being β-4GT, β-5GT and one of the rings in the crystallographic repeat of 3GT and of α-5GT. It therefore appears that for mGT, m-odd, both rings cannot simultaneously meet this packing requirement, and also that the larger separation for β-4GT might contribute to the instability of this phase. Despite the very short separation between one pair of adjacent rings in β-5GT there are no bad contacts between them.

With each a-axis direction in the crystal lattice there is associated a set of ring planes. If the sets corresponding to adjacent a-axis directions separated by a b-axis are considered, for good packing the planes of each set might be expected to interleave. The ratio of the distance between neighbouring rings of adjacent sets to that between neighbouring rings of the same set (r) is also given in Table 6 and might be expected to be close to 0·5. The deviations are considerable but, with the exceptions of 6GT and one ring in each phase of 5GT, all values lie between 0·25 and 0·75. This again indicates difficulties in simultaneously packing both rings satisfactorily with m-odd. The result for 6GT is surprising, and might be due to poor accuracy in θ which has already been discussed. On the other hand, the values for 2GT, β-4GT, β-6GT (the extended conformations with m-even) lie on a definite sequence. The low value for β-4GT might be associated with the instability of this phase, and that for β-6GT with the polymorphism in the chain packing of this material.

If adjacent benzene rings along the b-axis were packed in the same manner for all materials then $\alpha - \phi$ should be constant. For m-even it is (see Table 6), and does not differ significantly from 76°. For these materials the hydrogen atom H'_1 (see Fig. 5) on one ring interleaves between H_1 and H'_2 on the adjacent one. For m-odd $\alpha - \phi$ is close to either 60° or 90° and in both cases the hydrogen atoms butt rather than interleave. This, together with the poor interleaving of planes, results in some poor separations between atoms of adjacent rings along the b-axis and along the diagonal of the ab plane. Thus again the suggestion emerges that good packing of the benzene rings cannot be achieved with m-odd.

The above discussion indicates that the chain packing is controlled either by the benzene rings or by the terephthaloyl residues, and it is instructive to compare the packings described above with those observed in terephthalic acid[43] and in benzene.[44,45]

Successive molecules of terephthalic acid are hydrogen bonded by the carbonyl groups and, with a suitable transformation of the published unit cell, form chains along the c-axis, the normals to the planes of the benzene rings lying close to the a-axis. With this unit cell, $a' = 3\cdot 56$ Å, $b' = 5\cdot 19$ Å and $\gamma' = 98\cdot 6°$, values which are similar to but slightly smaller than those for the polymers (see Table 3). The area of this projection of the unit cell is $18\cdot 3$ Å^2 compared with values of a little over 20 for the polymers. This indicates that the packing is similar, but closer.

θ, ϕ, ψ and the other packing parameters are given in Table 6. C_3C_4 lies almost exactly along the c-axis, which is within $1°$ of lying in the plane of the benzene ring. The plane is more severely inclined in the polymers, presumably because of the constraint imposed by the remainder of the chain, and this probably causes the looser packing which is evident from the relative magnitudes of q in Table 6. However, interleaving of the planes of adjacent rows is no better than in the polymer, and, from the value of $\alpha - \phi$ the hydrogen atoms on adjacent molecules will butt rather than interleave.

Two polymorphs of terephthalic acid were found, but the structural details given for the second form are insufficiently precise for a detailed comparison. However, the similarities noted above are sufficient to indicate that the packing of the terephthaloyl residues exerts a very strong influence on the crystal structure and so the polymorphism of the acid and the polymer might be related. Certainly, if this polymorphism is to be explained by energy calculations, that which is associated with molecular packing cannot be neglected.

With benzene a packing arrangement is observed which is completely different from that described above. Two forms have been reported. At -3 °C crystallisation is in an orthorhombic cell containing four molecules in space group *Pbca*. Each ring is surrounded by neighbours whose planes are inclined to it. Under high pressure it crystallises in a monoclinic cell containing two molecules in space group $P2_1/c$. In this case the packing in the ac plane ($\beta \neq 90°$) does consist of neighbouring rings with parallel planes, but the second molecule in the cell lies in the bc plane with its plane inclined to the other. Thus molecular packing in the polymer is controlled by the entire terephthaloyl residue, not just the benzene ring.

3.7 Tilting of Molecular Chain Axes from the Fibre Axis

As stated in Section 2.1, for all members of this series crystallisation is such that the axis of the molecular chain is inclined by a few degrees to that of the fibre by rotation about an axis lying in a particular direction in the a^*b^* plane. As illustrated in Fig. 3 this causes some reflections to be displaced above and others below their mean layer line positions. Specification of this tilt requires two parameters which are given in Table 6. The tilt azimuth (μ) is the angle measured from the tilt axis to the line of intersection of the a^*b^* and ac planes and it is positive if made in a clockwise direction looking from positive c (i.e. the reference line and sign convention are the same as for θ). The tilt angle (η) is the angle of rotation about this axis (from coincidence of fibre and chain axes) and it is positive if it is clockwise viewed from the a^* positive region of the lattice.

The values given in Table 6 have, in most cases, been taken from the references giving the cell parameters, and were obtained as described in Section 2.1. Only for β-6GT have the published values been obtained as a result of a refinement procedure minimising discrepancies between the observed and calculated displacements. This procedure has also been used to refine the values published for 2GT and β-5GT and it is these refined values which are given. The shifts obtained on refinement indicate probable error of about 10° and 1° in the unrefined values of tilt azimuth and angle respectively. The diffraction patterns of α- and γ-6GT showed no evidence of tilting.

In order to determine the cause of tilting its correlation with three structural parameters, also shown in Table 6, has been studied. If μ is compared with θ, it will be seen that there is no consistency in the difference between their values for the different materials, and so the orientation of the terephthaloyl unit plays no part in producing tilt. Additional confirmation of this conclusion is provided by 3GT and 5GT where there are two terephthaloyl units of different orientation within the unit cell. If it was caused by some growth habit of the crystal, the plane which remains vertical during tilting might be expected to be the same crystallographic plane for all materials. The ratio k/h for this plane is given in Table 6 and varies appreciably for the different members of the series indicating that this is unlikely to be the cause.

The line of steepest slope in the ab plane of the unit cell has also been determined and is given in Table 6. (This is denoted by σ, measured from the line of intersection of the a^*b^* and ac planes, with the same sign convention as used for μ and θ.) The inclination of the normal to the ab plane to the c-axis (ν) is also given. With the exception of 2GT and 3GT, $\mu - \sigma$ is seen to lie between 0° and 30° and so is the most consistent of the parameters considered (especially when probable errors in μ are considered). The reason for this behaviour is not clear; intuitively tilting might be expected to reduce the inclination of the ab face. If this was so $\mu - \sigma$ would be near $-90°$ with η negative. With $\mu - \sigma$ positive and η negative, the inclination will be increased. Also, the tilt angle is of a very different magnitude from the inclination of the ab plane (ν).

Thus there is not a consistent pattern of behaviour throughout the series and so the cause of tilting cannot be ascertained, although the most likely relationship appears to be with the shape of the unit cell.

4 SUMMARY

If structural parameters are to be determined from the X-ray diffraction patterns of crystalline polymers then great care is necessary, both in the preparation of a suitable test-piece and in the recording of the X-ray diffraction pattern. Even when these are optimised, the probable errors are likely to be about 0·05 Å in the lengths of cell edges and 0·7° in cell angles, leading to an error of about 2% in cell volume. If this is then used to determine the crystallinity by density an error of about 20% is probable. Because of the small number of unambiguous reflections it is essential to assume that the bond lengths, angles and certain parts of the conformation are the same as those observed in closely related low molecular weight compounds. Even so, because of the poor and uncertain accuracy with which the intensities of the diffraction spots may be measured, conformational parameters can contain significant and unquantifiable errors.

Intercomparison of a large number of independent structural studies on members of the mGT series of polymers shows that the preferred structure is that in which the terephthaloyl units are approximately planar and the conformation angles of the methylene sequence close to either *trans-* or *gauche-*. The conformation angle connecting these two parts is about 20° out of plane. However, the packing of the terephthaloyl units plays an important role and the structures of the various compounds appear to be a compromise between the need to adopt a conformation similar to that just described and the need to achieve satisfactory packing. The conflict between these two requirements probably leads to different structures of similar energy, and is thus a possible cause of the allomorphism which is observed.

When m is odd, there are two terephthaloyl units in the cell, and it seems that satisfactory packing of both of these and a suitable chain conformation cannot be simultaneously achieved. This leads to the low melting points and slow crystallisation of these compounds.

REFERENCES

1. Mencik, Z. (1975). *J. Pol. Sci. (Phys. Edn)*, **13**, 2173.
2. Yokouchi, M., Sakakibara, Y., Chatani, Y., Tadokoro, H., Tanaka, T. and Yoda, K. (1976). *Macromolecules*, **9**, 266.
3. Hall, I. H. and Pass, M. G. (1976). *Polymer*, **17**, 807.
4. Poulin-Dandurand, S., Perez, S., Revoi, J-F. and Brisse, F. (1979). *Polymer*, **20**, 419.
5. Desborough, I. J., Hall, I. H. and Neisser, J. Z. (1979). *Polymer*, **20**, 545.
6. Hall, I. H. and Rammo, N. N. (1978). *J. Pol. Sci. (Lett.)*, **16**, 2189.
7. Hall. I. H., Pass, M. G. and Rammo, N. N. (1978). *J. Pol. Sci. (Phys. Edn)*, **16**, 1409.
8. Hall, I. H. and Ibrahim, B. A. (1980). *J. Pol. Sci. (Phys. Edn)*, **18**, 183.
9. Ibrahim, B. A. (1980). Ph.D. Thesis, Victoria University of Manchester.
10. Brisse, F. and Marchessault, R. H. (1980). In *Fibre Diffraction Methods* (A. D. French and K. H. Gardner (Eds)) ACS Symposium Series No. 141, p. 267.
11. Bateman, J., Richards, R. E., Farrow, G. and Ward, I. M. (1960). *Polymer*, **1**, 63.
12. Goodman, I. (1962). *Angew. Chem.*, **74**, 606.
13. Desborough, I. J. and Hall, I. H. (1977). *Polymer*, **18**, 825.
14. Joly, A. M., Nemoz, G., Douillard, A. and Vallet, G. (1975). *Die Makromol. Chem.*, **176**, 479.
15. Daubeny, R. de P., Bunn, C. W. and Brown, C. J. (1954). *Proc. Roy. Soc.*, **A226**, 531.
16. Heffelfinger, C. J. and Schmidt, P. G. (1965). *J. Appl. Pol. Sci.*, **9**, 2661.
17. Fakirov, S., Fischer, E. W. and Schmidt, G. F. (1975). *Die Makromol. Chem.*, **176**, 2459.
18. Arnott, S. (1965). *Polymer*, **6**, 478.

19. Cella, A. J., Lee, B. and Hughes, R. E., (1970). *Acta Cryst.*, **A26,** 118.
20. Neisser, J. Z. (1980). Ph.D. Thesis, Victoria University of Manchester.
21. Atkins, E. D. T. and Meader, D. (1980). In *Fibre Diffraction Methods* (A. D. French and K. H. Gardner (Eds)) ACS Symposium Series No. 141, p. 113.
22. Hall, I. H. and Pass, M. G. (1975). *J. Appl. Cryst.*, **8,** 60.
23. Perez, S. and Brisse, F. (1975). *Acta Cryst.*, **B31,** 2746.
24. Perez, S. and Brisse, F. (1975). *Canad. J. Chem.*, **53,** 3551.
25. Perez, S. and Brisse, F. (1976). *Acta Cryst.*, **B32,** 470.
26. Perez, S. and Brisse, F. (1976). *Acta Cryst.*, **B32,** 1518.
27. Brisse, F. and Perez, S. (1976). *Acta Cryst.*, **B32,** 2110.
28. Perez, S. and Brisse, F. (1977). *Acta Cryst.*, **B33,** 1673.
29. Perez, S. and Brisse, F. (1977). *Acta Cryst.*, **B33,** 3259.
30. MacGillavry, C. H., Reick, G. D. and Lonsdale, K. (Eds) (1962). *International Tables for X-ray Crystallography*, Vol. III, Kynoch Press, Birmingham.
31. Williams, D. E. (1969). *Acta Cryst.*, **A25,** 464.
32. Zugenmeier, P. and Sarko, A. (1972). *Acta Cryst.*, **B28,** 3158.
33. Hall., I. H. (1980). In *Fibre Diffraction Methods* (A. D. French and K. H. Gardner (Eds)) ACS Symposium Series No. 141, p. 335.
34. Stambaugh, B., Koenig, J. L. and Lando, J. B. (1979). *J. Pol. Sci. (Phys. Edn)*, **17,** 1503.
35. Pilling, D. E., Cruickshank, D. W. J., Bujosa, A., Lovell., F. M. and Truter, M. R. (1961). *Computing Methods and the Phase Problem in X-ray Analysis*, Pergamon, Oxford.
36. Tomashpol'skii Ya.Ya. and Markova, G. S. (1964). *Pol. Sci. USSR*, **6,** 316.
37. Tadokoro, H., Private communication.
38. Bornschlegl, E. and Bonart, R. (1980). *Colloid and Pol. Sci.*, **258,** 319.
39. Hall, I. H., Unpublished work.
40. Hamilton, W. C. (1965). *Acta Cryst.*, **18,** 502.
41. Sundararajan, P. R., Labrie, P. and Marchessault, R. H. (1975). *Canad. J. Chem.*, **53,** 3557.
42. Turner-Jones, A. and Bunn, C. W. (1962). *Acta Cryst.*, **15,** 105.
43. Bailey, M. and Brown, C. J. (1967). *Acta Cryst.*, **B22,** 387.
44. Cox, E. G., Cruickshank, D. W. J. and Smith, J. A. S. (1958). *Proc. Roy. Soc.*, **A247,** 1.
45. Fourme, R., Andre, D. and Renauld, M. (1971). *Acta Cryst.*, **B27,** 1275.
46. Arnott, S. and Wonacott, A. J. (1966). *Polymer*, **7,** 157.
47. Campbell-Smith, P. J. and Arnott, S. (1978). *Acta Cryst.*, **B34,** 3.

Chapter 3

TRANSMISSION ELECTRON MICROSCOPY OF POLYMERS

EDWIN L. THOMAS

Polymer Science and Engineering Department, University of Massachusetts, Amherst, USA

1 INTRODUCTION

Electron microscopy investigations have provided much of the detailed information on polymer morphology. Whereas scattering methods are statistical averages over the bulk sample, electron microscopy permits the examination of features from the micron to the atomic scale. It is the aim of this chapter to outline the central techniques and the theoretical basis for interpreting the images derived from each technique, including scanning transmission electron microscopy (STEM) as well as conventional transmission electron microscopy (CTEM). The usefulness and limitations of the various techniques will be documented with illustrations taken from the polymer literature (influenced of course by the author's bias to his own work and interests!).

In general the practice of polymer microscopy (as compared to that of metals and ceramics) is hindered by the radiation sensitivity of the specimens and the difficulty of obtaining thin samples with sufficient contrast from bulk material. It is further hindered by some practitioners who display insufficient regard for the influence of radiation damage and electron optical conditions on their resultant images and on an overly subjective approach to image interpretation. This chapter begins with a short section on instrumentation, followed by an overview of radiation damage problems and finally the main section, on image interpretation. The theory of phase contrast imaging is extensively developed, but it appears not to be generally known to polymer microscopists and 3 examples of image misinterpretation provide ample justification for including the details. Lattice imaging and moiré pattern (indirect lattice imaging) formation are also illustrated. The discussion of amplitude contrast imaging contains a section on mass thickness contrast with emphasis on quantitative mass thickness determination and on the relatively new staining techniques which provide a means of investigating lamellar structures in bulk polymers. The section on diffraction contrast deals extensively with dark field microscopy of defects in single crystals and the imaging of crystallites in fibres.

2 MICROSCOPY TECHNIQUES

There exist many excellent references on electron microscope theory and practice. Their emphasis is largely towards metals and biological materials and towards amplitude contrast (diffraction contrast for metals and mass thickness contrast for stained biological sections). For phase contrast imaging, the book by Misell[1] is excellent, and for an introduction to STEM, the book edited by Hren et al.[2] is very useful.

To aid in our critical discussion of polymer microstructural studies it is first necessary to define briefly the standard transmission electron optical techniques. Figure 1 shows schematically the normal situation of an electron beam incident at 90° to a thin specimen. Bright field (BF) imaging utilises the electrons scattered in the forward direction, which are selected by the objective aperture, whereas for dark field (DF) imaging the objective aperture is set to collect a portion of the scattered electrons. DF imaging by simply displacing the objective aperture off the optic axis is very convenient, but for high resolution the resultant image shift and image aberration of the displaced aperture technique can be prevented by using tilted incident illumination, which is now standard on all modern CTEMs (see Fig. 1b). BF imaging is certainly the most often used mode of microscope operation, whereas DF provides a higher contrast image though of weaker intensity, which is strongly dependent on the sample–electron beam geometry and on the size and location of the objective aperture in the scattering pattern.

In addition to imaging capabilities, electron diffraction (see Section 3.8) is readily accomplished and indeed for crystalline materials studied by diffraction contrast, correct image interpretation requires a properly oriented diffraction pattern to specify the diffraction condi-

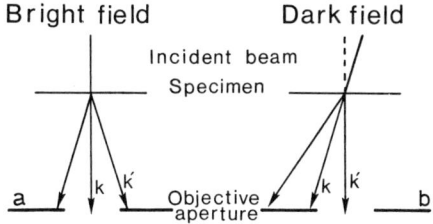

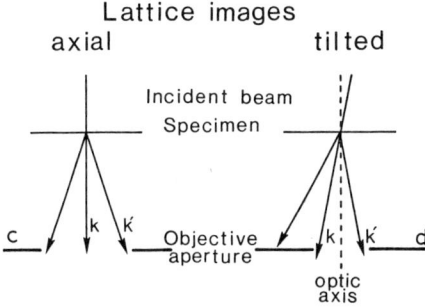

Fig. 1. Optical configuration in the electron microscope for different imaging conditions.

tions. Crystal structure studies are also possible using electron diffraction; this is particularly valuable for polymer crystals which cannot be easily grown to a size sufficient for standard X-ray methods.

2.1 CTEM and STEM Imaging

It is presumed that the operational principles of CTEM are known to the reader as descriptions are readily available in a large number of texts. However, STEM is a relatively new technique with few applications as yet to polymer systems and this section presents a brief introduction to this technique. Because of the higher collection efficiency of scattered electrons[3] and the precise control of beam position afforded by STEM, applications of STEM to polymer morphology will certainly increase. Specific comparison between STEM and CTEM depends on the type of incident illumination, mode of image formation, type of image contrast and image resolution desired. Because the

STEM image is collected point by point, the various kinds of transmitted electron signals can be processed in many possible ways, permitting, for example, selected energy loss images for chemical mapping, elastic to inelastic scattering ratio images for atomic number contrast, etc.[2,6] Our approach here will be restricted to applications of STEM DF imaging of radiation sensitive crystalline polymers.

Figure 2 shows schematically the basic components of a STEM. The central feature is a highly focused beam of small diameter (10 to 60 Å for a CTEM with a scanning attachment and as low as 2 Å for a dedicated STEM) which is sequentially scanned over the specimen as in a SEM. In order to form the image, the transmitted electrons are detected by a photomultiplier tube (PMT) and the amplified signal is scanned across a cathode ray tube (CRT) which is synchronised with the scan coil of the incident illumination. Image contrast arises from variations of the transmitted intensity. The magnification is set by the ratio of the area of the scan on the CRT to the area of the scan on the specimen. The BF image is recorded by an on-axis disc detector and a second annular detector outside the BF detector may be used with a second viewing CRT to provide simultaneous BF and DF images. If all but one of the diffracted beams are restricted by the selected area diffraction (SAD) aperture from reaching the PMT detector, the conventional single beam DF image is recorded. Various types of multiple beam DF images are possible with STEM. Indeed, with an array of

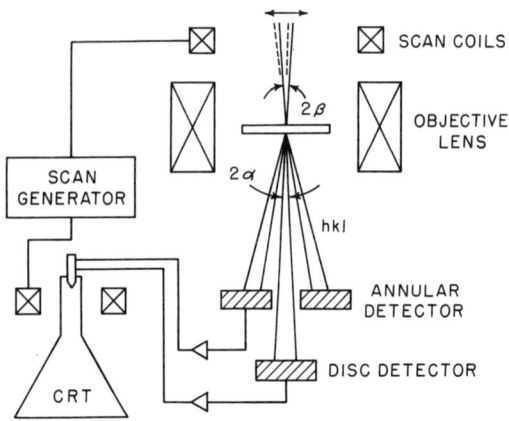

Fig. 2. Schematic illustration of the basic components of a STEM.

detectors, each coupled to its own amplifier, optimum use may be made of all the simultaneous DF images. Because of electronic image intensification, control of illumination location and magnification independent focus, STEM is superior to CTEM for the systematic observation of diffraction contrast features and diffraction patterns from radiation sensitive crystalline polymers.

2.2 Special Concerns for Polymer Specimens

The chief limitation to the use of electron microscopy for the study of crystalline polymers is the radiation damage produced by the electron beam. Therefore, in any electron microscopy study it is vital to take steps to both *recognise* the effects of radiation damage and to *minimise* radiation damage to the specimen. The excellent review articles on radiation damage by Grubb†[4] and Isaacson[6] should be required reading before beginning any polymer microscopy investigation. Many papers have appeared detailing the specific effects of dose, dose rate, temperature, voltage, vacuum and specimen environment on the resulting electron-beam–specimen interactions. The main issue for the morphologist to consider is how to obtain the maximum amount of microstructural information from the specimen. The electron dose (ϕ_{max}) that can be used to extract information before the sample is severely damaged (the chief contribution to the image becoming essentially artefact diffraction contrast) depends on the radiation physics and chemistry taking place in the sample.[7] The maximum level of damage which can be tolerated depends on what type of information is desired from the specimen. For example, diffraction contrast images and diffraction patterns depend on the crystallinity of the specimen. Because the long range crystalline order of the sample is destroyed with increasing electron dose, only a limited number of scattered electrons can be used to obtain crystallographic information. The crystalline lifetime is normally taken as the electron dose required to change the diffraction pattern from crystalline reflections into broad amorphous haloes. Use of this criterion to determine the maximum specimen dose ϕ_{max} would however allow a large number of electrons scattered from a partially damaged crystal to contribute to the image. It is therefore best to assess the effect of beam dose directly on the image by comparing successive low magnification (hence relatively low dose) DF

---

† Grubb has also recently written a review chapter on the electron microscopy of crystalline polymers.[5]

(or BF) micrographs taken at different fractions of the crystal lifetime for an area of the crystal containing a prominent diffraction contrast feature. Figures 3a and 3b are successive dark field micrographs of a PE single crystal containing (100) microsectors taken at identical microscope conditions, showing the effect of radiation damage on the image. The horizontal black–white line pairs (300 Å wide microsector boundaries) visible in Fig. 3a are hardly recognisable in Fig. 3b. Crosslinks due to radiation damage set up strains in the lattice causing the lattice to distort which results in a black–white patchwork in the dark field image. With continued irradiation all diffraction contrast features become fainter and finally disappear altogether. The nonplanar crystals may also shift and tilt during irradiation owing to the dimensional changes from crosslinking.[4] Occasionally fine sets of 'irradiation lines'[8] also appear, which have been previously misinterpreted as due to slip traces[9] or as evidence for a mosaic block structure,[10] (see Section 3.10).

From results similar to Fig. 3 it was determined that PE single crystal DF micrographs show an unacceptable amount of artefact diffraction contrast due to radiation damage when taken with electrons scattered after doses greater than $0·6\phi_{max}$. This dose for PE single crystals at 100 KeV and room temperature is 3×10^{16} electrons cm^{-2}.

In addition to loss of crystallinity, radiation damage can result in large mass loss and dimensional changes with resultant distortion of the object. While the state of the specimen with respect to diffraction contrast may be determined by inspection of the fading and broadening of the crystalline reflections in the electron diffraction pattern,† changes in the overall morphology in routine BF investigations can be overlooked because the rather high doses employed may cause changes long before the operator has adjusted the microscope for a particular area. Figure 4a is a 'normal' appearing BF micrograph, similar to many in the literature, of lamellae in a PE spherulite. Figure 4b shows the actual morphology, preserved by means of a special staining technique developed by Kanig (see Section 3.6). Establishing a ϕ_{max} criterion based on other than crystalline order will be necessary for systems which undergo large mass loss and dimensional changes during irradiation. The safest approach is to use a low magnification series of micrographs at increasing levels of total dose in order to assess possible

† This is not the case for samples, however, which damage primarily by scission mass loss. Here the diffraction spots lose intensity but remain sharp.

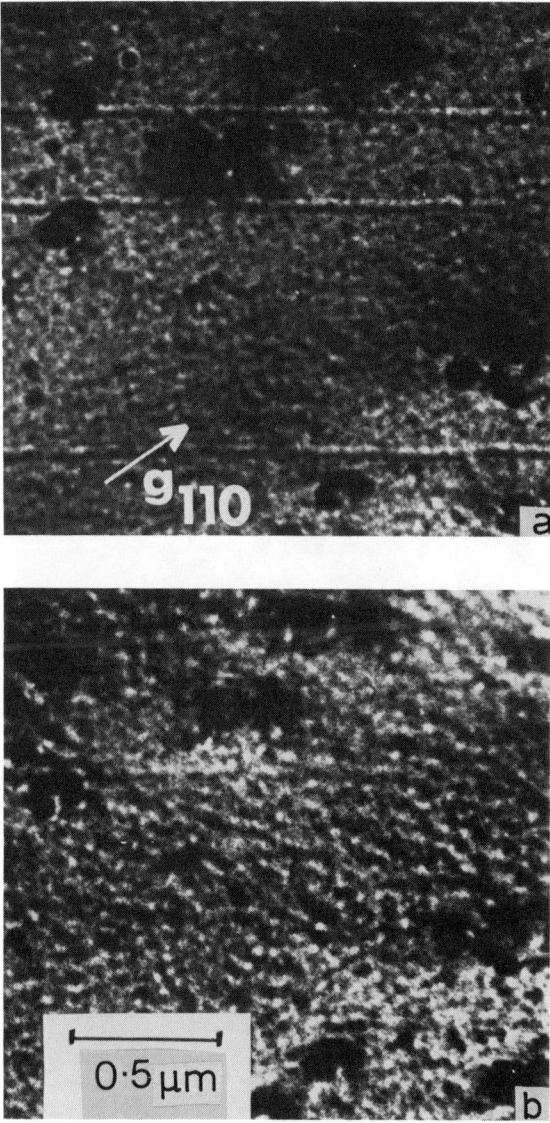

Fig. 3. Illustrating the effect of radiation damage on the image of crystalline polymers. (a) 110 DF micrograph taken using electrons from time 0 to time $0\cdot4\,\phi_{max}$ (b) 110 DF micrograph of the same area as (a) taken using electrons from time $0\cdot4$ to $0\cdot8\,\phi_{max}$.

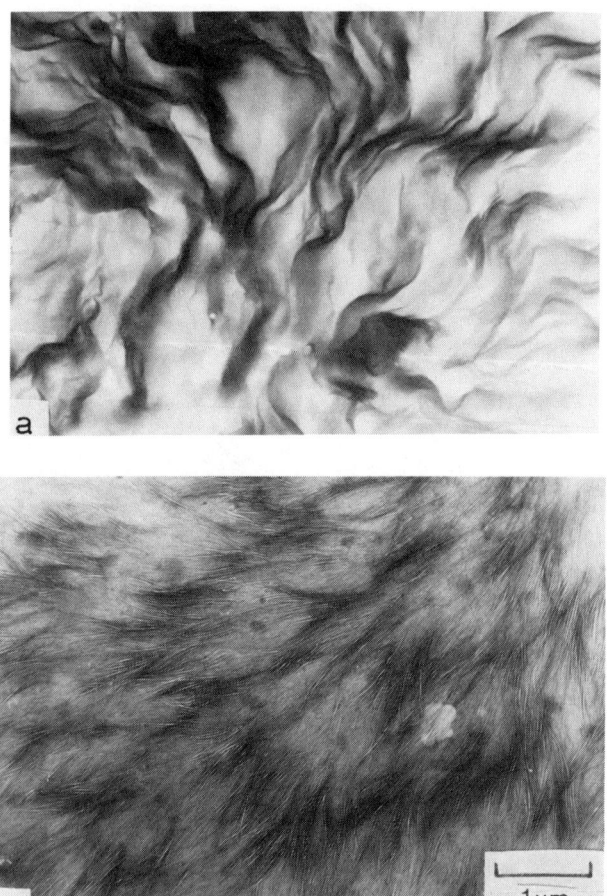

Fig. 4. BF micrographs of PE spherulites. (a) Apparent lamellar morphology from radiation damage unstained film. (b) Actual lamellar morphology visualised and preserved using the Kanig method of staining[74] (Reprinted by permission from Kanig, G. (1973). *Kolloid Z. und Z. Polymere*, **251,** 782. Copyright by Dr. Dietrich Steinkopff Verlag, Darmstadt, Germany).

beam-induced image features and to select the ϕ_{max} appropriate for the particular sample and type of structural information desired.

Once a suitable figure for ϕ_{max} is available it is straightforward to calculate the optimum working magnification and the attainable resolution.

The image resolution δ is related to ϕ_{max} by the equation:

$$\delta = \frac{SNR}{[f\phi_{max}]^{1/2}C} \quad (1)$$

where SNR is the signal to noise ratio sufficient to detect a signal in a noisy background (it is usually taken as at least 5[16]), f is the utilisation efficiency (i.e. the fraction of the electrons passing through the specimen which contribute to the image), and C is the contrast.[17] Both f and C are dependent on the imaging mode. Because f and C are coupled as $f^{1/2}C$ the most efficient use of the transmitted electrons (BF) does not necessarily provide the highest resolution.[4,11]

The optimum electron dose for recording, ϕ_{record}, is set by signal/noise considerations and the photographic plate density. The optimum magnification M_{opt} is given by:

$$M_{opt} \simeq \left[\frac{\phi_{max}}{I_0\left(\frac{\phi_{focus}}{I_{000}} + \frac{\phi_{record}}{I_{hkl}}\right)}\right]^{1/2} \quad (2)$$

where I_0 is incident beam intensity, I_{000} is bright field intensity, I_{hkl} is diffracted beam intensity, and ϕ_{focus} is minimum dose on phosphor screen or image intensifier for bright field focusing of image.

The optimum working magnification at 100 KeV for a 120 Å thick PE lamella for bright field is $M \sim 16\,000\times$ and for (110) dark field $M \sim 4000\times$ using an image intensifier for focusing ($\phi_{focus} \ll \phi_{record}$). The specimen detail resolvable is in the order of 50 Å (based on a signal to noise ratio of 7 and a photographic plate resolution of 50 line pairs/mm) for both bright and dark field since the resolution depends inversely on the contrast ($C_{BF} \sim 40\%$, $C_{DF} \sim 100\%$).[11] If n photographs of an area are desired, then the working magnification must be lowered by a factor $n^{-1/2}$. White[12] has recorded up to 14 successive DF micrographs and 2 diffraction patterns of PE single crystals at 1000×.

By employing higher accelerating voltages or by specimen cooling, ϕ_{max} may be increased.[4,6] High voltage does not result in any net improvement because although ϕ_{max} increases due to the decrease in inelastic scattering, the elastic diffracted intensity per unit incident dose decreases by the same amount.[13] Specimen cooling to cryotemperatures can result in improvement in radiation lifetime due to reduction of motion of the molecular fragments caused by ionisation of the specimen. Until recently only modest improvements (factors $\leq 4\times$)

were obtained using cryotemperatures, with a general trend towards greater improvements at lower temperatures. However, improvement factors of up to 330× for a variety of organic specimens have been claimed with a cryomicroscope equipped with superconducting lenses.[14,15] More careful radiation lifetime experiments by Dietrich's group in Munich and by other groups now however find improvements of 10–20× rather than several hundred. Certainly further development of commercial cryostages is warranted and application of cryomicroscopy to polymer morphology studies presents exciting possibilities for significant gains in microstructural information.

The only way of improving resolution at a given specimen temperature is to increase f. A STEM equipped with an annular detector can collect nearly all electrons scattered outside the central beam. Optimum information extraction is achieved if all this signal is transferred without loss to the recording medium and if the changes in the sample during the focusing, area selection, and choice of diffraction optics steps are negligible in comparison to the radiation damage which occurs during recording. The inherent image intensification, control of illumination location and magnification independent focus capabilities of STEM permit very convenient and precise focusing, area selection and choice of diffraction optics without significant radiation damage to the area of interest. The usual approach for CTEM is however, to focus on an area which is then damaged, followed by translation to an adjacent undamaged area for subsequent recording. The chief drawback of the CTEM method is the low yield of useful pictures (imprecise focus and uninteresting object features).

Image intensification can be provided by the electronic contrast and brightness amplifiers of the STEM detection system or by image intensifiers for CTEM.[11] The electronically manipulated image does not, of course, contain more information but is merely brighter than the unintensified image. The lower limit at which an ideal image intensification system can be used for focusing or area selection is determined by the statistical electron beam noise to about $1 \times 10^{-14}\,A\,cm^{-2}$. This is a factor 100–200× lower than the necessary current density required for minimum microscope phosphor screen brightness for unaided focusing/area selection by the dark adapted eye.[18,19] For STEM, only the specimen region viewed on the CRT is radiation damaged, consequently use of the selected area mode (a variable size reduced raster of the CRT) and variable beam scan speeds are quite useful as discussed in the following example. The specimen is first observed with a rapid scan at low magnification (hence low dose rate), thus coarse focus and area selection are accomplished with only slight sample damage. A low quality BF micrograph of the area is taken for reference. At this point the selected area mode is used to observe an area from a region that does not contain the precise feature of interest but is sufficiently close so that focusing on the observed region will give adequate focus for the feature of interest. Now since focus is independent of magnification in STEM (which is not the case for CTEM), the image is focused for high resolution at a high magnification in the selected area mode with a slow scan speed, to improve *SNR* for precise focus, damaging (severely) only a very small area. The magnification is then reduced and the full CRT scan is used to record a high resolution image from the nearly undamaged and in-focus selected area.

3 IMAGE INTERPRETATION—CONTRAST MECHANISMS

The incident electrons, after interacting with the specimen, carry information in both the amplitude and phase of their wave function. The usual way to extract information from the electrons is to use the squared amplitude of the wave function to produce *amplitude contrast* (mass thickness contrast or diffraction contrast). The phase part of the electron wave function may also be used to produce changes in intensity (*phase contrast*). In order to produce a change in intensity from phase differences in the electron wave functions by the object, the phase of a portion of the electron waves must be further shifted by the microscope system. In a light microscope, this shift can be produced by a phase plate that uniformly changes the phase of selected photons (Zernike contrast). In an electron microscope this phase shift is produced by spherical aberration and defocusing the objective lens. However, defocusing shifts the phases of the electrons in a complicated manner and makes image interpretation more complex.

For polymer specimens, which may consist of complex arrays of quite small structural units, it should be emphasised that the image is a two-dimensional projection of the three-dimensional specimen. Thus, for phase contrast the *projected phase change* determines the image, while for amplitude contrast imaging of noncrystalline samples, it is the *projected mass thickness* distribution which is imaged. Figure 5 shows two possible geometries of structural units in a matrix. If, as in Fig. 5a,

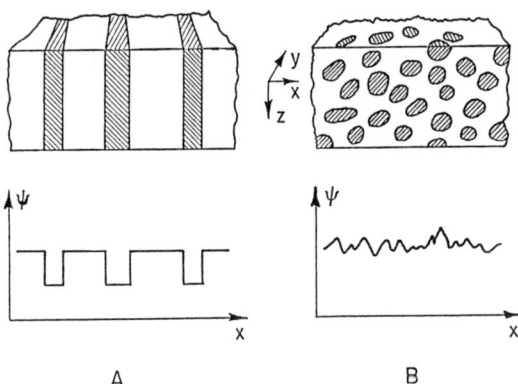

Fig. 5. Two possible geometries of domains in a matrix. The projected potential (phase contrast) or projected mass-thickness (amplitude contrast) of configuration (a) is interpretable while that of (b) is indistinguishable from random fluctuations.

the structural unit size and the film thickness are similar, the projected object function is uncomplicated and correct interpretation of the image is possible. In Fig. 5b, however, the units are randomly distributed and much smaller than the film thickness so that the resulting projection will be complex and hard to distinguish from an otherwise homogeneous film with small random fluctuations (of thickness, density or composition). For this reason very thin specimens must be prepared to study small scale structures. Moreover, determination of the volume fraction of an image feature requires careful consideration of the size of the feature with respect to the film thickness (for an excellent discussion of stereology see the book by Underwood[20]).

3.1 Phase Contrast Imaging

Because of a succession of misinterpretations of images dominated by defocus phase contrast effects in the polymer literature, a short discussion of the theory of phase contrast imaging is given. The theory is well known but unfortunately not widely known. This section presents a brief outline of the salient features of the transfer theory of imaging with regards to polymer specimens. More extensive treatments of the transfer theory of imaging may be found in a number of texts[21,22] and in recent papers where the theory has been applied to polymers.[23–27]

Thin (<1000 Å thick) polymer specimens do not significantly alter the amplitude of the electron waves, so that the object function consists essentially only of a phase term: $\exp[i\psi(\mathbf{r})]$. The object function is directly proportional to the mean inner potential $\bar{\phi}(\mathbf{r})$ of the specimen:

$$\psi(\mathbf{r}) = \frac{\pi}{\lambda V_0} t\bar{\phi}(\mathbf{r}) \tag{3}$$

where λ is the wavelength of the electron, V_0 the accelerating voltage and t the specimen thickness. The inner potential can be further expressed in terms of sample density ρ (g cm^{-3}) and f_i, the electron scattering factor at zero angle of the ith atom of the basic structural unit of molecular weight M(g):

$$\bar{\phi} = 69 \frac{\rho}{M} \sum_i f_i \tag{4}$$

Equations (3) and (4) thus permit calculation of the projected phase shift at any point of the exit face of the object.

When the phase modulations (more precisely when the relative phase changes at the exit face of the object) are small compared to unity (the so-called weak phase object) the microscope works as a linear imaging system so that a simple linear relation exists (for axial BF) between the Fourier transform (F) of the image intensity $I(\mathbf{r})$ and the Fourier transform of the object function $\psi(\mathbf{r})$:

$$F[I_{\mathrm{BF}}(\mathbf{r})] = \delta(\mathbf{K}) - 2T(\mathbf{K})[F\psi(\mathbf{r})] \tag{5}$$

For this case the image intensity is just the inverse Fourier transform of a delta function minus a term involving the Fourier transform of the object function modulated by the microscope transfer function $T(\mathbf{K})$, where $\mathbf{K}$ is the reciprocal space scattering vector and $T(\mathbf{K})$ is given by:

$$T(\mathbf{K}) = A(\mathbf{K}) \sin(-\chi(\mathbf{K})) \tag{6}$$

Here $A(\mathbf{K})$ is equal to one inside the objective aperture and zero outside, and the phase shift of the microscope optics is given by:

$$\chi(\mathbf{K}) = \pi\lambda(\Delta Z)\mathbf{K}^2 + \frac{\pi}{2} C_s \lambda^3 \mathbf{K}^4 \tag{7}$$

where ΔZ and C_s are the defocus and spherical aberration coefficient of the objective lens. The BF image contrast is:

$$\frac{I_{\mathrm{BF}}(\mathbf{r}) - I_{\mathrm{AVG}}}{I_{\mathrm{AVG}}} = F^{-1}[2A(\mathbf{K}) \sin \chi(\mathbf{K}) F\psi(\mathbf{r})] \tag{8}$$

The effect of the microscope transfer function is to modulate object frequencies in the image. Sin $\chi(\mathbf{K})$ can be adjusted by means of objective lens defocus in order to highlight selected frequencies corresponding to object details of interest. However, because sin $\chi(\mathbf{K})$ alternatively enhances or suppresses the contribution of various object spatial frequencies to the final image, a weak phase object containing random density fluctuations (amorphous carbon or polymer films are good examples) will exhibit a 'salt and pepper' structure due to the contrast reversals. The size of this phase contrast structure is dependent on both the projected potential of the specimen and the microscope optical conditions employed. The actual microscope defocus can be measured by taking an optical transform of the image as shown in Fig. 6. The position of the maxima and minima in the optical transform are determined by the $\sin^2 \chi(\mathbf{K})$ function. The largest phase contrast structures in the image correspond to the first maximum of $\sin^2 \chi(\mathbf{K})$. Thus from eqn. (7) a defocus value of ΔZ yields a phase contrast structure in the image of approximately $(2\lambda \Delta Z)^{1/2}$ (for example, a defocus of 3 μm produces an apparent domain structure of about 50 Å). For polymers which have been stained using high atomic number elements the amplitude contrast under defocus conditions will be modulated by the phase contrast, causing a confusion in the small scale structure. Figure 6 is a through focus series of a thin amorphous carbon film with many osmium particles on its surface. At approximately zero defocus sin $\chi(\mathbf{K})$ is nearly zero for the small object frequencies so that the phase contrast contribution from the amorphous carbon support film on the >10 Å scale is minimal. This permits a reasonably accurate determination of the osmium particles' size and shape from their residual amplitude contrast (Fig. 6c). As the microscope is defocused, it becomes more and more difficult to distinguish the true size and shape of the small particles owing to the increased phase contrast contributions from both the substrate and the particles. The contribution of phase contrast increases with the amount of objective lens defocus and is already comparable to the amplitude contrast of the particles at a defocus of about 1 μm. The antisymmetry of sin $\chi(\mathbf{K})$ with defocus (eqn. (7)) leads to a reversal in contrast of the phase contrast structure of the carbon support (see arrowed regions in Figs 6a and 6e). Only by taking a through focus series of micrographs is it possible to distinguish phase contrast from amplitude contrast. For these reasons, the small scale (<100 Å) dark regions in a BF image of unspecified defocus (of which a great number have been published for

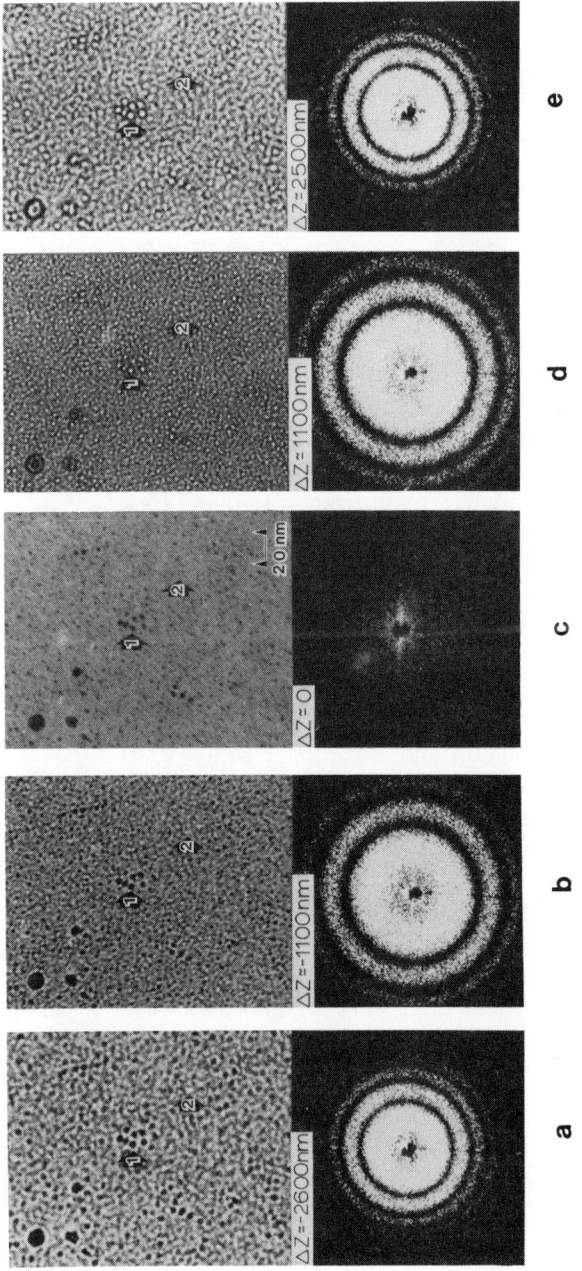

Fig. 6. BF through-focus series of osmium particles on the surface of an evaporated carbon film. Defocus is (a) $\Delta Z = -26\,000$ Å, (b) $\Delta Z = -11\,000$ Å, (c) $\Delta Z \simeq 0$ Å, (d) $\Delta Z = +11\,000$ Å, (e) $\Delta Z = +25\,000$ Å as measured from the optical transforms. Arrow 1 shows small particles changing size and contrast with defocus and arrow 2 points to featureless regions of the carbon film which develop structure under defocus conditions.

a wide variety of polymer systems) simply cannot be cited as regions of higher electron density. As there still appear[28] publications purporting to visualise various complex small scale micromorphologies by defocus high resolution BF imaging without careful attention to the transfer theory it seems worthwhile to re-emphasise the importance of phase contrast in images by recounting the previous main misunderstandings and their subsequent correct explanations.

Schoon and Kretschmer[29,30] were the first to attempt to obtain direct information on the form of the polymer molecule by high resolution BF imaging of thin polymer films. They observed 30–70 Å regions in SiO_2 replicas and uranyl acetate stained films of polystyrene, polyisoprene, PMMA, polyvinylalcohol, and styrene–butadiene random copolymers, which they inferred were the basic subunits of single molecules ('perlschnurartige Aufbau von Polymeremolekülen aus primären Struktureinheiten'[30]). They explored molecular weight effects and mechanical degradation by ultrasonication and found no change in the images, which they interpreted as proof of the fundamental nature of the 'string of pearls' structure. Zingsheim and Bachmann[31] repeated the work of Schoon and showed that for in-focus micrographs it was not possible to distinguish the image of a polymer film from an evaporated amorphous carbon support film. Their defocused micrographs showed structures for both polymer and blank support films which were identical to those previously published[29,30]. Utilising optical transforms, Zingsheim and Bachmann demonstrated the relation between the observed object spatial frequencies and the value of objective lens defocus and made clear the connection to the then (1971) emerging transfer theory.

In 1967, Yeh and Geil made note of the high resolution BF contrast features in nominally homogeneous, unstained amorphous polymer films and advanced the nodule hypothesis, which explained the dark regions as due to diffraction contrast from ordered regions of parallel packed chains.[32,33] After initial work on amorphous PET and PS, they concentrated on crystallisation from initially amorphous materials.[34] Other workers continued to find nodular structures in virtually every nominally amorphous system investigated, causing a general re-examination of the Flory concept of randomly coiled chains in the condensed amorphous phase. When a large variety and number of careful structural investigations[35] could find no evidence for the nodules, including the new isotopic substitution neutron scattering results which were in rather excellent agreement with Flory's ideas, the microscopy experiments were repeated by Uhlmann and coworkers and the BF defocusing artefacts rediscovered.[36] However, there existed additional evidence from electron microscopy for the ordered structure of nodules from DF observations where bright speckles in the same size range (25–100 Å) were observed.[32,33] The interpretation of this DF speckle as regions of local order in the noncrystalline polymer specimen was based on the idea that for a truly amorphous specimen the DF image intensity would be completely uniform with no contrast. Therefore the observation of the bright speckle in the DF image (usually formed by selection of the scattered electrons from the first diffuse halo) was presumed to be due to regions in the specimen sufficiently ordered to diffract coherently. Thomas and Roche[25] carried out several systematic DF experiments in order to clarify the nature of the DF speckle. The behaviour of the speckle with respect to changes in objective lens defocus and objective aperture size and location in the scattering pattern were shown to be entirely consistent with a statistical image modulated by the transfer function of the microscope for DF, rather than an ordered domain image interpretation.

The third and most recent use of the same BF defocus artefacts is in the study of microphase separated polymer systems. The story is the same; putative identification of the small 25–100 Å dark regions in the high resolution BF image (of both unstained and stained films) as electron dense regions representing the more dense phase separated regions (hard segment domains in urethanes,[37,38] ionic clusters in ionomers[39]), and a mixed micellar phase for polystyrene–polybutadiene triblock copolymers.[40] In this case however, bulk scattering, DSC, dynamic mechanical, IR and other experiments definitely support a phase separated microstructure, so the deductions from the micrographs are not at odds with other established facts. However, it is still not possible to interpret casually the small scale (<100 Å) dark regions in BF micrographs of unknown defocus of microphase separated polymers unambiguously as the higher density domain structures in the polymer film.

In general, to decide whether an observed image feature is related to a real structural unit in the object or is simply due to the microscope spatial frequency filtering of a homogeneous array of scatterers (with random fluctuations) is quite difficult, since images of completely different object models are qualitatively indistinguishable (recall Fig. 5b). Faithful image reconstruction of the projected object structure by phase contrast imaging only occurs when the transfer function is near unity and the same for all the important object frequencies. For irregular domain structures, a single micrograph does not reliably reconstruct the projected potential variations of the object. The projection to two dimensions means that the interpretation of the properly reconstructed image will be ambiguous unless the film thickness is of the order of the domain size.

There have been successful uses of phase contrast imaging—but necessarily for much simpler systems. Petermann and Gleiter[41] have investigated thin lamellar films of several semicrystalline polymers and enhanced the image contrast between crystalline lamellae and noncrystalline regions by objective lens defocus. Handlin[12] has recently employed phase contrast to image polystyrene/polyisoprene lamellae without OsO_4 staining. But even such simple one-dimensional systems still require caution. Some typical cases of misinterpretation of the fine structure of cellulose and keratin fibres are given by Johnson and Crawford.[23] For useful deductions on lamellar domain shape and interface profile a systematic defocus series is required. A reconstructed image can then by made by appropriate filtering and two-dimensional Fourier processing (for details see Reference 1).

3.2 Lattice Images

Lattice imaging is a specialised phase contrast imaging technique for crystalline materials and has produced very important atomic scale information of crystal defects in nonpolymeric crystalline materials where the materials' resistance to electron irradiation permits high (~ 2 Å) resolution imaging.[43] Synthetic polymer lattice images have only been achieved for poly(*para*-xylene) (18 Å),[44] poly(*para*-phenylene terephthalamide) (PPTA, 6·4 Å and 4·3 Å)[45] and most recently poly(*para*-phenylene benzobisthiazole) (PBT, 12·4 Å and 5·9 Å).[46] The radiation stability necessary for such high resolution work is provided in these polymers by extensive electronic conjugation of the backbone, permitting structurally non-destructive modes of de-excitation of the excited states created by inelastic scattering of the incident electron beam. The reported crystal lifetimes for PPTA and for PBT are of the order of 1 C cm^{-2},[45,46] about a 100-fold increase over PE.[4]

A two beam lattice image can be formed by objective aperture selection of the forward scattered beam (k) and one Bragg diffracted beam (k') (see Fig. 1). When a value of objective lens defocus is chosen to permit good transfer of the spatial frequencies of the two beams, the two beams interfere producing a set of sinusoidal fringes, the period of which is the spacing of the crystal planes producing the diffracted beam. The position of the fringes is influenced by several factors[47] and bears no simple relation to the positions of the atomic planes in the crystal.

Dobb, Johnson and coworkers have produced elegant lattice resolution micrographs of PPTA fibres.[45] Figure 7 shows the 6·4 Å meridional (M) fringes and the 4·3 Å equatorial (E) fringes extending over regions 50 Å wide by 250 Å along the fibre axis. Such observations permit direct assessment of crystallite size and orientation for comparison to X-ray line broadening and orientation measurements. Lattice images also permit detection of defects in the crystal by the disturbance of the crystal periodicity. In the few published lattice images, the fringe regions are limited in extent to less than 300 Å in any one direction [due to small crystallite size or elastic bending of the lattice (see Section 3.7)]. As yet, no evidence of structural defects has been found. The possibility for direct imaging of, for example, chain end defects, certainly exists for these radiation resistant, stiff chain materials.

3.3 Moiré Patterns

Moiré fringe patterns are produced by the mutual interference of diffracted beams from two overlapping crystals differing either in relative orientation or lattice spacing. For the usual case of the boundary between the crystals being normal to the incident electron beam, a moiré fringe pattern of spacing D_d occurs from crystals of different lattice spacings d_1 and d_2:

$$D_d = \frac{1}{1/d_1 - 1/d_2} \tag{9}$$

and for crystals of the same spacing d but with a relative rotation of η about the incident beam direction a moiré fringe pattern has a spacing D_R given by:

$$D_R = d/\eta \tag{10}$$

White[48] has shown that for very thin crystals tilt moiré patterns can also occur because of an effective change in the lattice parameter of a thin crystal slightly misoriented from the Bragg condition. For all these types of patterns the fringes are oriented normal to the difference

Fig. 7. BF axial lattice fringe image of poly(*para*-phenylene terephthalamide) showing both equatorial 4·3 Å (regions labelled E) and meridional 6·4 Å (regions labelled M) spacings.[120] (Reprinted by permission from Dobb, M. G., Johnson, D. J. and Saville, B. P. (1980). *Phil. Trans. R. Soc.* (*London*), **A294,** 483. Copyright by the Royal Society, London, UK.)

diffraction vector, $\Delta\mathbf{g}$. A detailed discussion of each type and combination of types has been presented by White.[48]

The usefulness of moiré patterns derives from their sensitivity in revealing defects in either of the two contributing crystals. In 1959, Agar, Frank and Keller[49] showed the existence of dislocations in PE single crystals as evidenced by terminating moiré fringes. The fringes on one side of the image are shifted by the component of the Burger's vector (**b**) of the dislocation normal to the diffracting planes. By employing different sets of crystal planes to form the images, the total Burger's vector of the dislocation can be determined. Holland[50] used systematic DF moiré observations to characterise the dislocations in PE as being partial edge dislocations $\mathbf{b} = [a/2, b/2, 0]$ on {110} planes. This type of defect could be formed by termination of a {110} fold plane during growth. The change in spacing and orientation of moiré fringes on crossing a fold domain boundary led Bassett[51] to suggest that each fold domain possessed a slightly different, distorted orthorhombic unit cell. The values deduced for the rotation and lattice spacing change are indeed consistent with the measured values of the splitting of reflections in the ($hk0$) electron diffraction pattern of a single crystal with 4 simultaneously diffracting fold domains.[52,53]

Holland and Lindenmeyer[54] found that overlying PE lamellae occasionally exhibited interfacial dislocation networks instead of rotation moiré patterns. The spacing of the dislocation arrays was found to depend on the relative crystal misorientation in the same way as for moiré fringes but, unlike the straight moiré fringe patterns, the dislocation lines were curved and formed pseudohexagonal networks. It was suggested that the interfacial dislocations were due to interaction of the contacting lamellar surfaces such that correct crystalline registry occurred over areas bounded by sets of interfacial screw dislocations where the rotational mismatch between the crystals was accommodated. DF imaging determined that the dislocations were screw in character having Burger's vectors [100], [010] and ⟨110⟩. The detailed diffraction contrast behaviour of the dislocations is given by Holland and Lindenmeyer in a collaborative paper with Amelinckx.[55] Sadler and Keller[56,57] systematically explored the origin of these interfacial dislocation networks, especially as regards the need for low molecular weight polymer. They were able to rule out models based on fold–fold interactions and suggested that the penetration of molecular ends into a relatively thick surface region containing folds establishes crystallographic register. Niinomi *et al.*[58] in their examination of how the transition from moiré fringes to dislocation network depends on the angle of rotation, found an intermediate transition region in which the two crystals interact locally, forming rippled moiré patterns called 'deformation moiré'. This slight 'trembling nature' of certain moiré fringe patterns had been previously cited as evidence for a mosaic substructure of the lamellae.

In the mosaic block model of single crystals, adjacent 300 Å diameter blocks are assumed to be independent scatterers to account for the observed X-ray line broadening. The requirement of independent scattering can be met by a combination of tilt, rotation and translation of the blocks. Fischer and Lorenz[59] examined rotation moiré patterns and noted that a high density of terminating fringes from rotations and translations of blocks was not observed. They therefore supposed that block tilt must be responsible for small crystallite size and subsequent line broadening. The mosaic block model of Hosemann et al.[60] suggests the polymer crystal is composed of 300 Å blocks with tilts between neighbouring blocks of up to 11°. The boundaries between blocks are assumed to consist of two arrays of screw dislocations with Burger's vectors [001] and [$hk0$] (twist boundary). However, it must be stated that moiré fringe patterns are sensitive to details of crystal tilt and strain due to partial collapse of the pyramidal lamellae, as well as to specific lamellar surface interactions, so that subtle contrast effects in moiré patterns are difficult to interpret.[48] Furthermore, while [001] screw dislocations are invisible in ($hk0$) moiré DF, [$hk0$] screw dislocations should be detected and are not. Finally, considering the size of the supposed mosaic blocks and the large tilts and dislocations associated with them, high resolution DF diffraction contrast imaging should be able to reveal their presence directly. The DF results are, however, in direct conflict with the mosaic model (see Section 3.10).

3.4 Amplitude Contrast

Amplitude contrast arises from loss of electrons from the image by scattering outside the objective aperture. Crystals at the Bragg condition and regions of higher density or increased thickness scatter more and appear dark in the BF image. Methods to enhance the normally weak mass thickness contrast in polymer films include heavy atom staining, metal decoration, metal shadowing and replication. While interpretation of stained, decorated and shadowed specimen images (see Section 3.5) is reasonably straightforward at the level of resolution of lamellae and above, deductions concerning structural features approaching the scale of the grain size of the heavy atom clusters (10–30 Å) requires careful separation of the role of phase contrast from amplitude contrast as well as care in distinguishing genuine structural features of the stained (decorated, etc.) specimen from the fluctuations present in the heavy atom distribution itself. Medium resolution BF images will exhibit predominantly amplitude contrast if one operates near focus where $\sin \chi(\mathbf{K})$ is nearly zero over the range of object frequencies corresponding to large scale object structure (recall Fig. 6).

3.5 Mass Thickness Contrast

For samples which primarily scatter incoherently, that is, no significant diffraction occurs from crystalline regions (e.g. replicas, non-crystalline polymers or radiation damaged polymers) the fraction of the incident beam intensity I_0 which is collected by the objective aperture depends on the microscope optics through the choice of beam energy E_0, objective aperture size α, sample density ρ and thickness t as:

$$I(E_0, \alpha) = I_0 \exp\left(-S_p(E_0, \alpha)\rho t\right) \quad (11)$$

where S_p is the effective mass scattering cross section (including elastic, inelastic and plural scattering[61]) and ρt is the mass thickness of the specimen.

Bahr and Zeitler[62] employed a series of various sized polysytrene latex spheres as standards to first demonstrate this relationship in 1962. Quantitative BF microscopy for determination of particle mass without recourse to standards is now possible because of improved theoretical scattering calculations. The high collection efficiency of STEM annular DF for scattered electrons with direct coupling of the electronic signal to a computer[63] has been utilised recently for mass measurement of biological macromolecules. Misell and Burdett[64] have shown that S_p is nearly independent of atomic number so that the calculated theoretical values of S_p may be used for stained and unstained material. Thus for given electron optical conditions the transmitted intensity is related exponentially to the projected density–thickness variations. By employing theoretical values for S_p (see Reference 64) and measuring specimen thickness (a review of thickness measurement techniques is given in Reference 65) or by using standards, variations in sample density (and hence composition) may be determined. In addition, by measuring the average value of ρt and the area A of a particle, the particle mass m can be calculated ($m = A\rho t$).

Kramer and Lauterwasser[66] have recently employed quantitative microdensitometry to obtain measurement of mass thickness variations due to deformation in polystyrene films. Adams[67] has adapted this method using the electronic BF intensity signal in STEM (thus avoiding difficulties of photographic film linearity and microdensitometry) for deformation studies of PE spherulite films.

A series of papers[68-71] has appeared claiming to determine 'the instantaneous shape and segmental density of flexible chain macromolecules' via mass thickness contrast. These papers serve as a useful example in the (mis)application of mass thickness contrast microscopy.

The experiment consists of tagging polymer chains with a low concentration (about 2 mole %) of heavy atoms and dispersing these tagged chains at about 10% concentration in a matrix of identical, but untagged chains. After microtomy, the section is examined in BF to observe specific details of the actual conformation of individual chains (actually their 2-D projection) by the overall spatial variation of heavy atom enhanced contrast measured by microdensitometry. Both amorphous and crystalline polymers have been examined in this manner. In the case of crystalline polymers (PE tagged with bromine and isotactic polystyrene tagged with iodine) particular attention was paid to the influence of a lamellar morphology on the chain conformation. While the idea is attractive and addresses basic morphological questions of long standing interest, the results on crystalline polymers are derived from a naive and incorrect calculation of the expected image contrast and from micrographs which appear to feature chiefly radiation damaged lamellae and thickness variations from ultramicrotomy.

After the optical density of a region (typically 1500–4000 Å in lateral extent) purportedly containing a single, isolated tagged molecule and that of the matrix are determined by microdensitometry, the densities are converted to opacities which are then compared to those derived from a theoretical calculation. The theory employed assumes that the opacity is a constant times the scattering per unit volume, where the scattering per unit volume is taken as being proportional to the number of atoms of a given type times the square of their atomic number summed over all types of atoms in the basis volume. Nowhere are the important effects of specimen thickness, inelastic scattering, multiple scattering and electron optical parameters taken into account. The value of α is never mentioned. Given all this, the measured opacity ratio of tagged to untagged region and the calculated ratio approximately agree—which is taken as proof that the observed contrast variations represent visualisation of the shapes of individual tagged molecules. The correct theoretical BF contrast (using eqn. (11)) from a region containing a single polystyrene molecule with 1 in 61 carbon atoms tagged with an iodine label compared to the untagged matrix for a 1000 Å thick film is completely negligible (assuming polymer $\rho = 1$ g cm^{-3}, $t = 1000$ Å and $E_0 = 75$ KeV and $\alpha = 10$ mrad, for which $S_p = 4 \cdot 7 \times 10^{-3}$ m^2 mg^{64}). Moreover, even for the case where the region of tagged molecules is phase separated—i.e. all molecules in the 1000 Å thick region have the same 1:61 composition ratio, the contrast is still only 8%. To achieve the measured opacity ratio of 2 would require the density of the labelled region to be 2·5 times that of the matrix. This would correspond to a phase separated region of purely tagged molecules with average composition of approximately C_8H_6I, that is an iodine atom for every mer repeat along the chain! This would provide no information on the size and shape of the individual molecules (which would hardly be considered unperturbed at such an iodine content) but rather of the phase separated, iodine rich domains. While the paper does not comment on the measured or calculated opacity ratio for PE, the lower mass of the bromine label makes matters worse. Moreover, pendant halogen atoms are a poor choice for labelling since polymer molecules rapidly loose halogens upon irradiation.[6,7] The experimentally observed contrast must then be attributed to other sources such as thickness variations and dimensional changes from radiation damage. (Note the similarity of Figs 8 and 9 of reference 74 with Fig. 4a.)

It is possible to employ heavy atom tagging of individual molecules to learn about the positions of the labels. With highly specific labelling chemistry, the sequencing of the chain backbone is possible. Such work, however, requires isolated individual molecules on ultrathin, uniform, clean, low atomic number substrates and dedicated high resolution microscopy (see for example recent issues of J. Molecular Biology, J. Ultrastructural Research).

3.6 Enhancement Methods for Mass Thickness Contrast

Employment of lower beam energy and smaller objective apertures increases the fraction of electrons scattered outside the objective aperture and improves image contrast. For example, the calculated contrast for the 1 in 61 iodine tagged polystyrene chains increases approximately 2·5× to almost 19% on lowering the voltage from

75 KeV to 40 KeV and using a 5 mrad aperture instead of a 10 mrad one.

Staining methods are extensively used in cell biology because, like polymers, cell components are chemically very similar leading to little intrinsic image contrast. Staining methods for polymers are recently finding expanded usage, most notably because of the success of osmium tetraoxide (OsO_4) as a stain for double bonds[72,73] and chlorosulphonic acid–uranyl acetate (CSA–UA) as a stain for semicrystalline olefins[74] (see Fig. 4b). Kanig's CSA–UA method provides strong contrast between the impermeable crystalline lamellae and the stained noncrystalline interlayers of bulk crystallised material because the chlorosulphonic acid reacts selectively with the amorphous regions of the polymer, and the uranyl groups attach to the resulting acid groups. Additionally, because of chemical crosslinking in the noncrystalline regions, the material becomes quite stiff, improving microtomy and making the thin sections resistant to beam induced dimensional changes.

Voigt-Martin has applied Kanig's method to study lamellae thickness distributions quantitatively for a series of melt crystallised polyethylenes.[75,76] The long spacing and crystal thicknesses observed by electron microscopy of stained, microtomed sections were plotted as histograms and compared with the results obtained from the electron density correlation functions by Fourier inversion of SAXS curves. The results for minimum, maximum and average long spacing were in good agreement.[75] However, in a subsequent investigation,[76] it has been shown that when the lamellae thickness distributions are rather broad the agreement is substantially less, indicative of the different sensitivity of each of the techniques to the details of the lamellae thickness distribution.

Any substance which selectively increases the local scattering can be used as a stain, but contrast will be better the larger the mass of the staining atom and the higher the number of staining sites occupied per unit volume. Careful work is necessary to establish that the final distribution of heavy atoms which is imaged bears a specific relationship to the original specimen microstructure. Geometrical limitations restrict unambiguous interpretation to views of lamellae in the edge-on position.

Metal shadowing makes use of the variations in surface topography to enhance image contrast. The distribution of metal atoms also depends on the particular metal and evaporation conditions employed.

Apparent surface irregularities can be brought about just by the fluctuations in metal island build-up and one should always employ blanks if fine (<50 Å) detail is sought.[77] Length measurements (e.g. lamellar thickness) based on shadow length and shadowing angle should use latex spheres to determine the *local* shadowing angle. The lack of sensitivity of metal shadowing to specific polymer surface features has been noted by Wunderlich,[78] who mentions the similarity of BF images of metal shadowed ethylene oxide–polystyrene copolymer single crystals (polystyrene surface) and polyethylene single crystals.

Metal decoration is sensitive to specific metal atom–substrate interactions. Instead of a high melting point metal shadowed at a shallow angle, metal decoration employs a highly mobile species, such as gold, incident at 90° to the specimen surface. Bassett[79] originally developed this technique for ionic crystals and has successfully studied surface features of PE single crystals.[80] Spit[81] has discussed the influence of metal type, angle of incidence, etc., on the resultant metal atom distribution.

Replication is a technique long employed to study surface topography. Metal shadowing is used to contrast the topological variations in a plastic film previously in contact with the surface of interest. The drawback here is the additional information transfer step; major advantages are the view of interior as well as exterior sample microstructure and stability of the replica against beam damage. Detachment replication is also useful in obtaining thin specimens from bulk material by removing with the replica a thin layer of the polymer surface.[82,83] Various types of *etchants* have been used to treat free or fracture surfaces before replication in order to enhance contrast between crystalline lamellae and noncrystalline regions. Bassett *et al.*[84] have recently developed a permanganic etch technique which works well for several olefins. The etchant selectively removes disordered material and is even sensitive to crystallographic orientation. Such a combined surface etch and replication technique allows study of lamellar morphology of melt crystallised polymers without the projection limitations of Kanig's technique.

3.7 Diffraction Contrast

Diffraction contrast derives from scattering of the diffracted beams outside the objective aperture by crystalline regions in the sample and will therefore be limited by the radiation lifetime of the polymer. BF

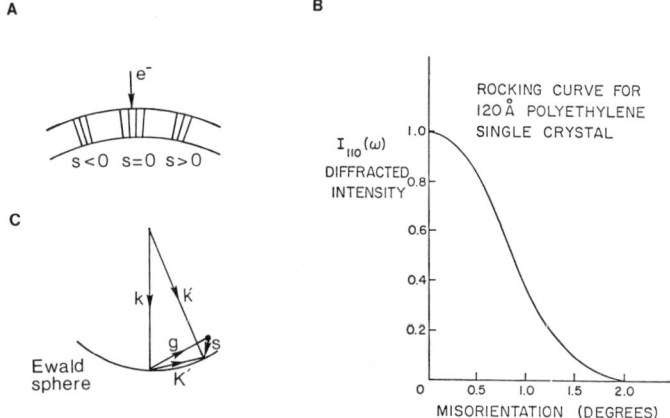

Fig. 8. (a) Schematic illustration of effect of bending on the Bragg condition ($s = 0$). (b) Rocking curve for diffracted intensity from (110) planes in a 120 Å thick polyethylene crystal at 100 KeV. (c) Definition of the deviation parameter s in terms of the reciprocal lattice vector **g** and the diffraction vector **K**.

diffraction contrast occurs in all regions of the specimen where the Bragg condition is satisfied. Single beam DF imaging uses Bragg diffracted electrons from one particular set of crystallographic planes. Contrast arises from the local deviation of the lattice planes from the Bragg angle (see Fig. 8a). The diffracted intensity varies as:

$$I(s) \simeq |F(hkl)|^2 \frac{\sin^2 \pi st}{\sin^2 \pi s} \qquad (12)$$

where $F(hkl)$ is the structure factor for the (hkl) reflection, t is the crystal thickness parallel to the optic axis and s is the deviation of the (hkl) planes from the Bragg condition ($s = 0$) (see Fig. 8b). The relationship of s and the diffraction vector, **g** and the Ewald sphere can be seen in Fig. 8c. Strain fields from defects present in the crystal will cause characteristic displacements of the lattice planes and affect the local diffracted intensity accordingly. Specimen texture can be similarly studied by the overall displacements of arrays of crystals from the Bragg condition. By using different sets of diffracting planes (different **g** vectors) to form DF images of the same region, it is possible to map out the strength and symmetry of the displacement field and understand details of the defects and textures present. The advantage of DF is the direct interpretation afforded diffraction contrast features in the image. The disadvantage is the much lower diffracted intensity compared to the transmitted beam intensity, requiring focusing to be done in BF and limiting the maximum useful DF magnification in radiation sensitive materials (see Section 2.2).

3.8 Micro-area Electron Diffraction

Electron diffraction can be used to identify regions of different orientation and crystal structure. Figure 9a shows schematically selected area diffraction (SAD) obtained in CTEM by using a large diameter, parallel beam of incident electrons on the sample with a field limiting aperture (the SAD aperture) placed in the first image plane of the intermediate lens. SAD permits examination of a much smaller scattering volume than is possible with X-ray techniques. Since large single crystals could not be grown, Claffey et al.[85] used electron diffraction to solve the crystal structure of alpha-poly(3,3-bis(chloromethyl)oxacyclobutane). Diffracted intensities from the dilute-solution grown crystals less than 100 Å thick were found to be approximately kine-

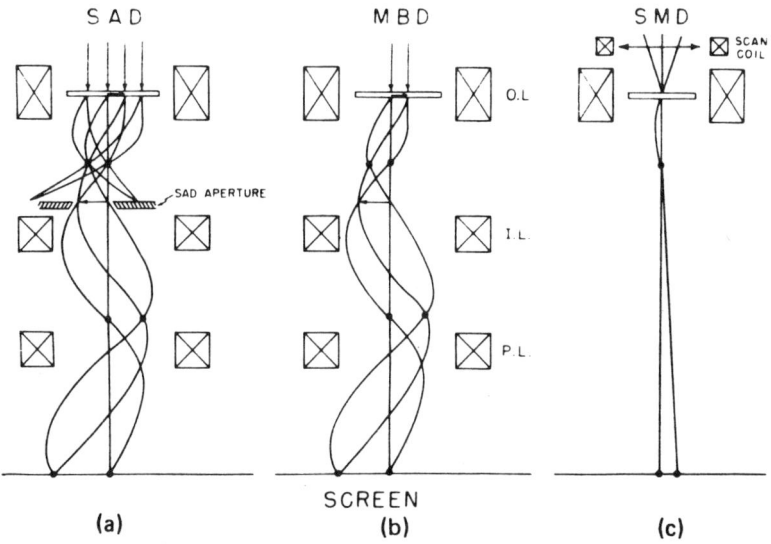

Fig. 9. Schematic ray diagrams of various diffraction geometries. (a) CTEM—selected area diffraction, (b) CTEM—microbeam diffraction, (c) STEM—convergent beam microdiffraction.

matic. Allen and Bevis[86] have used SAD for systematic studies of the stress induced phase transformation from orthorhombic to monoclinic crystal structure in individual fold domains of PE single crystals.

In SAD, the minimum sample area contributing to the diffraction pattern is determined by the size of the SAD aperture and spherical aberration of the objective lens (i.e. diffracted electrons from outside the area defined by the SAD aperture contribute to the diffraction pattern). Spherical aberration limits the minimum diffraction area to several thousand ångströms for CTEM. Moreover, the large incident beam size on the sample eliminates the possibility of recording successive patterns from undamaged adjacent areas for radiation sensitive materials. However, microbeam diffraction (MBD) can be employed using a strongly excited first condenser lens and a very small second condenser lens aperture (typically 20 μm) to illuminate the sample with a fine parallel beam of electrons (see Fig. 9b). Since no field limiting aperture is used, spherical aberration of the objective lens does not play a role and the minimum diffraction area is the incident beam diameter (c. 2500 Å). Because the incident beam only illuminates the area of interest, successive patterns may be obtained from adjacent areas.

The development of STEM with its fine electron probe has made possible diffraction from regions as small as 30 Å in metals, which is approximately the theoretical minimum crystal size for a meaningful diffraction pattern of about 5 times the unit cell size.[87] Figure 9c shows convergent (nonparallel) beam microdiffraction (MD) where the sample area producing the diffraction pattern is again determined by the incident beam diameter (variable down to ~10 Å in commercial non-dedicated STEM, i.e. a CTEM with a scanning attachment). The main obstacle to the successful application of MD to polymers is, of course, radiation damage. The minimum polymer sample size which can be used to form a useful diffraction pattern is limited by insufficient *SNR* statistics in the scattered peaks at low doses and destruction of the crystal by radiation damage at high doses. By systematically varying the diffraction area at fixed incident beam current, it is possible to determine the smallest area which yields a useful pattern. Vesely *et al.*[88] have reported MD from 1000 Å regions in 3000 to 7000 Å thick films of PE and Thomas *et al.*[89,90] have recorded patterns from 1000 Å regions in 120 Å thick PE single crystals and 1500 Å thick spherulitic films. Microbeam diffraction in combination with STEM DF has been used by Khoury[91] to determine variations of the tilt of the molecular chain axis with respect to the lamellar surface in PE single crystals. Roche et al.[92] employed microbeam diffraction and small angle light scattering to identify the molecular orientation differences between normal and abnormal spherulites of polybutylene terephthalate.

3.9 Defects in Crystals

Wunderlich[78] has classified polymer crystal defects into 9 categories and reviewed theoretical ideas and experimental investigations through 1972. The observation of defects in polymer crystals by diffraction contrast will be limited by the strength and extent of the strain field of the defect and the radiation damage limit of resolution for the particular polymer and particular choice of **g**. For example, the lower limit of resolution in a relatively undamaged PE crystal is estimated at about 50 Å. Thus, there may very well be present defects which, because of their insufficient effect on the diffracted intensity, will not be detectable. To the author's knowledge, there have been as yet no claims of experimental observation of point defects (e.g. chain disorder defects: chain ends, jogs, kinks, chain torsion, isolated folds inside the crystal, vacancies, interstitial or substitutional monomer or comonomer units) by diffraction contrast methods.

Because of their longer range characteristic displacement fields, dislocations, stacking faults and twins have been experimentally observed. To illustrate the application of diffraction contrast image interpretation of defects in crystalline polymers, three examples of increasing complexity are considered: mosaic blocks, stacking faults and screw dislocations.

3.10 Mosaic Blocks

X-ray diffraction peaks from crystalline polymers are normally quite broad. The peak width is usually attributed to small crystal size. The original fringed micelle model of crystalline polymer morphology incorporated the micelle regions as the small independent scattering entities. Since the discovery of chain folded lamellar crystals from dilute solution and a similar lamellar morphology of bulk crystallised polymers, for which the lateral lamellar sizes are in the range of tens of microns, a mosaic block substructure of the order of 300 Å in diameter has been assumed to account for the observed line broadening.[60,93,94] Early DF microscopy and electron diffraction studies of polymer single crystals grown from dilute solution found evidence to support the

X-ray diffraction results, but subsequent work indicates chain obliquity plays an important role in controlling the line width of diffraction peaks.

As mentioned in Section 3.3, Fischer and Lorenz[59] used the moiré technique to try to identify the blocks but concluded from the unperturbed nature of the moiré fringe patterns observed that no boundaries were present due to rotations and translations of the blocks. Their electron diffraction line width investigations however, indicated a 300 Å crystallite size for which they concluded the blocks are incoherent scatterers due to *tilt* about an axis normal to the chain axis. Thus the current mosaic block model depicts the lamellar crystal as being composed of mosaic blocks with an average lateral extent of 300 Å with tilts between neighbouring blocks of 0·6° to 11°. Considering the size of the mosaic blocks and the large tilts associated with them, DF microscopy should be able to reveal their presence. The rocking curve in Fig. 8b shows the expected variation of diffracted intensity as a function of tilt away from the Bragg condition. For tilts $\geqslant 2°$, the blocks will appear as 300 Å dark regions in a ($hk0$) DF micrograph. Qualitatively one would expect a black and white 'checkerboard' appearance in the DF image for the range of tilt indicated above.

Geil[10] reported direct DF observation of 250 Å blocks in polyoxymethylene (POM) single crystals, but did not assess the role of radiation damage on the DF image. As was noted in Section 2.2 (Fig. 3), radiation damage can set up strains in the crystal lattice causing the lattice to distort, resulting in the appearance of a black–white patchwork texture in the DF image during the irradiation, which confounds correct image analysis. Thomas *et al.*,[53] working with electron doses $\leqslant 0·6\phi_{max}$, obtained DF micrographs of PE single crystals showing nearly uniform contrast over large (micron scale) regions. The same investigation also examined the electron diffraction patterns using much lower beam divergence and radiation dose than earlier workers[59] and quantified the observation (noted ever since the discovery of polymer single crystals) that the reflections appear unusually sharp for a polymer. An effective block size of more than 2000 Å was obtained from the diffraction peak width. Dorset[95] has examined the electron diffraction pattern of paraffin crystals and found dynamical intensity effects indicating that the crystal cannot be considered a collection of small incoherent scattering units. Thus several careful microscopy studies (with special precautions against radiation damage) did not reveal evidence for the mosaic blocks in the DF image or in the diffraction pattern. It is indisputable that while there can be various extraneous sources which may cause DF image contrast and contribute to diffraction line broadening, there is nothing but the intrinsic perfection of the crystal which yields uniform DF image contrast and sharp reflections. It follows that solution grown single crystals can be highly perfect in the lateral direction. It remained to understand why the crystals produce broad X-ray diffraction lines.

X-ray measurements of Harrison et al.[96] on PE indicate that the reflections are broader if the solution grown crystals are dried down into a mat, as is usual for X-ray studies, than if they are studied in suspension. White[97] has argued on this basis that simple elastic bending of the lamellae could account for the observed X-ray line broadening. However, examination of line broadening from POM crystals yields a large lateral crystal size (~2000 Å) both in suspension and after sedimentation, so that elastic and plastic distortion, while undoubtedly contributing to line broadening, are alone not sufficient. Detailed consideration of the electron and X-ray diffraction physics revealed a simple explanation: X-ray powder diffraction from crystals where the effective crystal size is determined by the Scherrer equation is a measure of the perpendicular distance through the set of diffracting crystallographic planes *within* the crystal. For PE the chain axis is about 30° to the lamellar normal.[98] For *oblique* chains the lateral crystal size is thus given by $t/\sin w$, where t is the crystal thickness and w is the angle between the chain axis and the lamellar normal. For the case of electron diffraction there is not an assembly of crystals as in the powder diffraction geometry. Only one orientation is present and the intersection of the crystal shape transform and the Ewald sphere is only a point (hence large crystal size) regardless of chain obliquity. Such a simple geometrical argument also accounts nicely for the observed increase in crystallite (mosaic block) size with increase in lamella thickness.[99] Moreover, since SAXS and Raman scattering studies indicate that crystalline lamellae of melt crystallised materials also have oblique chains,[100] the interpretation of X-ray line broadening of bulk polymers should also consider the effect of chain obliqueness.

3.11 Stacking Faults

The new method of solid state polymerisation[101] presents exciting possibilities for the study of defects in polymer single crystals. This type of crystal can be grown as thin platelets as well as macroscopic (several millimetre) crystals and affords unique opportunities for study-

112   EDWIN L. THOMAS

ing defects because of the extended chain structure, high crystal perfection and reasonable radiation stability. Microscopic investigations on these interesting materials are just beginning. An initial analysis of stacking faults has been presented by Young and Petermann.[102] Figure 10 shows a strong variation of the deviation parameter of the crystal in the vicinity of a bend contour (region A) indicating the presence of defects which impose their own displacement field on the

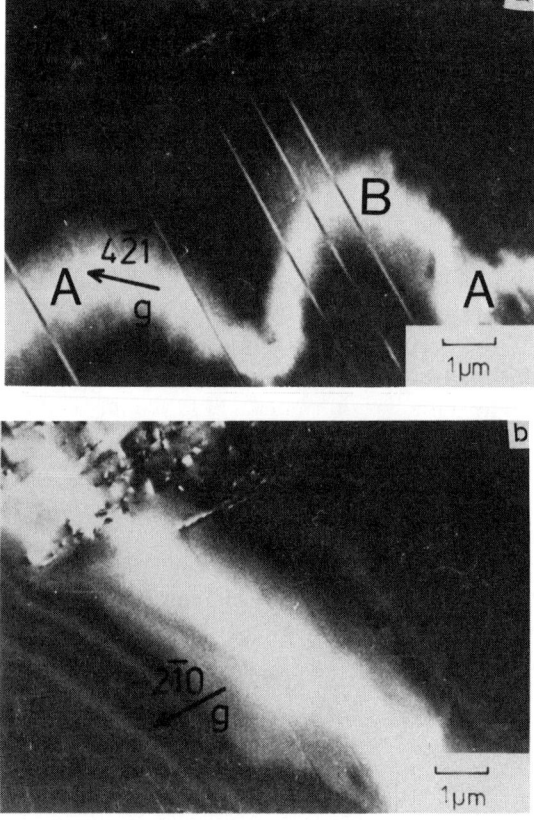

Fig. 10. (a) DF micrograph ($\mathbf{g} = 4\bar{2}1$) of stacking faults in PTS crystal. (b) DF micrograph of same region as (a) but with $\mathbf{g} = \bar{2}10$. (c) Schematic of 9 unit cells of a PTS crystal viewed parallel to the chain direction, showing the perfect crystal and a region containing a stacking fault (indicated by broken line). (Reprinted by permission from Young, R. J. and Petermann, J. (1981). *J. Mat. Sci.*, **16**, 1835. Copyright by Chapman and Hall, London, UK.)

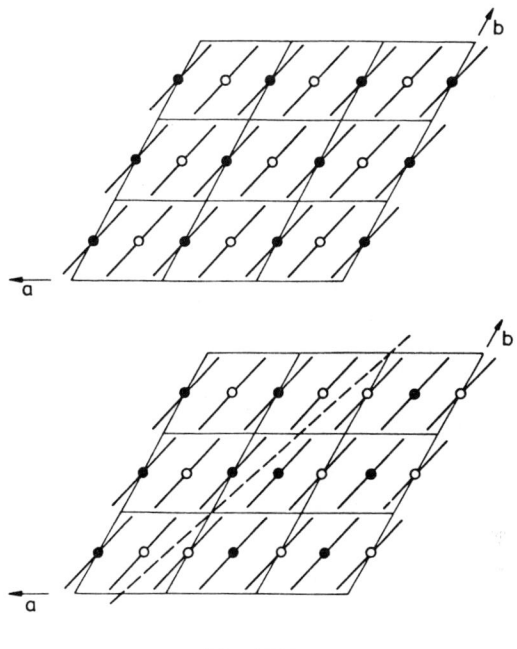

Fig. 10(c)

lattice (region B). In general, defect contrast will be strong or nonexistent depending on the phase shift caused by the defect region. The defect phase shift is given by $2\pi\mathbf{g}\cdot\mathbf{R}$, so that no visible contrast is produced by a displacement $\mathbf{R}$ in the image of planes (hkl) if the product $\mathbf{g}\cdot\mathbf{R}$ is an integer.

The displacement field of a stacking fault is particularly simple and systematic DF microscopy can identify such defects. Below a certain plane (called the fault plane) the entire crystal is translated by a constant amount with respect to the region above the fault plane. Young and Petermann examined solid state polymerised p-toluene sulphonate (PTS) polymer crystals and found regions of black and white fringe contrast which exhibited strong contrast for $\mathbf{g} = [4\bar{2}1]$ and weak contrast for $\mathbf{g} = [2\bar{1}0]$ (see Figs. 10a and 10b). Such $\mathbf{g}$ vector contrast dependence and other characteristic features such as fixed location of the specimen contrast on tilting, variation of number of fringes with change in s and sample thickness, and the trace of the boundary of the defect and the crystal basal habit plane were consistent with the interpretation of the contrast as due to stacking faults on

(210) planes with displacement vector $\mathbf{R} = \frac{1}{2}[1\bar{2}1]$. The stacking fault requires a translational shift of $a/2$, $-b$ normal to the chain axis and $c/2$ along the chain axis. Figure 10c shows schematically a crystal containing such a fault viewed along the chain axis. The large side groups of the PTS molecule lie approximately in the (210) plane and the shift is such to neither disrupt the molecular backbone or to unduly pack the relatively bulky side groups, therefore strongly reinforcing the deduction from microscopy.

3.12 Screw Dislocations

A considerably less straightforward situation regarding contrast interpretations for [001] screw dislocations in PE crystals exists and illustrates the difficulty and hazards of interpretation from a single DF image. In 1964, Keith and Passaglia[103] worked out the possible types of conventional dislocations which could be present in chain folded polymer crystals and noted that only [001] screw dislocations do not require bending of molecular chains for dislocation motion. [001] and [$hk0$] screw dislocations are supposed to comprise the twist boundaries between adjacent mosaic blocks.[60] It has been previously mentioned that this favored [001] type of dislocation would unfortunately remain invisible in moiré patterns. For the normal case of a PE lamella lying flat on a carbon support film, the crystal is oriented such that only ($hk0$) planes diffract and, by the $\mathbf{g} \cdot \mathbf{R}$ criterion, [001] screw dislocations will also be invisible for all possible $\mathbf{g}$'s (here $\mathbf{R} = \mathbf{b}\theta/2\pi$ where $\mathbf{b}$ is the Burger's vector of the dislocation ($\mathbf{b} = [00c]$) and θ is the angle of rotation about the dislocation line (the [001] direction)). However, in thin metal foils deliberate tilting of the sample to image screw dislocations end-on (the geometrical situation for PE lamellae) showed residual diffraction contrast. Tunstall et al.[104] suggested that the contrast might arise from surface relaxations and employed the equations of Eshelby and Stroh[105] for the displacement field of a screw dislocation near a surface and found good agreement of the experimental images and their two beam dynamical calculation. However, their contrast calculation was for thick, elastically isotropic foils (several $\xi\mathbf{g}$'s thick, $\xi\mathbf{g}$ being the extinction distance for the $\mathbf{g}$ reflection). Guan[106] repeated this calculation for very thin crystals (thickness of $1/7\ \xi\mathbf{g}$, i.e. appropriate for a 120 Å thick PE crystal viewed in (110) DF) with essentially the same result as for thick crystals (see Fig. 11a). The DF contrast is seen to be a symmetric pair of faint, dark lobes for the crystal set at $s = 0$. With increasing deviation of the crystal from

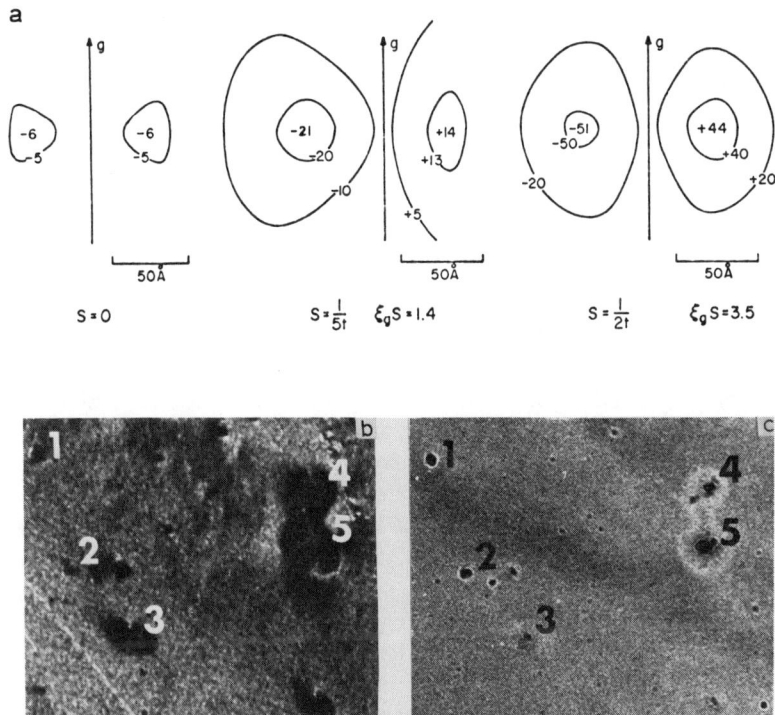

Fig. 11. (a) Two beam dynamical calculation of DF image of a screw dislocation viewed end-on in a PE single crystal. ($g = 110$, $\xi g = 890$ Å, foil thickness = 120 Å $\simeq \xi g/7$). (b) image ($g = 110$) showing double dark lobes with line of no contrast perpendicular to g. (c) BF image of same area as (b). Note the presence of dirt particles, numbered 1-5, some of which are associated with the strain contrast in (b).

the Bragg condition, the contrast increases substantially as a pair of lobes of opposite contrast. In all cases there is a line of no contrast (i.e. the intensity is the same as for the crystal far from the defect) *parallel* to the direction of g. The centre to centre spacing of the lobes is about 100 Å and the maximum size of each of the lobes (determined by minimally observable contrast level of 10%) is slightly less than 100 Å.

Petermann and Gleiter[107] reported diffraction contrast evidence for [001] screw dislocations in PE single crystals which had been placed on specially stepped carbon substrates in order to induce plastic deformation parallel to the chain axis and hence the production of [001] screw dislocations. The published evidence (Figure 1 of reference 107) is a (110) DF image showing many black–white lobe contrast regions of 200–400 Å centre to centre spacing with the line of no contrast *parallel* to g_{110}. Because the crystals are quite buckled, many bend contours are present and the correct effect of the deviation parameter on the image contrast is demonstrated. It must be noted however, that many published DF micrographs of polymer crystals[48,108–110] exhibit two lobes separated by a line of no contrast *perpendicular* to **g** (see Fig. 11b). By noting the orientation of the line of no contrast for several **g**'s, the symmetry of the strain field projected on the crystal face is determined to be radial. By comparing the areas indicated by the same numbers in Figs. 11b and 11c it is noted that the strain contrast in the DF image is associated with small dark regions in the BF image. These dark spots are interpreted as 'dirt' particles adhering to the crystal in solution or already present on the carbon support film, which then produce circular 'tent-like' bulges in the crystal upon sedimentation onto the support film (note the dark diffraction contrast in BF is 360° around particles 4 and 5). The local rotations of the crystal planes will have approximately radial symmetry around the particle giving rise to the line of no contrast perpendicular to **g**. The strain contrast in DF corresponds to the particles observed in BF in roughly 50% of the observations. This is to be expected since dirt particles which lie on top of the crystal surface do not cause any bulges.

Two features thus serve to distinguish the [001] screw dislocation image from that of bulges due to dirt particles—*orientation* of the diffraction contrast with respect to **g** (correct orientation of diffraction pattern and image requires calibration of the microscope image rotation) and *variable size* of the diffraction contrast region for the dirt particle interpretation (i.e. large dirt particles result in large regions of lobe contrast (see Fig. 11b)) versus *constant size* for the screw dislocation interpretation. A careful review of the diffraction contrast features in Fig. 1 of Reference 107 shows in addition to the contrast features attributed to screw dislocations, various larger (up to 1500 Å) regions of double lobe contrast (not referred to in the paper) with the line of no contrast parallel to the given **g**. It now remains to explain these features as due to a new defect similar to a [001] screw dislocation but producing much larger lobe contrast, or to suppose that the actual operating **g** vector is 1$\bar{1}$0 instead of 110 (a rotation of **g** of approximately 90°), for which all the observed contrast effects are simply attributed to elastic bending of the crystal about dirt particles of various sizes. Application of Ockham's razor favours the latter interpretation.

3.13 Fibres

Electron microscopy, along with X-ray scattering, has provided much of the information on the ultrastructure of fibres. Current interest in fibre science centres on the structural details of extremely high modulus, high strength fibres. Such fibres can be produced from various crystallisation–orientation processes from both flexible and rod-like chain molecules. Many models have been proposed describing the morphology of these fibres. The models represent the latest versions of the continuing evolution of the early ideas of Mark,[111] Staudinger[112] and (later) Kiessig[113] for the molecular structure of long chain fibres. The models, while all emphasising highly oriented chains, differ as to whether a given type of fibre consists of a discrete microfibrillar structure with crystallites connected axially in various ways or a homogeneous continuous crystalline material with various randomly distributed defects.

X-ray studies can be used to distinguish between alternative models (see Chapter 7). How can microscopy decide amongst the many models? BF imaging of fibres readily detects any microfibrillar substructure and permits direct measure of microfibre diameter, length and interconnectedness. The earliest evidence[114] for fibre axial periodicity, in fact predating SAXS,[113] comes from Kiessig's BF images of iodine stained fibres of polyamide, poly(vinyl alcohol) and cellulose. Because annealing drawn fibres results in substantial topological variations, replica methods and metal shadowing are useful.[115] Provided there exist regions of sufficient inner potential (density) difference, defocus phase contrast imaging may be able to show axial structure.[28]

DF imaging of fibres can directly measure the size of the diffracting crystalline regions (extent of lateral and axial coherence), and with much patience even provide the actual crystal size distributions[116] rather than a moment of the distribution as is the case for wide angle X-ray determinations. Moreover, positional variations (skin–core effects) in fibre morphology arising from processing conditions may be investigated. Of course, only those regions of the crystal oriented for

Bragg diffraction are imaged and dark regions can occur due to elastic bending, rotation, kinking, tilts, twins or local accumulation of defects as well as noncrystalline areas. Certainly when a long, uniformly diffracting region is observed, it is clear evidence for a large axial crystallite size[117] (see Fig. 12a). One difficulty is in ensuring that sample preparation steps necessary to obtain thin specimens for microscopy do not themselves cause a reduction in the crystal size. Microtomy, ultrasonic fragmentation and detachment replication are all far from 'gentle' and it is prudent to measure simultaneously crystallite size from unaltered bulk specimens by X-ray line broadening methods. A recent technique developed by Petermann[118] employs film drawing from a thin molten layer of polymer on either hot phosphoric acid or a heated glass surface. When the film is drawn off the heated surface it simultaneously undergoes high deformation and rapid cooling, resulting in a well oriented thin film suitable for direct microscopic examination, and by accumulating the film on a take-up roll sufficient sample can be made for WAXS, SAXS, DSC, etc. Micron long needle crystals (~200 Å diameter) have been observed in DF of both PE and isotatic polystyrene films prepared in this manner.[119]

In principle, systematic imaging using different $\mathbf{g}$ and s conditions can help to distinguish the various causes of the dark nondiffracting regions seen in a single DF image. However, for most polymers, small crystallite size and polymer radiation sensitivity severely limit this approach (as evidenced by the lack of published serial DF studies on fibres). A rather useful method in making a comparison between two DF micrographs is to employ a transparent sheet to mark the location and shape of features of interest (diffracting crystallites) and several prominent background features in the first micrograph which can then be overlaid on the second using the background features common to both for correct registration and accurate comparison. Experience indicates that radiation damage can cause unintentional changes in fibre diffraction so that a third micrograph with electron optical conditions identical to the first should be made if possible to assure that the changes in the second micrograph were due to the experimental design and not radiation damage. To achieve the best DF image contrast, the smallest available objective aperture should be used to select only the crystalline reflections of interest and to minimise the contribution of inelastic and amorphous region scattering to the image background.

Instead of a fibrillar structure, an unusual, radially oriented sheet structure is observed for poly(*para*-phenylene terephthalamide)

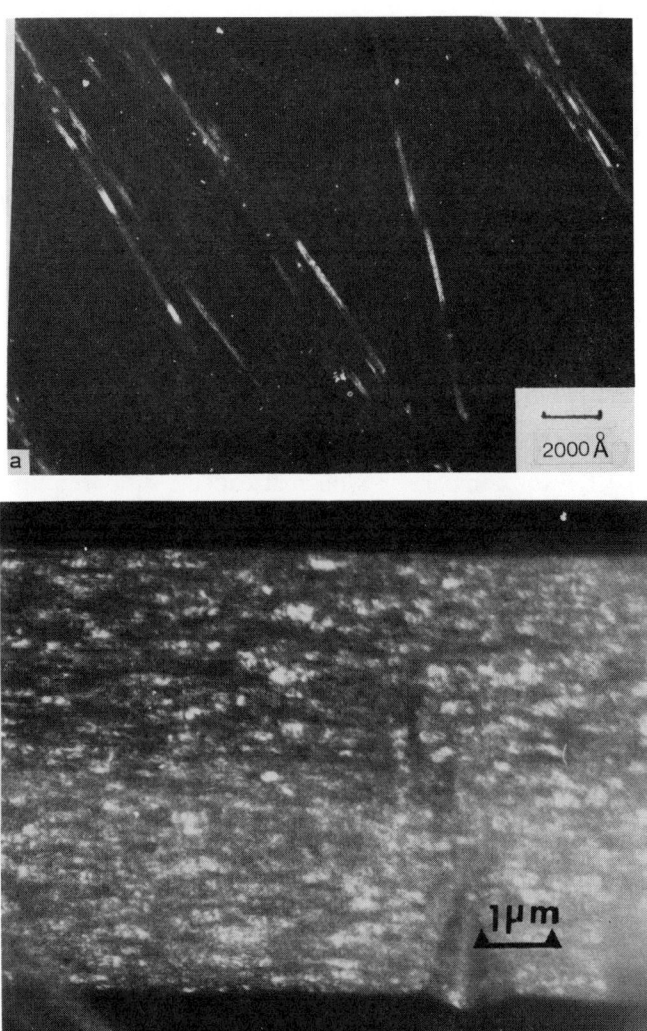

Fig. 12. (a) DF micrograph ($\mathbf{g} = \bar{1}01$, 101 and 002) of Valonia ventricosa (cellulose) fibrils showing long crystals[117] (Reprinted by permission from Bourret, A. Chanzy, H. and Lazaro, R. (1972). *Biopolymers*, **11,** 893. Copyright by John Wiley & Sons, New York, USA.) (b) DF micrograph ($\mathbf{g} = e_2 - e_4$) of poly(*para*-phenylene benzobisthiazole) film with texture of crystallites about the orientation axis.

(PPTA) fibres. Crystallite size in PPTA has been determined by measuring the coherently scattering regions in DF and by the extent of lattice fringe regions in axial BF (recall Fig. 7). The values of crystallite size by DF and BF imaging agree rather well with those measured by X-ray and electron diffraction line broadening measurements.[120] It is interesting to note that DF images of rigid molecule fibres and films can be remarkably similar in appearance to those of flexible chain fibres even though the molecules in the rod-like polymer are fully extended and there is no amorphous phase in the conventional sense.

SAD patterns from large regions usually exhibit apparent fibre symmetry, but microdiffraction patterns can display a local pseudo-single crystal texture.[121] Examination of DF images of thin areas (thickness of the order of the crystallite size) can be used to determine the population of diffracting crystallites per unit area in non-meridional DF for comparison with that expected for true fibre symmetry of the crystallites about the orientation direction. The fraction of diffracting crystals per unit area of fibre can be estimated as [(number of independent non-meridional reflections used to form the DF image) × (average angular half width of rocking curve for a crystallite of average diameter)]/360. For example, crystals with an average diameter of 120 Å have an angular half width of 2° (see Fig. 8b), hence for a single beam DF image less that 1% of the crystal population should be in the diffraction condition for true fibre symmetry. Rather than a small population of randomly separated diffracting individual crystallites, non-meridional DF images of PPTA fibres[45] and PBT fibres and films[121] reveal regions with upwards of 30% of the crystallites diffracting simultaneously (see Fig. 12b). Such a texture may arise from the extended chain nature of the molecules, the absence of noncrystalline regions and a ribbon rather than fibril morphology, imparting not only high axial orientation but substantial lateral alignment.

Solid state extruded PE fibres of high ($\geq 24\times$) extrusion draw ratio also show significant departure from true fibre symmetry in DF images.[122] In combined [110, 200] DF, many fibrils are non-diffracting except for sequences of 3–5 adjacent 350 Å long crystallites, each diffracting region being separated by an approximately 50–70 Å long non-diffracting region. Similar observations of 'colonies' of diffracting crystallites in drawn PE films were reported in 1962 by Scott in the first paper utilising DF for the study of the structure of microfibres.[123] The results for the solid state extruded PE fibres indicate that there is no lateral coherence between adjacent microfibrils but substantial axial coherence along a given microfibril.[124]

ACKNOWLEDGEMENTS

The support of the Alexander von Humboldt Foundation and the hospitality of the Institute of Macromolecular Chemistry, University of Freiburg are gratefully acknowledged.

REFERENCES

1. Misell, D. L. (1978). *Image Analysis, Enhancement and Interpretation*, North Holland, Amsterdam.
2. Hren, J. J., Goldstein, J. I. and Joy, D. C. (Eds) (1979). *Introduction to Analytical Electron Microscopy*, Plenum Press, New York.
3. Langmore, J., Wall, J. and Isaacson, M. S. (1973). *Optik*, **38**, 335.
4. Grubb, D. T. (1974). *J. Mat. Sci.*, **9**, 1715.
5. Grubb, D. T. (1982). *Developments in Crystalline Polymers—1*, Applied Science Publishers, London.
6. Isaacson, M. S. (1977). In *Principles and Techniques of Electron Microscopy*, Vol. 7 (M. Hayat (Ed.)) Van-Nostrand and Reinhold, New York.
7. Bahr, G. F., Johnson, F. B. and Zeitler, E. (1965). In *Quantitative Electron Microscopy* (G. F. Bahr and E. Zeitler (Eds)) Williams and Wilkins Co., Baltimore.
8. Petermann, J. and Gleiter, H. (1973). *Kolloid Z. u. Z. Polymere*, **251**, 850.
9. Neigisch, W. D. and Swan, P. R. (1960). *J. Appl. Phys.*, **31**, 1906.
10. Garber, C. A. and Geil, P. H. (1968). *Makro. Chem.*, **113**, 236.
11. Thomas, E. L. and Ast, D. G. (1974). *Polymer*, **15**, 37.
12. White, J. R. (1975). *Polymer*, **16**, 157.
13. Parsons, D. F. (1972). *J. Appl. Phys.*, **43**, 2885.
14. Dietrich, I., Formanek, H., Fox, F., Knapek, E. and Weyl, R. (1979). *Nature*, **277**, 380.
15. Knapek, E. and Dubochet, J. (1980). *J. Mol. Biol.*, **141**, 147.
16. Rose, A. (1948). *Adv. Electron.*, **1**, 131.
17. Isaacson, M. S., Johnson, D. and Crewe, A. (1973). *Rad. Res.*, **55**, 205.
18. Agar, A. (1957). *Brit. J. Appl. Phys.*, **8**, 410.
19. Herman, K. H., Krahl, D. and Rindfleisch, V. (1972). *Siemens Forsch. Entwickl. Ber.*, **1**, 167.
20. Underwood, E. E. (1970). *Quantitative Stereology*, Addison-Wesley, Reading, MA.
21. Hanszen, K. (1971). In *Advances in Optical and Electron Microscopy 4*, (R. Barer and V. E. Cosslett (Eds)) Academic Press, London.
22. Erickson, H. P. (1973). In *Advances in Optical and Electron Microscopy 5*, (R. Barer and V. E. Cosslett (Eds)) Academic Press, London.
23. Johnson, D. J. and Crawford, D. (1973). *J. Microscopy*, **98**, 313.
24. Christner, G. L. and Thomas, E. L. (1977). *J. Appl. Phys.*, **48**, 4063.
25. Thomas, E. L. and Roche, E. J. (1979). *Polymer*, **20**, 1413.
26. Roche, E. J. and Thomas, E. L. (1981). *Polymer*, **22**, 333.

27. Handlin, D. L., MacKnight, W. J. and Thomas, E. L. (1981). *Macromolecules*, **14,** 795.
28. Miles, M. J. and Petermann, J. (1979). *J. Macro. Sci. (Phys.)*, **B16,** 243.
29. Schoon, Th. G. F. and Kretschmer, R. (1964). *Kolloid Z. u. Z. Polymere*, **197,** 45–51 and 51–55.
30. Schoon, Th. G. F. and Kretschmer, R. (1966). *Kolloid Z. u. Z. Polymere*, **211,** 53.
31. Zingsheim, H. P. and Bachmann, L. (1971). *Kolloid Z. u. Z. Polymere*, **246,** 36.
32. Yeh, G. S. Y. and Geil, P. H. (1967). *J. Macro. Sci. (Phys.)*, **B1,** 235.
33. Yeh, G. S. Y. (1972). *Crit. Rev. Macromol. Chem.*, **1,** 173.
34. Miyaji, H. and Geil, P. H. (1981). *Polymer*, **22,** 701.
35. See special issue on 'Physical structure of the amorphous state,' *J. Macro. Sci. (Phys.)*, **B12,** pp. 1–301 (1976)
36. Uhlmann, D. R., Renninger, A. L., Kritchevsky, G. and Van der Sande, J. (1976). *J. Macro Sci. (Phys.)*, **B12,** 153.
37. Koutsky, J. A., Hein, N. V. and Cooper, S. L. (1970). *J. Pol. Sci. (Lett.)*, **8,** 353.
38. Wilkes, G. L., Samuels S. L., and Crystal, R. J. (1974). *J. Macro. Sci. (Phys.)*, **B10,** 203.
39. Marx, C. L., Koutsky, J. A. and Cooper, S. L. (1971). *J. Pol. Sci. (Lett.)*, **9,** 167.
40. Beamish, A., Goldberg, R. A. and Hourston, D. J. (1977). *Polymer*, **18,** 49.
41. Petermann, J. and Gleiter, H. (1975). *Phil. Mag.*, **31,** 929.
42. Handlin, D. L. and Thomas, E. L. (1983). *Macromolecules*, **16,** 1514.
43. Kihlborg, L. (Ed.) (1979). *Direct Imaging of Atoms in Crystals and Molecules*, Proceedings of the 47th Nobel Symposium, Royal Swedish Academy, Stockholm.
44. Keller, A. (1969). *Kolloid Z. u. Z. Polymere*, **231,** 389.
45. Dobb, M. G., Johnson, D. J. and Saville, B. P. (1977). *J. Pol. Sci. (Pol. Symp.)*, **58,** 237.
46. Shimamura, K., Minter, J. R. and Thomas, E. L. (1983). *J. Mat. Sci. (Lett.)*, **18,** 54.
47. Hirsch, P., Howie, A., Nicholson, R., Pashley, D. and Whelan, M. (1965). *Electron Microscopy of Thin Crystals*, Butterworths, London.
48. White, J. R. (1974). *J. Pol. Sci. (Phys. Edn)*, **12,** 2375.
49. Agar, A., Frank, F. C. and Keller, A. (1959). *Phil. Mag.*, **4,** 32.
50. Holland, V. F. (1964). *J. Appl. Phys.*, **35,** 3235.
51. Bassett, D. C. (1964). *Phil. Mag.*, **10,** 595.
52. Keller, A. (1972). *Rep. Prog. Phys.*, **31,** 623.
53. Thomas, E. L., Kramer, E. J. and Sass, S. L. (1974). *J. Pol. Sci. (Phys. Edn)*, **12,** 1015.
54. Holland, V. F. and Lindenmeyer, P. H. (1965). *J. Appl. Phys.*, **36,** 3049.
55. Holland, V. F., Lindenmeyer, P. H., Trivedi, R. and Amelinckx, S. (1965). *Phys. Stat. Solidi*, **10,** 543.
56. Sadler, D. M. and Keller, A. (1970). *Kolloid Z. u. Z. Polymere*, **239,** 641.
57. Sadler, D. M. and Keller, A. (1970). *Kolloid Z. u. Z. Polymere*, **242,** 1081.

58. Niinomi, M., Abe, K. and Takayanagi, M. (1968). *J. Macro. Sci. (Phys.)*, **B2**, 649.
59. Fischer, E. W. and Lorenz, R. (1963). *Kolloid Z. u. Z. Polymere*, **189**, 97.
60. Hosemann, R., Wilke, W. and Balta Calleja, F. J. (1966). *Acta Cryst.*, **21**, 118.
61. Smith, G. H. and Burge, R. E. (1963). *Proc. Phys. Soc.*, **81**, 612.
62. Bahr, G. F. and Zeitler, E. (1962). *J. Appl. Phys.*, **33**, 847.
63. Engel, A. (1978). *Ultramicroscopy*, **3**, 273.
64. Misell, D. L. and Burdett, I. D. (1977). *J. Microscopy*, **109**, 171.
65. Flood, R. (1980). *Proceedings of SEM*, **1**, 183.
66. Lauterwasser, B. D. and Kramer, E. J. (1979). *Phil. Mag.*, **A39**, 469.
67. Adams, W. W. (1984). Ph.D. Thesis, University of Massachusetts.
68. Aharoni, S. M. (1978). *Polymer*, **19**, 401.
69. Aharoni, S. M. (1978). *J. Macro. Sci. (Phys.)*, **15**, 635.
70. Aharoni, S. M. (1978). *Macromolecules*, **11**, 677.
71. Aharoni, S. M., Kramer, V. and Vernick, D. A. (1979). *Macromolecules*, **12**, 265.
72. Andrews, E. H. (1964). *Proc. Roy. Soc. (London)*, **A277**, 562.
73. Kato, K. (1965). *J. Electron Microsc.*, **14**, 219.
74. Kanig, G. (1973). *Kolloid Z. u. Z. Polymere*, **251**, 782.
75. Voigt-Martin, I. G. (1980). *J. Pol. Sci. (Phys. Edn)*, **18**, 1375.
76. Voigt-Martin, I. G., Fischer, E. W. and Mandelkern, L. (1980). *J. Pol. Sci.*, **18**, 2347.
77. Schimmel, G. (1969). *Elektronenmikroskopische Methodik*, Springer Verlag, Berlin.
78. Wunderlich, B. (1973). *Macromolecular Physics*, Vol. 1, Academic Press, New York.
79. Bassett, G. A. (1958). *Phil. Mag.*, **3**, 1042.
80. Bassett, G. A., Blundell, D. J. and Keller, A. (1967). *J. Macro. Sci. (Phys.)*, **B1**, 161.
81. Spit, B. J. (1968). *J. Macro. Sci. (Phys.)*, **B2**, 45.
82. Bassett, D. C. (1961). *Phil. Mag.*, **6**, 103.
83. Geil, P. H. (1963). *Polymer Single Crystals*, Wiley Interscience, New York.
84. Olley, R. H., Hodge, A. M. and Bassett, D. C. (1979). *J. Pol. Sci. (Phys. Edn)*, **17**, 627.
85. Claffey, W., Gardner, K., Blackwell, J., Lando, J. and Geil, P. H. (1974). *Phil. Mag.*, **30**, 1223.
86. Allen, P. and Bevis, M. (1974). *Proc. Roy. Soc. (London)*, **A341**, 75.
87. Geiss, R. H. (1976). In *Developments in Electron Microscopy and Analysis*, (J. Venables (Ed.)) Academic Press, New York.
88. Low, A., Vesely, D., Allan, P. and Bevis, M. (1978). *J. Mat. Sci.*, **13**, 711.
89. Sherman, E. S. and Thomas, E. L. (1979). *J. Mat. Sci.*, **14**, 1109.
90. Sherman, E. S., Adams, W. W. and Thomas, E. L. (1981). *J. Mat. Sci.*, **16**, 1.
91. Khoury, F. (1979). Faraday Discussions of the Chemical Society, No. 68, *Organization of Macromolecules in the Condensed Phase*, The Faraday Division, Royal Society of Chemistry, London, p. 404.

92. Roche, E. J., Stein, R. S. and Thomas, E. L. (1980). *J. Pol. Sci. (Phys. Edn)*, **18,** 1145.
93. Thielke, H. G. and Billmeyer, F. W. (1964). *J. Pol. Sci.*, **A2,** 2947.
94. Stuart, H. A. (Ed.) (1955). *Die Physik der Hochpolymerer*, Vol. 3, Springer Verlag, Berlin.
95. Dorset, D. L. (1979). *J. Pol. Sci. (Phys. Edn)*, **17,** 1797.
96. Harrison, I. R. and Runt, J. (1976). *J. Pol. Sci. (Phys. Edn)*, **14,** 317.
97. White, J. R. (1978). *J. Pol. Sci. (Phys. Edn)*, **16,** 387.
98. Bassett, D. C., Frank, F. C. and Keller, A. (1963). *Phil. Mag.*, **8,** 1753.
99. Harrison, I. R., Keller, A., Sadler, D. M. and Thomas, E. L. (1976). *Polymer*, **17,** 736.
100. Hsu, S. L. and Krimm, S. (1977). *J. Appl. Phys.*, **48,** 10.
101. Wegner, G. (1969). *Z. Naturforsch.*, **24,** 824.
102. Young, R. J. and Petermann, J. (1981). *J. Mat. Sci.*, **16,** 1835.
103. Keith, H. D. and Passaglia, E. (1964). *J. Res. NBS*, **68A,** 513.
104. Tunstall, W., Hirsch, P. B. and Steeds, J. (1964). *Phil. Mag.*, **9,** 99.
105. Eshelby, J. D. and Stroh, A. N. (1951). *Phil. Mag.*, **42,** 1401.
106. Guan, D. Y. (1973). Unpublished results.
107. Petermann, J. and Gleiter, H. (1972). *Phil. Mag.*, **25,** 813.
108. Thomas, E. L., Sass, S. L. and Kramer, E. J. (1974). *Phil. Mag.*, **30,** 335.
109. Bassett, D. C. (1964). *Phil. Mag.*, **10,** 595.
110. White, J. R. (1974). *J. Mat. Sci.*, **9,** 1860.
111. Meyer, K. H. and Mark, H. F. (1930). *Der Aufbau der Hochpolymeren Naturstoffe*, Akademische Verlagsgesellschaft, Leipzig.
112. Staudinger, H. (1932). *Die Hochmolekularen Organischen Verbindungen*, Springer Verlag, Berlin.
113. Hess, K. and Kiessig, H. (1943). *Naturwissenschaften*, **31,** 171.
114. Hess, K. and Kiessig, H. (1944). *Z. Physik. Chem.*, **A193,** 196.
115. Peterlin, A. (1967). In *Man Made Fibers Science and Technology*, Vol. 1, (H. Mark, S. Atlas and E. Cernia (Eds)) Interscience, New York.
116. Grubb, D. T. and Hill, M. J. (1980). *J. Crystal Growth*, **48,** 321.
117. Bourret, A., Chanzy, H. and Lazaro, R. (1972). *Biopolymers*, **11,** 893.
118. Petermann, J. and Gohil, R. M. (1979). *J. Mat. Sci.*, **14,** 2260.
119. Petermann, J., Miles, M. and Gleiter, H. (1979). *J. Pol. Sci. (Phys. Edn)*, **17,** 55.
120. Dobb, M. G., Johnson, D. J. and Saville, B. P. (1980). *Phil. Trans. R. Soc. (London)*, **A294,** 483.
121. Minter, J. M., Shimamura, K. and Thomas, E. L. (1981). *J. Mat. Sci.*, **16,** 3303.
122. Sherman, E. S., Porter, R. S. and Thomas, E. L. (1982). *Polymer*, **23,** 1069.
123. Scott, R. G. (1962). *J. Pol. Sci.*, **57,** 405.
124. Clements, J., Jakeways, R. and Ward, I. M. (1978). *Polymer*, **19,** 639.

Chapter 4

NEUTRON SCATTERING BY CRYSTALLINE POLYMERS: MOLECULAR CONFORMATIONS AND THEIR INTERPRETATION

D. M. SADLER

H. H. Wills Physics Laboratory, University of Bristol, UK

1 INTRODUCTION

1.1 Aims and Scope

What are the shapes of molecules inside polymer crystals, and how do the crystals form that way? Neutron scattering, thanks to its ability to 'see' individual molecules, has already made considerable progress in answering the first if not the second of these questions.

The technique relies on the different neutron scattering from protons (^{1}H nucleus) and deuterons (^{2}H). A solid solution of ^{1}H and ^{2}H polymer molecules is chemically almost homogeneous, yet the neutron scattering contains a contribution which depends only on *intra*molecular separations. This means that one molecule can be imaged indirectly if not directly in isolation from the surrounding molecules. Figure 1 shows the sort of conformation we wish to test and illustrates in 'cartoon' form how, without the isotope contrast shown in the diagram by filled and open circles, it would not be possible to discern the passage of one molecule. Figure 2^1 shows a useful representation of conformations in crystal lamellae, in a projection along the chain axis direction in the lattice. In this case the large dots describe the positions of the stems of one molecule. A 'stem' is that sequence of monomers necessary to cross the crystal lattice of a lamella.

The theme of this book is how to assess the degree of certainty of current models and explanations. There are several levels: the preci-

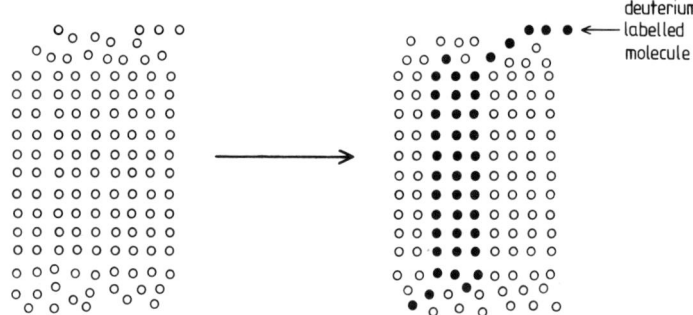

Fig. 1. Schematic representation of a cross-section of a polymer crystal, the circles representing monomers: (left) no labelling, (right) with labelling.

sion and accuracy of data, the inherent difficulties of reconstructing the conformation of molecules from scattering in the absence of an imaging system, and how to interpret the conformation in terms of crystallisation mechanisms. The first does not represent a serious limitation to date. The emphasis in discussing the second level will be: which features of the conformations can be tested most effectively against the neutron scattering results? Finally, and most interestingly, it will be shown that the conformations which have been proposed cannot be considered independently of the ideas on crystallisation which they presume, either explicitly or implicitly.

In order to serve best these aims it has been necessary to be selective. While the description of at least the outline of results for a range of crystalline polymers is extensive, the detailed discussion of conformations is restricted mainly to polyethylene. The contents are arranged in the following manner:

1. Introduction to the theories and evidence relevant to lamellar crystal growth.
2. Scattering theory, with an emphasis on an explanation (2.4) of the isotope labelling technique in terms of 'partial coherence' and scattering from anisotropic samples (2.7).
3. Measurements: experimental methods and treatment of data.
4. An account of results of neutron scattering on crystalline polymers, listed according to polymer and sample type.

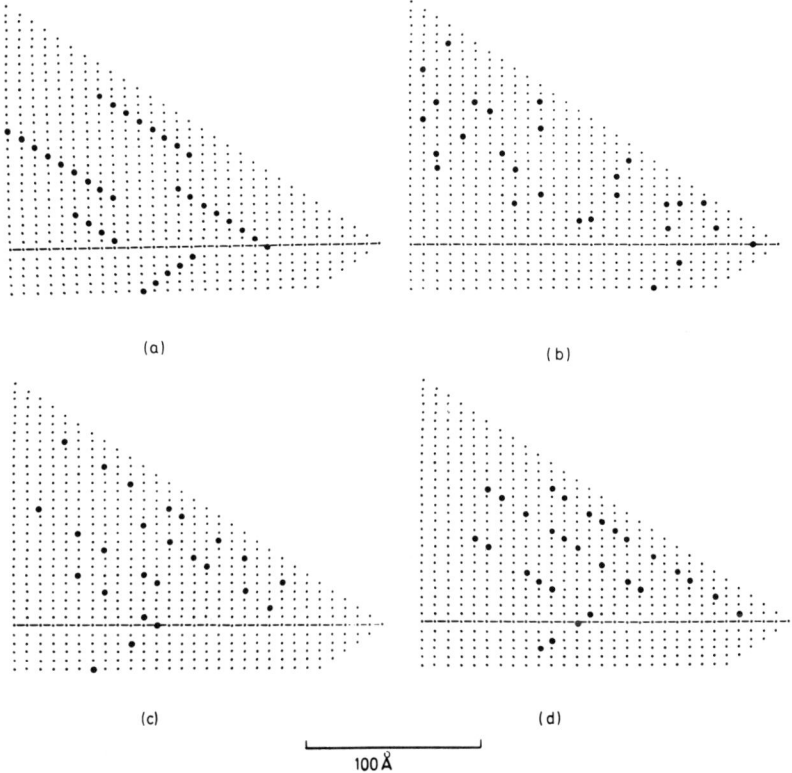

Fig. 2. Representation of the positions of intersections of stems (i.e. chain sequences passing through the crystalline region) and a (001) plane. The small points represent PE chains, the large points DPE. The examples of chain trajectory are: (a) adjacently re-entrant folding; (b) randomly re-entrant folding; (c) and (d) other forms of folding restricted to (110) planes (c) large gaps between stems, (d) small gaps. The broken line indicates a fold sector boundary.

5. Interpretation of results on polyethylene in terms of molecular conformation, including a recent interpretation of their significance in terms of the (variable) structure of the crystal growth face.
6. Deformation studies.
7. Concluding remarks.

1.2 Mechanisms of Crystallisation

Many of the debates on how best to approach an understanding of crystallisation turn on the relative weight to be given to equilibrium or kinetic approaches. Inspiration is drawn either from (equilibrium) theories of polymer liquids and solutions,[2] or on the other hand from theories based on the growth of crystals which are not polymeric.[3] Since crystals only reach equilibrium in exceptional circumstances, a crystal growth approach naturally emphasises kinetic effects.

An equilibrium approach may be itself subdivided according to whether (a) the conformation in the crystals is in some sense an equilibrium one or (b) it is thought that the (equilibrium) conformation prior to crystallisation is largely retained during the crystallisation process. Models based on (a) and (b) may be superficially very similar, but they imply quite different concepts of how polymers crystallise.

Kinetic approaches are potentially even more diverse, since it is necessary to specify not only the end state (the conformation after crystallisation) but also the sequence of transition states which lead to it. It has been customary to suppose a secondary nucleation mode of crystal growth, which is based on the crystal growth surface being smooth on a molecular scale except during relatively rare events when a new growth layer is being nucleated.[3,4] The model does not severely restrict the types of conformation: although folding the chain adjacently (Figs 1 and 2) simplifies the calculations, at least some degree of non-adjacency could be incorporated in this approach.

Again, a kinetic approach may be adopted which is based on a rough or 'crenellated' growth surface on the lamellar edges. The roughness may be purely kinetic in origin[5] or may be primarily an equilibrium phenomenon.[6]

One is led to the conclusion that there are not just two or three clearly defined conformations which follow unambiguously from the choice of a type of theory of crystallisation. It is true that to some extent the scattering data can be used to 'test' which type of theory is most promising, but it is also necessary to examine the theories constantly and search for how they may be adapted in the light of new results. Put another way, we are not asking which immutable theory is 'correct': we are looking for new insights.

1.3 The Starting Point

It would be desirable to distil the work of the last decade so as to provide a core of solid conclusions which cannot seriously be questioned when constructing models for neutron scattering. Companion chapters in this volume are very relevant to this, but it is useful to make a list which is intended as an *aide-mémoire*. Several reviews exist[7-11] which describe the material more fully.

1. Lamellar crystals[12] are characteristic of the crystallisation of regular polymers from the coiled state (either liquid or solution). This excludes crystallisation of proteins, crystallisation under flow[13] or stress when coils are extended, and crystallisation when the liquid crystalline state is involved.[14] The chains are usually much longer than the lamellar thickness, and are at an angle to the lamellar normals in the range 0°–30°. For monolayer crystals it follows that the molecules fold to and fro across the crystals. (In this chapter the words 'fold' and 'folding' do not imply a type of fold, e.g. 'regular', 'tight', or 'random'.) Folding in this sense is also the norm for multilayered crystals, unless the molecule expands during crystallisation to a size much larger than the thickness of the lamella. This point is amplified in Section 5.2.
2. The lamellar thickness varies approximately as the inverse of the supercooling,[15,4] though for supercoolings greater than about 50 °C a plateau or constant thickness is reached.[16] There can be effects specific to long monomer repeat distances.[17]
3. The outlines of the lamellae sometimes follow low index crystallographic directions (usually a close packed direction such as ⟨110⟩ in polyethylene), but they can be curved or irregular, especially for crystallisation from the liquid polymer (melt). A catalogue of morphologies is given in Tables III.3 and III.5 of Reference 11 for melt and solution crystallisation respectively. The simple straight facets are normally taken as evidence for singular surfaces (in the sense discussed in Reference 18), e.g. as expected for molecularly smooth surfaces. No account has been taken until recently[6] of growth faces which show characteristic curvature (e.g. Reference 19).
4. In several detailed studies it has been shown that when simple straight facets exist they can lead to 'sectors' which are physically distinct according to the orientation of the facet.[20] For example, the lozenge shaped crystals of polyethylene are divided into four sectors: two grow using the (110) faces and two using ($\bar{1}$10) faces. They differ by the direction of the normal to the fold surface with respect to the unit cell,[20] by unit cell distortions[21] and by their

deformation properties.[22] The existence of sectors is explained by a preferential fold direction parallel to the growth face.

5. The crystallinity as assessed by the usual methods (e.g. density, heat of fusion) has a maximum of about 80–90%, though it can be much less.[23] It is accepted that in general the disordered regions are on the fold surfaces of the lamellae, although in some low crystallinity polymers there may be regions of uncrystallised polymer between the crystals.

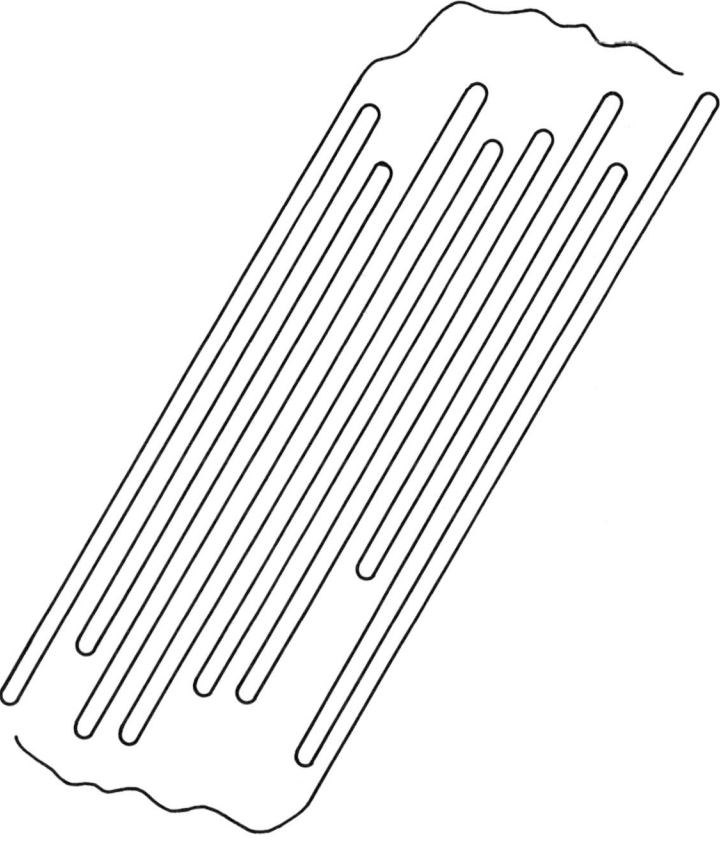

Fig. 3. Schematic representation (in cross-section) of some of the features of folding in polyethylene solution grown crystals, with both fluctuations in fold heights and some loose loops giving rise to disorder. Strictly adjacency for folding is shown for clarity (cf. later in the review).

NEUTRON SCATTERING BY CRYSTALLINE POLYMERS 131

6. A (fold) surface free energy is derived from melting points.[24]
7. Small-angle X-ray scattering shows that the surface disordered regions are similar in width and overall density to that expected for amorphous polymer.[25]

Many observations have been made on selected systems and may not necessarily be relevant to others. In particular, solution grown crystals (SX) are much more amenable to study, but conclusions based on them may not be valid for melt grown crystals (MX).

1.4 Fold Surface Structure

Neutron scattering, while it is sensitive to the separations between stems of the same molecule, is much less informative on the nature of chain packing in the surface regions. Most of the studies on this subject have been on solution grown crystals of PE as a model system. Some of the conclusions are worth bearing in mind.

Figure 3 shows a model for most of the surface disorder which explains many disparate observations. A significant source of disorder is the fluctuating height of the fold with respect to the outer surface of the lamella which was predicted theoretically.[26] The results which favour, or are in agreement with, this model are based on electron microscopy,[27] degradation studies,[28,29] Raman scattering[30–32,7] and X-ray diffraction line widths.[33]

It is perhaps not surprising that a simple two phase crystal/amorphous model should not be adequate: as the chains emerge from a crystalline towards an amorphous region many reasons exist[34] for supposing that the transition from order to disorder should not be abrupt.

2 SCATTERING THEORY

2.1 General

Neutron scattering is used to distinguish one molecule from its neighbours in the way that one would distinguish one woollen thread in a tangle from neighbours of a different colour (see Fig. 1). It is common to compare a mixture of molecules of different isotopes with a solution, with (usually) the hydrogenous (H) polymer as solvent and the deuterated (D) as solute. For this reason the deuterium concentration is usually less than a few percent. As the following sections make clear, a better analogy is with spin incoherence. Whatever the relative concentrations of H and D, the scattering is a superposition of that expected for the structure with H and D nuclei replaced with an 'average' nucleus, together with a signal related to the isotope differences (Section 2.4).

All the methods of calculation in this chapter stem from one basic equation. The difficulties in interpretation come not from uncertainties in the scattering laws, but from the basic fact that the intensities and not the amplitudes of waves are measured. This means that it is not, in general, possible to do a reverse Fourier transform since part of the information necessary to do so is missing (the crystallographers' 'phase problem').

What follows is an abbreviated summary of the relevant theory. Its main purpose is to indicate in outline how the basic equations can be traced to what is commonly known about diffraction at undergraduate level. Several textbooks and reviews (e.g. References 35 and 36) give the general theory of neutron scattering. Jacrot[37] gives a useful review of the method as applied to biological systems. This has many affinities to its use for synthetic macromolecules.

2.2 Scattering from Atoms

For neutrons (or X-rays) the individual scattering process from nuclei (or electrons) is sufficiently weak that the incoming wave is perturbed very little. The approximation due to Born is then very accurate: the amplitude of scattering is calculated as the sum of amplitudes from spherical wavelets originating from every part of the material which interacts with the incoming waves. For X-rays the scattering is related to the electron density, which in general varies smoothly in space. For neutrons most of the scattering is due to the nuclei, which act as point scatterers since their dimensions are very small compared with the neutron wavelengths. (Neutrons can also be scattered by unpaired electrons which produce net magnetic moments, but this is not as yet relevant to polymer science.) The scattering length, b, is sufficient to characterise the wavelet originating from one isolated nucleus. b^2 can be defined as the probability of an incoming neutron being scattered into one steradian per unit flux of neutrons (flux is number per area of beam). The value of b depends both on the isotope and on the way the neutron spin combines with any spin the nucleus may have. Thus a chemically homogeneous system will often have a heterogeneity as concerns neutron scattering because of the presence of several isotopes and because nuclear spins are uncorrelated except under exceptional circumstances. It is customary to quote atomic scattering characteristics so as to allow for the spin heterogeneity. The 'coherent scattering length' for one isotope only is the average b value (usually of the order of 0.5×10^{-12} cm). The incoherent scattering is zero when the nuclear spin is zero, and in other cases increases with the spin dependence of b. The coherent contribution has all the interference properties found with X-rays, for example, while for the incoherent contribution the intensities from different atoms are simply added, with no interference effects. As noted above the theory for random mixing of nuclear spins has analogies with that for non-random mixing of isotopes.

Hydrogen is not only the most plentiful atom in most polymers, but it also has highly atypical neutron scattering properties. Firstly ^{1}H has a very high incoherent cross section, which is very relevant to the spectroscopic use of (inelastic) neutron scattering (for a review see Reference 38). The structural work of the type discussed in this chapter is based on the other characteristic of hydrogen: the very large difference between the coherent scattering length of ^{1}H (b_H is -0.374×10^{-12} cm) from that of the other hydrogen isotopes (b_D, for deuterium, is $+0.667 \times 10^{-12}$ cm). Unfortunately, the chemical effects of changing the hydrogen mass by two, although small, are not completely negligible. For example, the melting temperature of ice is increased by 4 °C on deuteration, and that of polyethylene decreased by 4 °C.

The dimensions of b^2 are area, and the scattering is often characterised by the differential cross section, which for one nucleus is simply b^2. ('Cross section' can also refer to the total scattering integrated over all solid angles, i.e. $4\pi b^2$ for one nucleus). Experimentally, the differential cross section can be quoted per volume of sample (the 'macroscopic' cross section $d\Sigma/d\Omega$). Other conventions exist for expressing the intensity of scattering (see below). For example, the scattering may be expressed as a multiple of that expected for one nucleus. This is similar to the convention used in X-ray scattering[39] where the intensity is often expressed as a multiple of that from one electron.

Inelastic effects are normally neglected in the limit of small q (defined in Section 2.3). This is reasonable since they would only be important if there were movements which were highly correlated over large distances. The incoherent scattering does show inelastic effects which can be important when these are being subtracted (see Section 3). In particular, the physical state of the hydrogen atoms can influence the incoherent scattering even at small angles. (This has been observed for liquid polymers at different temperatures (Dr A. Maconnachie, private communication).)

2.3 Scattering and Structure

The coherent scattering from neighbouring nuclei will give interference effects, and this is how in principle structure information is available.

There will be a phase difference of $(\mathbf{k}-\mathbf{k}_0) \cdot \mathbf{r}_{ij}$ for the scattering between two nuclei separated in space by $\mathbf{r}_{ij}$, where $\mathbf{k}_0$ and $\mathbf{k}$ are vectors parallel to the incident and scattered neutron directions respectively. We consider elastic scattering only, so that $\mathbf{k}$ and $\mathbf{k}_0$ are equal in magnitude ($k = 2\pi/\lambda$). The scattering angle 2θ is commonly expressed in terms of:

$$\mathbf{q} \equiv \mathbf{k} - \mathbf{k}_0$$

hence:

$$q = 4\pi \sin \theta / \lambda$$

where 2θ is the scattering angle and λ the wavelength. A second convention uses $\mathbf{s} \equiv \mathbf{q}/2\pi$. (In this case $1/s$ is the 'equivalent Bragg spacing'.)

In order to calculate an intensity which can be compared to experiment it is necessary first to calculate the scattering for one 'assembly', which may be chosen for convenience (e.g. one molecule, one crystallite or even the whole sample). It is useful to choose an assembly such that the scattering from different assemblies add incoherently (i.e. the intensity contributions from the N assemblies per volume of sample are additive). The total amplitude of scattering is given by:

$$A(\mathbf{q}) = \sum_i b_i \exp(i\mathbf{q} \cdot \mathbf{r}) \quad (1)$$

where the exponential term arises from the phase differences, the origin is chosen arbitrarily, and:

$$d\Sigma/d\Omega = N |A(\mathbf{q})|^2 / V \quad (2)$$

where N is the number of assemblies and V the sample volume.

Most of the equations which are relevant to structure studies are elaborations and extensions of eqn. (1), whose identity as a Fourier transform (F) is more clearly apparent when a density function $\rho(\mathbf{r})$ is used, which represents the total $\sum b_i$ per unit volume at the position $\mathbf{r}$:

$$A(\mathbf{q}) = \int \rho(\mathbf{r}) \exp(i\mathbf{q} \cdot \mathbf{r}) \, d^3r \tag{3}$$

i.e.

$$A(\mathbf{q}) = F[\rho(\mathbf{r})]$$

Several Fourier transform theorems are useful:

$$\rho(\mathbf{r}) = \rho_1(\mathbf{r}) * \rho_2(\mathbf{r})$$

where the symbol $*$ signifies a convolution, then:

$$A(\mathbf{q}) = F[\rho_1(\mathbf{r})] \times F[\rho_2(\mathbf{r})]$$

For example, ρ_1 could signify the density inside one unit cell of a crystal, and ρ_2 a lattice of points representing the positions of a set of unit cells. The convolution theorem also leads to:

$$|A(\mathbf{q})|^2 = F(\rho(\mathbf{r}) * \rho(\mathbf{r})) = F[\gamma(\mathbf{r})]$$

This equation is of special interest since the 'density correlation function' $\gamma(\mathbf{r})$† can in principle be measured by calculating from experimental results:

$$\gamma(\mathbf{r}) = F|A(\mathbf{q})|^2$$

$\gamma(\mathbf{r})$ is related to the Patterson function used by crystallographers.

In many experiments the assemblies take up many orientations within a sample. This could be allowed for by making the requisite average of eqn. (3), but for an isotropic sample the averaging over orientations is best done before making the Fourier transform. The result, due to Debye, is:

$$\langle |A(\mathbf{q})|^2 \rangle = \sum_i b_i^2 + \sum_i \sum_{\substack{j \\ i \neq j}} \frac{b_i b_j \sin(qr_{ij})}{(qr_{ij})} \tag{4}$$

The magnitudes of all vectors $\mathbf{r}_{ij}$ between scattering centres is included in eqn. (4), and as one may expect there is no direct information in the intensity curve about the relative orientation of two

† There is further discussion of the correlation function in Section 3 of Chapter 5.

vectors $\mathbf{r}_{ij}$ and $\mathbf{r}_{kl}$. With continuous notation, eqn. (4) becomes:

$$\langle |A(\mathbf{q})|^2 \rangle = \int \gamma(r) \sin(qr)/(qr) \, dr \qquad (5)$$

2.4 Scattering from Mixed Systems

For a chemically and isotopically homogeneous material, the only fluctuations in scattering power across the sample are due to spin orientations (see above): the result is a scattering signal essentially the same as for X-rays, with an incoherent background. A random mixing of isotopes (e.g. as a result of natural isotope abundance) would simply add to the spin incoherent scattering. Non-random mixing of isotopes such that the isotope is the same *within* each molecule is the key contribution of neutron scattering to polymer science, since it can separate intra- from intermolecular interference contributions to the scattering. Its first application was to amorphous polymers and solutions (see for example References 40–43).

Consider a macroscopic assembly of identical molecules which are either fully hydrogenous (Isotope 1) or fully deuterated (Isotope 2). Quite generally:

$$A(\mathbf{q}) = \sum_i \sum_j b_i A_i A'_j$$

where

$$A_i = \exp(i\mathbf{R}_i \cdot \mathbf{q}) \quad \text{and} \quad A'_j = \exp(i\mathbf{r}_{ji} \cdot \mathbf{q})$$

where $\mathbf{R}_i$ is the position of the centre of gravity of molecule i and $\mathbf{r}_j$ is the displacement of atom j within that molecule. The hydrogen scattering length b_i of the ith atom can be written as $\bar{b} + \delta b_i$ where $\bar{b}$ is the average over the two isotopes. Hence

$$A(\mathbf{q}) = \sum_i \sum_j (\bar{b} + \delta b_i) A_i A'_j + \sum_i \sum_l b_l A_i A''_l$$

where b_l represents the scattering lengths of atoms other than hydrogen, and A_l is the corresponding scattering amplitude.

$$A(\mathbf{q}) = \left[\sum_i \sum_j \bar{b} A_i A'_j + \sum_i \sum_l b_l A_i A''_l \right] + \sum_i \sum_j A_i A'_j \delta b_i \qquad (6)$$

The measured intensity is obtained from the modulus square of $A(\mathbf{q})$, and the crux of the calculation is to identify those contributions which, over a sufficiently large number of molecules, will average to zero. The modulus square of the sum of the first and second terms in eqn. (6) (in square brackets) is closely related to the scattering expected from density fluctuations in a pure material. If $|A_0(\mathbf{q}, b_k)|^2$ is the scattering for pure isotope k, then the density fluctuation term will be $|A_0(\mathbf{q}, \bar{b})|^2$. Hence:

$$A(\mathbf{q}) = A_0(\mathbf{q}, \bar{b}) + \sum_i \sum_j A_i A'_j \, \delta b_i \tag{7}$$

A_0 is zero at small q for an incompressible homogeneous fluid in equilibrium. This can often simplify the equations for solutions whose compressibility can be ignored, but this is not relevant to crystalline polymers.[44] In taking the modulus squared of $A(\mathbf{q})$ (eqn. (7)) the cross term averages to zero over the whole assembly *if* the isotopic mixing is random. The randomness must fulfil two conditions:

(a) the value of $A_i A_{i'}$ is not correlated with whether molecule i and molecule i' are of the same or different isotopes
(b) there is no correlation between isotope fluctuations (in the second term of eqn. (6)) and density fluctuations (in the first term, in square brackets).

The modulus square of the second term in eqn. (7) then represents what is of special interest in these experiments—the scattering function for one molecule:

$$n^2 P(\mathbf{q}) \equiv \left| \sum_j A'_j \right|^2$$

where n is the number of hydrogens per molecule. The function $P(\mathbf{q})$ is normalised so as to go to unity as $\mathbf{q} \to 0$. Just as in the basic equation of Debye (eqn. (4)) there will be a series of cross terms of the type $A_i A^*_{i'} A'_j A'_{j'} \, \delta b_i \, \delta b_{i'}$, where A^*_i is a complex conjugate. In the ensemble average over all the sample, there will be many pairs of molecules which contribute the same factor $A_i A^*_{i'} A'_j A'_{j'}$. As long as the mixing is random (condition (a) above) the sum of all such terms with $i \neq i'$ will be zero, since δb_i is defined so as to sum to zero. Hence eqn. (6) becomes:

$$|A(\mathbf{q})|^2 = |A_0(\mathbf{q}, \bar{b})|^2 + \sum_k \delta b_k^2 (n^2 P(\mathbf{q})) N_k \tag{8}$$

where the sum over k extends over the different isotopes present (usually 2) and N_k is the number of molecules present of isotope k.

For several years after the use of the technique for polymers, experiments were carried out with one isotopic species dilute ($\approx 1-3\%$ deuterated D) compared to the other ($\approx 99-97\%$ hydrogenous H). In that case $\delta b_H = 0$, $\delta b_D \simeq b_D - b_H$, and $P(\mathbf{q})$ refers only to the D species. In addition $|A_0(\mathbf{q}, b_H)|^2$ is usually very small indeed at small $\mathbf{q}$, so that eqn. (8) becomes:

$$|A(\mathbf{q})|^2 = N_D n^2 P(\mathbf{q})(b_D - b_H)^2 \qquad (9)$$

In addition it has been recognised recently[45–48] that equations equivalent to eqn. (8) enable any concentrations to be used, the result remaining simple as long as the two isotopic species are nearly equivalent. For example, for two isotopes:

$$|A(\mathbf{q})|^2 = |A_0(\mathbf{q}, \bar{b})|^2 + n^2 P(\mathbf{q})\{N_H(b_H - \bar{b})^2 + N_D(b_D - \bar{b})^2\}$$
$$= |A_0(\mathbf{q}, \bar{b})|^2 + (N_D + N_H)n^2 P(\mathbf{q})(\overline{b^2} - \bar{b}^2) \qquad (10)$$

This result resembles closely the result of calculating the effect of spin incoherence[35,36] which differs from eqns (8)–(10) only in the term containing $P(\mathbf{q})$. For spin incoherence there are no remaining systematic interference terms for different nuclei, and eqn. (10) applies with $P(\mathbf{q}) = 1$ and $n = 1$. ($P(\mathbf{q})$ is in this case a 'form factor' for an individual nucleus, which, for elastic scattering, is independent of angle.) In effect, the use of isotope mixtures introduces a special sort of incoherence which removes the effect of interference *between* molecules and makes accessible the molecular scattering factor.

For a three component system which includes a solvent, an interchain contribution persists,[45,46] but the inter- and intrachain contributions can be separated since they depend on the second and first power of the D concentration in the polymer respectively.[46]

Equation (8) can be rewritten to allow for different conventions for the normalisation of the intensities. This is essentially a trivial operation but it can cause some confusion in the literature. In the derivation of eqn. (9) the 'assembly' considered is a macroscopic sample, so that the result gives $V\,d\Sigma/d\Omega$. The term $(\overline{b^2} - \bar{b}^2)$ can be written as $c_D c_H (b_D - b_H)^2$ where c is the volume fraction of D and H. Hence:

$$V\,d\Sigma/d\Omega = |A(\mathbf{q})|^2 = |A_0(\mathbf{q}, \bar{b})|^2 + N(nP(\mathbf{q}))c_D c_H (b_D - b_H)^2 \qquad (11)$$

where N is now the number of hydrogens in the sample. The quantity $nP(\mathbf{q})$ is a normalised intensity $I(\mathbf{q})$ which does not require a knowledge of n (n is often one of the less precisely known quantities in an experiment).

2.5 Interpretation of the Molecular Scattering Factor

Given that $I(\mathbf{q})$ can be measured, how can structures be derived from it? In principle a Fourier transform would yield a correlation function $\gamma(\mathbf{r})$, which for a homopolymer can be interpreted as a histogram of the probabilities of any given separation $\mathbf{r}$ within the molecule. In practice this route is not often taken, but it is worth bearing in mind $\gamma(\mathbf{r})$, since it represents the maximum information in real space which is in principle available from the scattering. When it comes to judging the uniqueness of an interpretation it is sometimes possible to exclude some models because they would have a correlation function incompatible with the measurements. On the other hand, some models can coincidentally have remarkably similar scattering. For example, a Gaussian coil and a thin sheet both have scattering patterns of the form (constant) $\times q^{-2}$ over certain ranges of q values. Such cases as this do not represent a serious problem since a wealth of other techniques are usually available to distinguish such different models.

2.5.1 Guinier Range of q

In the limit of the scattering as $q \to 0$ a series expansion of intensity from eqn. (4) in powers of q is possible, the usual versions being due to Guinier[39] and Zimm[49] respectively:

$$I(q) = I(0) \exp(-R_g^2 q^2/3) \quad (12)$$

$$I(0)/I(q) = 1 + R_g^2 q^2/3 \quad (13)$$

where R_g is the radius of gyration $= \langle x^2 \rangle + \langle y^2 \rangle + \langle z^2 \rangle$ for a homopolymer, where x, y and z are coordinates with respect to the 'centre of gravity' of the molecule.

Both versions can be derived from eqn. (4), where the $\sin(qr_{ij})$ term is expanded by Taylor's theorem. In the limit as $q \to 0$ these are equivalent, but in order to obtain a sensible measurement of a change in $I(q)$ it is necessary to use q values up to about R_g^{-1}, in which case the exactness of fit and the value of R_g obtained depends on which is used. For example, if intensities are calculated from eqn. (12) and plotted according to eqn. (13) a least squares fit gives a 10% difference between the R_g value used initially and derived from the plot. In effect, eqns (12) and (13) are slightly model dependent, the former being a better approximation for near spherical objects[39] (e.g. water soluble proteins) and the latter for Gaussian coils (for which it was originally derived[49]). Clearly, if q values larger than R_g^{-1} are used, the R_g values derived are even more model dependent. This may not however be a serious problem, since one is often more interested in trends in R_g values (e.g. with molecular weight or with crystallisation conditions) than in the absolute values. The use of these equations at relatively high q must always be carried out with caution.

A necessary condition for the validity of plotting according to eqns (12) or (13) is that there should be no disagreement between the calibrated value of $I(0)$ and the value of n known from the measurement of molecular weights (e.g. by gel permeation chromatography). If there is a disagreement this implies non-random mixing of isotopes (see below) or the existence of an additional source of intensity.

2.5.2 Intermediate Range of q

As the q values are increased beyond the so-called 'Guinier range', all the interpretations have been made by calculating intensities from models and comparing with measurements. There is no well accepted nomenclature for q ranges: 'Intermediate' refers to the q range being between the Guinier range and the crystallographic range, or wide angles. In polyethylene, the crystallographic range starts at the 110 peak $q/2\pi = 1/4 \cdot 4 \text{ Å}^{-1}$). The intermediate range can itself be subdivided:

(a) Intermediate q up to about $0 \cdot 1 \text{ Å}^{-1}$. In this range the atomic detail within the monomer unit need not be considered, and for polyethylene groups of about six monomers can be considered to scatter as points.

(b) Intermediate q with $0 \cdot 1 < q < 1 \text{ Å}^{-1}$.

Several methods of calculation of $I(q)$ can be distinguished:

1. The simplest makes use of several analytic expressions[39] for the scattering from thin sheets:

$$I(q) = 2\pi n_A C_1(q)/q^2 \qquad (14)$$

where n_A is the number of hydrogens per area of sheet or thin rods:

$$I(q) = \pi n_L/(q)C_2(q) \qquad (15)$$

where n_L is the number of hydrogens per length of rod. For these to be used it is not necessary for the whole molecule to be a sheet or rod; a molecule may consist of a sequence of such units joined in such a way that the interference between them is negligible over certain ranges of q. The functions $C(q)$ are unity if the thickness of the rods or sheets is very small compared with $1/q$. The use of eqns (14) and (15) can be extended to larger q by calculating the ratio $C(q)$ of intensities for the actual structure expected (e.g. an extended straight sequence of chain) to that of the infinitely thin structure. If R_0 is the root mean square of the distance of hydrogens from the axis of the molecule, then for $q < R_0^{-1}$ and $q \gg (\text{rod length})^{-1}$

$$C_2(q) = \exp\left(-\tfrac{1}{2}R_0^2 q^2\right) \tag{16}$$

similarly for the analogous case of sheets:

$$C_1(q) = \exp\left(-D_0^2 q^2\right)$$

where D_0 is the root mean square distance of hydrogens from the sheet centre.

2. The most complete calculations use 'Monte Carlo' methods based on eqn. (4) or (5). Computer programs are used to generate sets of x, y and z coordinates for each scattering unit (e.g. monomer). The particular advantage these methods have over analytic calculations is the ability to cope with complex models including disorder. Random number generators are used in conjunction with rules prescribed by the model. For example, the chain beyond the end of a stem is often 'generated' in the same way as for liquid polymers. Of course, a model complete with all the spatial coordinates is necessary, and usually an average is made over a number of computer generated structures. It is usual, in order to reduce the calculation to manageable proportions, to choose a group of monomers as the individual scattering unit, and for any given structure to select randomly only a proportion of the units in order to calculate $\gamma(r)$.

3. For polymers of high crystallinity, calculations can be made using the 'stem' sequences in the crystals. These calculations can be further simplified by noting that within one crystallite the structure is a convolution of a single stem with a two dimensional structure describing the position of the stems. The normalised intensity $(q \gg (\text{rod length})^{-1})$ for a pair of stems separated by a distance R is:

$$I(q) = \pi n_L/(2q)(2 + J_0(q, R))C_2(q)$$

where $J_0(q, R)$ is a zeroth order Bessel function, which often occurs in calculations of scattering from structures with a cylindrical character. For a group of n' stems:

$$I(q) = \pi n_L J(qn') \left\{ n' + \sum_{\substack{i \\ i \neq j}}^{n'} \sum_{j}^{n'} J_0(q, R_{ij}) \right\} C_2(q) \quad (17)$$

The advantages of this third method are that it can cope with structures involving disorder, is fairly rapid, and involves only those physical features which have a large influence on the scattering (notably the values of the distances R_{ij}).

2.6 Non-random Mixing

This can have several consequences, the details of which depend on the model chosen.[50-52] If zones of different isotope concentration exist, $|A_0(\mathbf{q}, \bar{b})|^2$ in eqn. (8) should be replaced by an average over the local isotope average b.[52] In general this average will always be greater than for one uniform value of b.

Another consequence which will probably be more important is the modification to eqn. (7), since the double sum

$$\sum_i \sum_{i' \neq i} A_i A_{i'} A'_j A'_{j'} \, \delta b_i \, \delta b_{i'}$$

will no longer average to zero. If there is an increased probability of isotopes being the same for shorter intermolecular distances $\mathbf{R}_i - \mathbf{R}_{i'}$ than for larger ones then this term can be written approximately as:

$$\left| \sum_j A'_j \right|^2 \overline{\delta b^2} \int G(\mathbf{R}_{ii'}) \exp(i\mathbf{R}_{ii'} \cdot \mathbf{q}) \, d^3 R \quad (18)$$

where $G(\mathbf{R})$ describes the actual probability of finding a molecule of like isotope minus the average probability. An additional intensity is indeed found when the isotope mixtures of polyethylene are crystallised slowly, but it is only significant at q values from the lowest normally measured (5×10^{-3} Å^{-1}) to about 5×10^{-2} Å^{-1}. Note that the intensity contribution from non-random mixing (the 'segregation' signal) contains the single chain scattering factor $|\sum A'_j|^2$. The intensity can in effect be thought of as arising from a density function which is a convolution of the chain structure and a concentration function $c(\mathbf{R})$ which describes the centre of gravity of the molecules. The integral is simply the Fourier transform of the correlation function $G(\mathbf{R})$ of $c(\mathbf{R})$.

The function $c(\mathbf{R})$ is interpreted below in terms of selective fractionation. For this purpose it is useful to use a two-phase scattering analysis much as given by Porod[39] or Debye.[53] The latter can be used to derive a characteristic length for the correlation function, which in turn can give a measure of the average 'chord length' of one or other of the phases.

It is instructive to realise that the familiar Zimm plot[49] for solutions (of chemically different solute and solvent) also has a term which arises from $G(\mathbf{R})$ not being flat. Zimm has shown[54] that the geometrical interpretation (where each molecule is imagined with an 'exclusion zone' for other solute polymer molecules) is equivalent to the more usual thermodynamic one for the purpose of calculating the intensity in the limit of $q = 0$.

2.7 Analysis of Anisotropic Scattering

For many polymeric materials anisotropy can be induced by deformation, and the anisotropy will usually be evident in the molecular conformation. Hence neutron scattering will play an increasingly important part in the understanding of deformation processes in polymers. For crystalline polymers there is, in addition, a distinct situation which does not arise for most solution, amorphous or gel states (block copolymers being a notable exception). Samples can often be prepared with oriented 'textures' (e.g. with fibre-like symmetry) even though the mode of crystallisation may be either identical or nearly identical to that in unoriented samples. For example, crystals grown from solution naturally orient during sedimentation, but the molecular conformation will be the same as for samples produced as unoriented powders.

There is as yet no well established notation for anisotropic scattering so it is important to establish the usage in each individual application. In general there are two sets of coordinates. One relates to the geometry of the sample and detector (e.g. in simple cases the fibre axis is in the plane of the detector). The second relates to the crystallite where the molecule resides. A simple case would be a highly drawn sample where there will be a narrow distribution of c-axis orientations around the fibre direction (c is normally along the chain or stem direction in the crystals). For solution grown crystal mats the relevant 'local' coordinate axis is not along c but along the lamellar normals n. In general x, y and z will be used to denote the 'local' coordinate system. For simple cases, this will coincide with an experimentally defined direction, e.g. z being along the macroscopic fibre direction.

Equation (4), for spherically averaged scattering, has so far been the basis of detailed analyses of conformation, but is of course not relevant to fibre and other textures. From eqn. (1) the normalised intensity may be written in terms of the Fourier transform T of the scattering length density distribution (for one molecule):

$$I(q_{x'}, q_{z'}) = \langle T|(q_x, q_y, q_z)|^2 \rangle \qquad (19)$$

Where x' and z' are the equatorial and meridional directions for the sample. From the properties of the Fourier transforms it follows that the intensity depends only on the projection of the scattering length density on to the vector $\mathbf{q}$ for the measurement direction. In the Guinier regime of $\mathbf{q}$ therefore it is not R_g which now gives the intensity (eqns (12) and (13) derived from eqn. (4)) but a dimension related to the projection. For example, for a homopolymer,

$$R_z \equiv \langle z^2 \rangle$$

where z is measured from the centre of gravity of the molecule.

For fibre symmetry it is not possible to measure R_x separately from R_y. It is convenient to quote a combined dimension such as:

$$R_{xy} \equiv \tfrac{1}{2}(\langle x^2 \rangle + \langle y^2 \rangle)$$

which would be equal to R_x if $R_x = R_y$. Equation (13) then has its counterparts in equations such as:

$$I(0,0)/I(0, q_{z'}) = 1 + (q_{z'}^2 R_z^2) \qquad (20)$$
$$I(0,0)/I(0, q_{x'}) = 1 + (q_{x'}^2 R_{xy}^2) \qquad (21)$$

2.7.1 The General Case: Moderate Anisotropy

This section refers to the situation when all three R values can correspond to the Guinier regime. It has generally been realised that for fibre symmetry two R values are sufficient to define the scattering. This has recently been shown independently by two articles in *J. Appl. Cryst.*[55,56] (The reason why two R values are sufficient is for the same reason that two absorption coefficients are sufficient to describe the linear dichroism of a sample with fibre symmetry.) Reference 55 includes the complications of having a range of z directions in the sample (partial orientation only) and of having the sample fibre axis out of the plane of the detector.

If, experimentally, values of R are obtained as a function of the angle α between the measurement direction and the fibre axis then:

$$R^2 = \langle a \rangle + \langle b \rangle \cos^2 \alpha \qquad (22)$$

Thus a plot of R^2 against $\cos^2 \alpha$ will give a straight line from which $\langle a \rangle$ and $\langle b \rangle$ may be determined. From there, the molecular dimensions can be calculated using:

$$R_z^2 = \langle a \rangle + \langle b \rangle \{1 + 1/\langle P_2(\gamma) \rangle\}/3 \qquad (23)$$

and

$$R_{xy}^2 = \langle a \rangle + \langle b \rangle \{1 - 1/\langle P_2(\gamma) \rangle\}/3 \qquad (24)$$

where:

$$P_2(\gamma) = \tfrac{1}{2}(3 \cos^2 \gamma - 1)$$

where γ is the angle between a local direction z and the fibre axis. $\langle P_2(\gamma) \rangle$ is a measure of the degree of local axis alignments, equal to zero for isotropy and unity for perfect alignment. It is readily seen that for simple cases such as $\alpha = 0$ or $\alpha = \pi/2$ these formulae give the expected simple results such as in eqns (20) and (21).

In order to optimise the measurement of anisotropy it is necessary to impose on the data the requirement that the 'forward scattering' $I(0, 0)$ does not depend on α.

2.7.2 High Molecular Anisotropy

If the molecules are highly extended (e.g. during deformation) they may be long enough to pose technical problems for small-angle spectrometers. For example, a fully extended polyethylene chain could have a length of 10^4 Å, yet data is only readily available down to $q \simeq 0.003$ Å^{-1}, so that the 'Guinier regime' will not be attainable. This difficulty can be circumvented if at the same time the dimension in the equatorial plane is small (clearly this would be the case for a fully extended molecule). In this case the $q_{z'}$ dependence of $I(q_{x'}, q_{z'})$ is very similar to that of $I(0, q_{z'})$. With two-dimensional position sensitive detectors $I(q_{x'}, q_{z'})$ can be measured down to $q_{z'} = 0$, and so the upper limit to measurable R_z values does not come from the usual limitation of small q intensities being obscured by the region around the main beam. It arises because the $q_{z'}$ dependence of $I(q_{x'}, q_{z'})$ is extremely high for high R_z values, and may be 'smeared' by the combined effects of finite beam collimation and detector size. This has the effects of limiting the maximum R_z value which can be attained, and also of decreasing the maximum $I(q_{x'}, q_{z'})$ value which is measured on the detector. This latter effect can produce some misleading results if comparisons are made with the expected 'forward scattering' $I(0, 0)$.

3 MEASUREMENTS

3.1 Instruments

A brief description will be given of the D11 diffractometer at the High Flux Reactor (Institut Laue Langevin, Grenoble). Much more detailed descriptions are given elsewhere.[36,57] Several instruments of the same type exist.

The initial intensity distribution with wavelength is determined by the reactor and the cold source, and has a maximum in the region 5–10 Å. A relatively generous 'slice' of this distribution is taken by a set of rotating helical slots which select neutrons according to velocity. (Normally $\Delta\lambda/\lambda$ is 8%.) The maximum specimen diameter is about 10 mm, and the divergence at the samples can be varied. An area detector contains boron in gaseous form (BF_3) so that the neutrons produce α particles whose charge makes them readily detectable. Its efficiency is approximately 50%, and varies slightly over the 64×64 array of elements, each of which is 10 mm square. The count rate ('intensity') is given in arbitrary units by the total number of neutrons counted divided by the counts recorded during the measurement by a beam monitor. Statistical errors in intensities can always be calculated. The specimen to detector distance can be chosen in the range 1–40 m, and the beam divergence is normally adjusted to give a beam width at the detector position of approximately 3·5 cm (full width, half height), though sometimes less.

A number of other instruments exist, for most of which the neutron flux is lower and such large specimen to detector distances are not available. These are often well suited to measurements of diffuse intensities at relatively large q. For some experiments the use of crystal monochromators rather than velocity selectors is important if crystal reflections are to be resolved.

3.2 Data Reduction

The geometry of the D11 apparatus conforms reasonably well to 'point collimation' in the sense used in X-ray scattering. Methods of correcting for collimation distortions and wavelength distributions exist (e.g.

Reference 58); this is usually most important when sharp maxima or minima are being studied (e.g. Reference 37) or for very high R_z values (see Section 2.7.2). In many diffractometers with one-dimensional detecting systems, the geometry corresponds to 'slit collimation', so that at least some collimation distortion is to be expected as angle is reduced.

A multidetector system requires a correction for any variations in detector efficiency which there may be. For D11 this is normally done by using a spectrum from H_2O, which is almost entirely incoherent and hence (at small q) independent of angle.

For the D11 multidetector system the intensities in the 4096 counter positions must be averaged in some way so as to be readily analysed.[59] For isotropic samples it is normal to average over all those cells whose centres lie on an annulus 10 mm wide centred on the beam position. For anisotropic samples a variety of geometries can be used, for example, annuli can be divided into annular sections, or strips can be taken across the detector. In any case it is desirable to use methods which exploit the measured intensities all over the detectors which are available and not just over strips along principal directions.

3.3 Subtractions

A measurement of $I(q)$ requires the recording of count rate versus angle for: (a) the isotope blend, (b) equivalent polymer samples of one (or both) pure isotopes, (c) the empty specimen container, (d) the container with a neutron absorbing piece of cadmium in place of the specimen, (e) an isotropic scatterer (usually H_2O) and (f) the container only for (e). Figure 4 shows examples of these. It is also necessary to measure transmission t of the specimens. The methods of algebraic manipulation in order to extract $I(q)$ are described in detail elsewhere.[59] The parasitic intensities from the diffractometer ((c) and (d)) must be subtracted, with due allowance for that intensity contribution which is attenuated when the sample is in place. It is necessary to subtract incoherent scattering, and sometimes a coherent contribution $|A_0(\mathbf{q}, \bar{b})|^2$ (eqn. (8)). These can be measured from (b). The incoherent scattering is largely controlled by the content of ^1H in the specimens, and it is preferable that this should be the same in (a) as in (b) (the transmissions are then very similar and contributions (c) and (d) almost cancel). If the ^1H contents are different, an adjustment must be made which takes account of the dependence of the incoherent scattering on the ^1H content (only at small ^1H content is this linear).

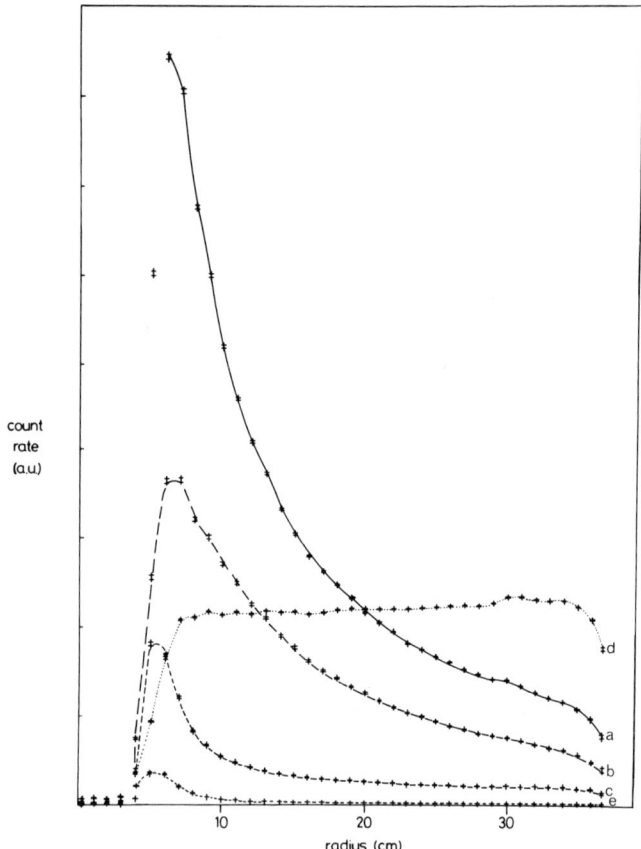

Fig. 4. Examples of plots of count rate versus radius on the detector: (a) isotopic blend (high molecular weight DPE), (b) isotopic blend (low molecular weight DPE), (c) HPE, (d) H_2O and (e) empty cell.

Subtraction of a coherent contribution is usually only a small correction. At very small q, with high ^1H it is usual to assume:

$$|A_0(\mathbf{q}, \bar{b})|^2 \simeq |A_0(\mathbf{q}, b_H)|^2$$

If it were certain that what is observed for ^1H polymers is due, for example, to voids or amorphous/crystal contrast, a rescaling should be applied which goes as the square of the scattering length density.

However, it is possible that the very small-angle scattering has a contribution from impurities (Summerfield, private communication). This is reasonable, since for polymers synthesised with Ziegler catalysts, some catalyst particles usually remain with the polymer.[60]

For measurements at large q it is desirable to minimise the ^{1}H content, so as to reduce the incoherent background and improve the signal to noise ratio. For large D concentrations $|A_0(\mathbf{q}, \bar{b})|^2$ from the crystal/amorphous density difference masks the single chain scattering, but this decreases sharply with q just as does the small-angle X-ray scattering. As a consequence, it is necessary to use different H/D ratios for different ranges in q: preponderantly H for low q and largely D for large q. For large q the background signals to be subtracted vary slowly with q, whereas $I(q)$ usually decreases quite quickly with q. In practice it is usually necessary to check that, at a q value greater than where $I(q)$ is calculated, the signal from the sample matches the signal calculated as a 'background'. For example, this has been done at $q = 1 \cdot 2$ Å^{-1} for $I(q)$ data in the range $0 \cdot 17$–$0 \cdot 36$ Å^{-1}.[51] When data are required up to about $q \simeq 1$ Å^{-1} it is necessary to use q values between those for the principal Bragg maxima (for polymers the intense peaks tend to be at $q \simeq 1 \cdot 4$ Å^{-1}). Whichever procedure is used, $I(q)$ values at the highest q are the least reliable. Unfortunately the methods used for subtractions are not always described in publications. The difficulties in achieving the optimum background subtractions are not necessarily the limiting factor in experiments, since calculations for models for $q \gtrsim 1$ Å^{-1} are also non-trivial (see $C_1(q)$ and $C_2(q)$ above). An accurate determination of the structure in the crystal unit cell is necessary—of average hydrogen positions and also, ideally, of the variability in their positions. Such detailed definition of structure is impractical for the amorphous regions and the transition zones between amorphous and crystalline regions.

3.4 Calibration of Intensities

A relatively simple procedure[37,61] can calibrate the macroscopic cross section $d\Sigma/d\Omega$. This is based on the true absorption being very small. Hence all the neutrons that are removed from the incident beam must be scattered. The integrated cross section is then known:

$$\int (d\Sigma/d\Omega)\, d\Omega = a(1-t)/V$$

where a is the sample area and t is the measured transmission. If the scattering is isotropic then its measurement at any angle must give the integrated cross section divided by 4π. Once the (differential) cross section for one specimen is calibrated in this way, cross sections for others are readily obtained by comparison. In this way the beam flux and various geometrical factors do not need to be measured directly in order to calibrate the intensity. The *precision* of this procedure should be only a few percent. The accuracy is not this good however since even for H_2O the scattering is not truly isotropic, for example because of inelastic effects which increase as wavelength decreases,[62,63] and multiple scattering.[64]

In order to change the calibration of $d\Sigma/d\Omega$ for the coherent scattering, in order to calculate $I(q)$, all that is necessary is a knowledge of the isotope concentrations. In order to calculate $P(q)$ a knowledge of molecular weights is also required.

3.5 Concluding Remarks on Techniques

One of the purposes of this book is to assess the degree of uncertainty of conclusions based on particular techniques. Where are the sources of experimental error in these scattering experiments and what are their consequences?

Neutron scattering equipment is in general built and maintained to high standards. In addition, because so many measurements are carried out on the same apparatus, there is an unusually high degree of informal contact about experimental methods among the community of users. If we are to explain why somewhat different interpretations are being placed on neutron scattering experiments, it seems unlikely that the reason is large experimental errors which are involved in collecting count rate data versus angle. Indeed, differing interpretations are often based on the very same measurements. Nor are the uncertainties in background subtractions particularly serious: in many cases the actual intensities being subtracted are small, and in others one or more samples can be expected to give particularly simple results, and can be used as checks of self-consistency.

It seems that to judge the uniqueness of conclusions we must look to the methods used to interpret $I(q)$ curves, and the words used to describe the models which are proposed. This is what will concern the bulk of the rest of this chapter.

4 EXPERIMENTAL CONDITIONS AND RESULTS

As a prelude to a more detailed discussion of interpretation, the scattering experiments will be described according to polymer, crystallisation conditions and range(s) of q used.

4.1 Polyethylene

4.1.1 General
This has long been the most widely used polymer for structural studies of crystalline polymers. The chemical structure is relatively simple, high crystallinities are usual, only one crystal modification is normally present, and the morphology is systematically lamellar unless conditions are specially chosen to produce fibres, gels or extended chain crystals. The very fact that polyethylene has been so intensively studied is very relevant to neutron scattering, since the technique relies on this other information being available. To some extent these advantages of using polyethylene offset the major disadvantage: not only do the two isotopic species have different melting points, but unless crystallisation is rapid the mixing of the two is no longer homogeneous after crystallisation.

4.1.2 Segregation
Inhomogeneous mixing leads to a significant term $G(\mathbf{R})$ and a very dramatic increase in intensity at small q according to eqn. (18). In the context of crystalline polymer conformations this topic is of somewhat specialised interest, but segregation is a common phenomenon and a clear understanding would be useful. Its existence was tested by comparing $I(0)$ from eqns (12) or (13) with the molecular weight as measured independently.[65] The precise form of the inhomogeneity has itself been the subject of debate. The author's view[51] is that fluctuations in isotope concentration over the samples result from the known propensity for polyethylene to fractionate. Previous results on fractionation[66,67] are the consequence of differences in crystallisation rate according to molecular weight, but there is no reason why it should not occur according to isotope. Fractionation leads to inhomogeneity in molecular type over the sample. In terms of Section 2.6 there are fluctuations in the density function $c(\mathbf{R})$ describing the position of molecular centres of gravity, leading to a correlation function $G(\mathbf{R})$

which is not flat. In principle one could measure $G(\mathbf{R})$ by dividing the intensity for a sample by the molecular form factor $nP(q)$ (supposing the latter were known). It would of course be very difficult to obtain from this a unique function $c(\mathbf{R})$ since the morphology of the lamellae, of stacks of twisting lamellae, and of spherulites is very complex.[68,69] Two level concentration models for $c(\mathbf{R})$ have been used[51,70,71] as being the first reasonable step at analysis. For SX (solution grown crystals)[51] with q in the plane of the lamellae, the intensity contribution due to $G(\mathbf{R})$ can be approximated by Porod's Law (intensity $\propto q^{-4}$). This was interpreted in terms of a 'picture frame' structure for $c(\mathbf{R})$ consisting of zones of different concentrations which are large in the plane of the lamellae (10^4 Å) (though much less at right angles). For (unoriented) MX (melt grown crystals) the q dependence is less and recent analyses,[70] still using a two level concentration $c(\mathbf{R})$, suggests dimensions of the zones of a few hundred ångströms. In terms of fractionation, this would imply neighbouring lamellae with differing isotope concentrations. This is not unlikely in view of microscopy results[68,69] which show that 'dominant' lamellae grow first with others growing later in the interstices. The characterisation of $c(\mathbf{R})$ by one distance and a two level concentration can of course only be a convenient oversimplification. In addition to fractionation being interlamellar, it will no doubt be 'interfibrillar' and interspherulitic, depending on such factors as crystallisation speed and self diffusion constants.

Other work on segregation[72] has concluded that molecular centres of gravity are always random, or very nearly so. It is not clear whether fractionation (differential crystallisation rates) could give an adequate description of this 'paraclustered' structure. Cross linking by irradiation prior to slow crystallisation was firstly shown to reduce the segregation signal,[51] as would be expected for non-random centres of gravity. Later work was interpreted differently.[73] Irradation is probably not sufficiently well understood in its own right for clear cut conclusions to be drawn. Experiments on annealing of SX[55] (Section 5.4) show that changes in chain conformation *can* occur without a segregation signal appearing. This is consistent with the segregation signal requiring movements of the centres of gravity.

4.1.3 Melt Grown Crystals

R_g values have been measured as a function of molecular weight for quenched melts.[65,74] R_g is proportional to $M_w^{1/2}$, where M_w is the weight average molecular weight, except for low molecular weights where R_g is higher than expected on this basis (see Fig. 5(b)). In the range of $M_w^{1/2}$ where the R_g dependence is linear, measurements on melts have been published[75] and show the same dependence of R_g on M_w. The departures from the $M_w^{1/2}$ behaviour have been associated with the stem length. A semi-empirical equation enables an extrapolation to be made to the situation of one stem only. The length of the stem obtained in this way is larger (294 Å) than the Bragg spacing observed by small-angle X-ray scattering (178 Å) for the samples used for the neutron scattering.

Experiments have recently been reported for a PE-like polymer: hydrogenated polybutadiene.[76] This polymer has 18 ethyl branches per 1000 main chain carbon atoms, and for this reason has lower crystal-

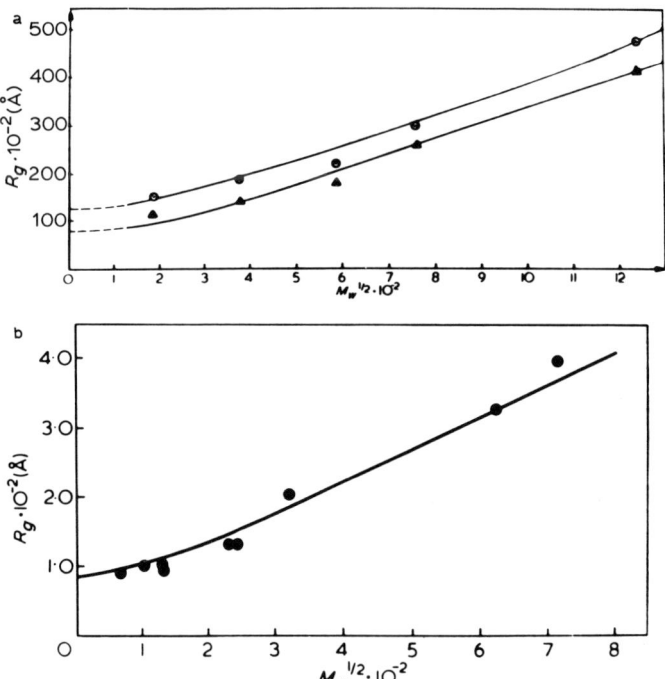

Fig. 5. (a) The radius of gyration of polypropylene in the semi-crystalline state produced by *quenching* (▲), and *slow cooled* (○) as a function of the square root of molecular weight M_w. (b) The radius of gyration of polyethylene in the semi-crystallised state produced by quenching. Taken from Reference 74.

linity and melting point than linear PE which usually has far fewer branches. However, it has the advantage of no isotopic segregation. The R_g values are the same in the melt, quench crystallisation and crystallisation with a moderately slow cooling rate (1 °C min^{-1}). The molecular weights were greater than the range where, for linear PE, crystallisation affected R_g values.

Intensities in the intermediate ranges of q are (as for smaller q) similar in MX to intensities either measured or expected for the molten polymer. The q ranges and molecular weights are as follows:

M_w 78 000	0·01–0·12 Å^{-1}	(Reference 77)
M_w 5200, 10 200, 31 000, 97 000 and 270 000	0·17–0·36 Å^{-1}	(Reference 51)
M_w 97 000	0·01–0·04 Å^{-1}	(Reference 78)
M_w 53 000	0·15–0·9 Å^{-1}	(References 79,80)

For hydrogenated polybutadiene (both quenched and cooled at 1° min^{-1}) the intensities are very similar for $0·02 < q < 0·12$ Å^{-1} as for linear PE (Crist, Graessley and Wignall, to be published).

4.1.4 Solution Grown Crystals

For crystallisation at 70°C from xylene, sedimented crystal mats give chain dimensions which increase much less steeply with M_w than with $M_w^{1/2}$ (Fig. 6).[81] Values of molecular dimensions along and perpendicular to the lamellae normals can be obtained from oriented mats.[55] For these sorts of samples the anisotropy is usually small but can be detected nevertheless by plots according to eqn. (22), (Fig. 7). The maximum in $\cos^2 \alpha$ of only 0·25 corresponds to the 'fibre-axis' normal to the mat being at 30° to the beam direction. The R_z values are 26 Å and 81 Å for as-grown (thickness 105 Å) and heat treated (annealed, thickness 210 Å) crystals respectively. In the former case the imprecision is at least 17% as a result, in effect, of extrapolating to $\cos^2 \alpha = 1$ in Fig. 7. These results agree with each molecule being confined to one lamella. The R_{xy} results were 54 Å and 56 Å showing no evidence of the annealing on the lateral dimensions. The results in Fig. 6 are dominated by R_{xy} since they were performed with $\cos^2 \alpha = 0$, and are not, strictly, 'R_g' values as shown on the label of the ordinate.

The scattering at intermediate q is very different from melt grown crystals, being much higher at low q and lower at high q. The References 51 and 78–80 given in Section 4.1.3 include work on SX as well as MX. The q ranges and molecular weights for other work are as follows:

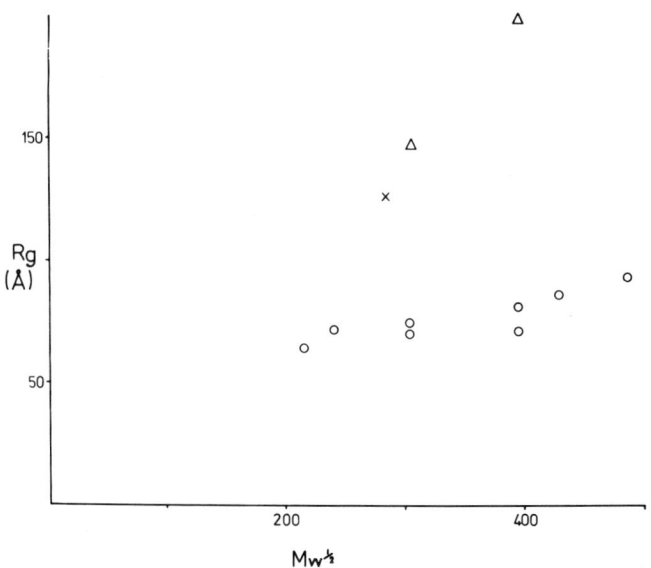

Fig. 6. 'R_g' values derived for solution grown crystals of polyethylene,[51,81] preferentially oriented with their normals along the beam direction ($\alpha = 0$). With the more recent terminology,[55] the ordinate is $\langle A \rangle$ which, for perfect orientation, is $3^{1/2} R_{xy}$. Triangles and the cross refer to molten samples.[81,75]

M_w 100 000	0·01–0·05 Å^{-1}	(Reference (1))
M_w 53 000	0·15–1·5 Å^{-1}	(Reference (82))
M_w 13 000–300 000	0·01–0·9 Å^{-1}	(Reference (83)).

4.2 Isotactic Polypropylene (Melt Crystallisation)

Although the melting point does depend on isotope, there is no evidence of inhomogeneous mixing of the isotopic species (solution crystallisation has not been studied however). For fast crystallisation R_g values are the same as for the melt (see Fig. 5), over the complete range of molecular weight.[84] The proportionality $R_g \sim M_w^{1/2}$ holds for higher molecular weights but not accurately for low.

A range of other conditions have been used: isothermal crystallisation, annealing and 'seed crystallisation'.[84,86] In the latter case cycling in temperature is used so as to crystallise on to crystalline fragments remaining after partial melting. The SAXS periodicity and crystallinity increase goes in the sequence quenched to seed crystallised. For annealing the R_g values are the same as for quenching or as for the melt. For isothermal and seeded crystallisation R_g values are higher,

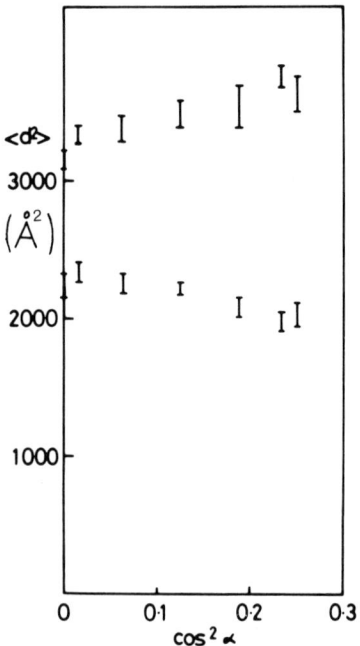

Fig. 7. Values of dimension squared ($\langle d^2 \rangle$) derived from Zimm plots for different inclinations α of the symmetry axis of the sample to the beam direction. For perfect crystallite alignment in the sample, $\langle d^2 \rangle$ would be R_z^2 for $\alpha = \pi/2$ and R_{xy}^2 for $\alpha = 0$. The lower points refer to crystals grown at 70 °C from xylene solution, the upper to such crystals that have been annealed in the dry state to 123 °C for one hour. The DPE (present as 3%) has molecular weight $M_w = 90\,000$. The degree of alignment of the lamellae to the symmetry direction of the sample is given by $\langle P_2(\gamma) \rangle$ as 0·57.

especially at lower molecular weights. As a consequence, the length which is derived for a single stem (see Section 4.1.3) is greater. This correlates with the increase in SAXS periodicities for these sample types. The stem lengths are systematically greater than the periodicity by a factor of about 2.

Intensities have been measured in the q range 0·1–0·8 Å^{-1} for quenched and annealed polypropylene.[87]

4.3 Isotactic Polystyrene

Polystyrene has a maximum crystallinity not much greater than 50%, and melt grown crystals are limited to about 35%. It seems likely that there are often amorphous zones within the samples, even though the degree of stereoregularity is usually very high. The polymer has however several important advantages for neutron scattering work, notably a high degree of compatibility between isotopic species. It is possible to quench the molten polymer to a glassy structure.[88]

Intensities at intermediate q, in the range 0.04–0.1 Å^{-1} at least, are not very sensitive to the type of sample preparation,[89–91] either in their absolute values or their q dependence. The conformation can be shown to be different however, from intensities at smaller q, including the Guinier range. The R_g values in particular are sensitive to whether the material is amorphous, melt crystallised (in several ways) or solution crystallised.

An ingenious way of changing the system is to use a range of H matrix materials: atactic will of course ensure an amorphous structure, but even using isotactic with high and low molecular weight has been found to induce different conformations.[92]

Figure 8 shows some examples of R_g values versus molecular weight.[89] Results for the amorphous structure (using atactic matrix) are also included in the figure. The various behaviours can be summarised

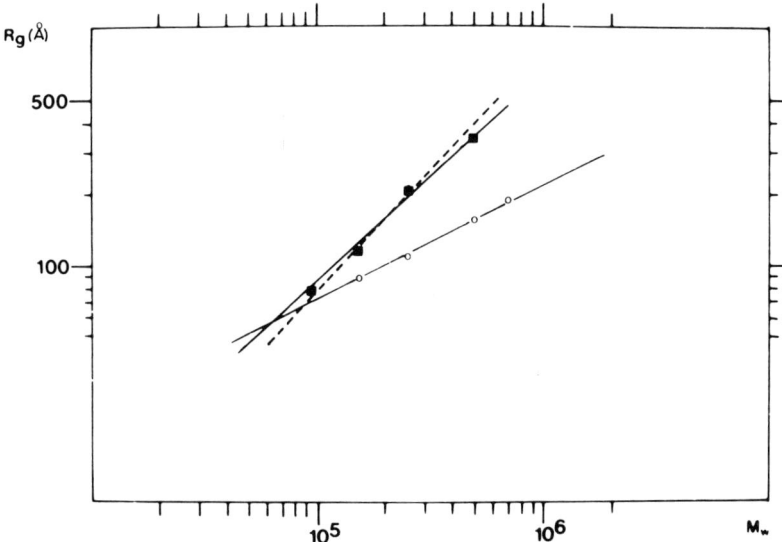

Fig. 8. Plot of R_g versus log M_w for solution grown crystals of iPS (■).[88] Also shown are results for amorphous samples moulded at 250 °C (○).[89]

by noting the power ν which controls the R_g behaviour of the type $R_g \sim M_w^\nu$. For amorphous samples, and those crystallised from the melt at relatively high supercoolings (crystallisation temperatures 140 °C) and high molecular weight matrix ($M_w = 1.7 \times 10^6$, high viscosity), ν is 0·5 (the Gaussian coil result). For a lower molecular weight matrix ($M_w = 4.2 \times 10^5$) and the same crystallisation temperature, ν is 0·63. For a higher crystallisation temperature (200 °C) and a matrix of M_w 8×10^5, ν is higher still at 0·77, and for solution grown crystals ν is quoted as 0·91.[89]

The time dependence was also investigated,[92,93] for a crystallisation temperature of 185 °C. For partly crystallised samples (e.g. crystallinity 16%) the R_g values are intermediate between those for the initial and final states.

4.4 Polyethylene Oxide

Some preliminary results have been published in reports to the Institut Laue Langevin.[94] For at least one type of melt crystallisation (long period 300 Å) the R_g values are similar to those measured in the melt. Intensities have also been measured to $q = 1$ Å^{-1}, which are similar but not equal to those measured for the melt. More recent results (E. W. Fischer, private communication) show substantial differences when crystallisation is carried out so as to give a long period of 430 Å.

5 INTERPRETATION OF RESULTS

5.1 Melt Crystallisation: General

Some general trends in the data have already been commented on and can be summarised as follows. For fast crystallisation from the melt, R_g values are similar after crystallisation to those before. ('Fast' in this context is relative to the rates of motion involving complete molecules.) When crystallisation is slow (iPS, PP) this is no longer the case, and molecular dimensions change significantly on crystallisation. On the other hand annealing to give a high crystallinity does not modify significantly the molecular dimensions for crystals whose initial crystallisation rate was high.

Intensities at intermediate q, which monitor more local details of the conformations, are also similar before and after fast crystallisation. This cannot mean that the conformations are similar, and a discussion of this point features prominently in what follows.

Although the experiments on polyethylene are limited severely by the fractionation problem, this polymer offers an example of very dramatic differences in intensity (e.g. by up to an order of magnitude) over the whole q range, according to whether crystallisation is from the melt or solution.

5.2 The Conformation for Quench Crystallisation of Polyethylene

This issue has been approached from utterly different points of view, yet the recent models which are now available in the literature differ more in their verbal descriptions than in the coordinates in space which specify the conformation. This is to the credit of the technique, since such a confluence of views had not been possible over the preceding two decades. Of course, the verbal descriptions are important, in particular because they embody explanations for conformations in addition to purely geometrical information.

The geometry of the chain in the crystals is as follows. Over the molecule as a whole the distribution over space is similar to that for the chain in the melt. On a local scale (c. 30 Å) stems from the same molecule fold back to give stem–stem separations in the range 4–15 Å (the distances between adjacent stems are 4·9 Å and 4·2 Å). Although not all these folds are adjacent (in the sense of the stem being in nearest neighbour positions) they could usefully be called 'tight' or 'nearby re-entrant'. In order to allow the chain to occupy a sufficiently large volume, it is necessary to suppose that the sequence of nearby folds is interrupted occasionally by long loops which may involve the molecule in changing lamellae. In some but not all models (see below) the sequences of nearby folds are specified as being straight.

The development of the models was as follows. Yoon and Flory[95,96] proposed a structure generated by Monte Carlo methods which has often been understood as being a purely random switchboard model. In fact, if the chains are allowed complete freedom when they emerge from the crystal to an amorphous layer, the R_g values are very high indeed. This extreme model can be compared to a random walk where the 'step length' is the length of one stem (about 200 Å). Rather than being given complete freedom, the chains as they emerge have a 70% chance of being confined to an interfacial layer 10 Å thick, and then re-entering the same lamella. They have a 30% chance of 'escaping' to a more truly amorphous layer, whereupon they may, according to chance, re-enter either the same or a different lamella. If the folds are predominantly adjacent it was shown that the intensity at $q \simeq 0\cdot1$ Å^{-1} is too high compared with experiment.

It was then shown[51] that in the region $q \simeq 0\cdot 3$ Å^{-1} the absolute values of the intensity and its q dependence were close to the scattering from random stems, at least in comparison with solution grown crystals. The 'interference' terms in eqn. (17) ($i \neq j$) were therefore contributing relatively little to the total scattering. Hence the folding was not systematically adjacent and folding was in some sense more 'random' than for solution grown crystals. The reasoning was that randomness would lead to terms with $i \neq j$ cancelling out on average. For adjacent re-entry this is certainly not the case. Subsequent calculations[97,98] showed that nearby re-entry with a range of stem separations (5–15 Å) introduced sufficient randomness to explain these data.

Intensities at the highest values of q (c. 1 Å^{-1}) were interpreted by a solidification model.[79,80] At first sight this implies that little molecular motion occurs on crystallisation (see below).

Guttman et al.[99] adopted the Monte Carlo approach and pointed out, for example, that the interfacial layer of Yoon and Flory was in fact resulting in a large proportion of folds which were relatively close if not adjacent. They also demonstrated that the interfacial layer could not accommodate conformations of the random folding type without unphysically high densities (e.g. twice the crystal density). The importance of space filling considerations was recognised by Frank,[100] who pointed out[101] that it amounts to considering *mutual* chain exclusion, whereas in the absence of a phase boundary it is usually *self* exclusion which has received intensive theoretical attention. If the boundary between crystal and amorphous phases is abrupt and can be described by a mathematical plane, then about 70% of the folds must be adjacent.[101] However, there is considerable evidence that the transition is more gradual and takes place over a distance of 10 Å or more (see Section 1.2) in which case fold patterns as shown in Fig. 9 are possible. These allow a certain degree of non-adjacent folds while avoiding space filling anomalies.[102] This idea is incorporated in some form or other in recent conformational models (see References 97–99 and 103).

The non-adjacency allowed by 'leapfrog' structures such as in Fig. 9 is not sufficient to allow an interface region which is amorphous in the conventional sense. Neutron scattering is not very sensitive to whether the folds are fairly taut as implied by the schematic (Fig. 9) or whether random chain excursions occur as part of the fold.[95,96] This issue is best resolved by packing calculations and by the experiments of the type listed in Section 1.

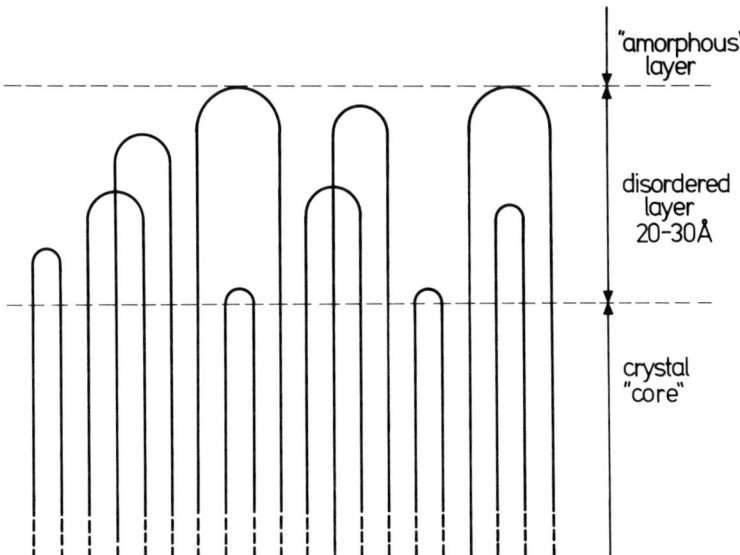

Fig. 9. Schematic two-dimensional diagram illustrating how the presence of folds at different depths below the lamellar surface leads to a transition layer, which allows some reduction in the restrictions imposed by mutual chain exclusion. A continuation of the all-trans (straight-stem) conformation into at least part of the transition layer is an essential feature of the model, though it is probably somewhat exaggerated in this illustration.

The type of nearby folding in the short sequence of stems (called a cluster or a subunit) has been considered in detail.[97,98] Figure 10 shows some examples. The insets to the diagrams show large spots whose area is proportional to the relative probabilities of re-entry at those positions, so that, for example, (a) defines 100% adjacent re-entry. Examples (d), (e) and (f) are all of the general type described as nearby re-entry. The calculations use eqn. (17) and plots are made not of $I(q)$ but of:

$$I_c(q) = I(q)/C_2(q)$$

$I_c(q)$ is a straight line through the origin for isolated stems. It is sometimes pointed out that this method of calculation (3 in Section 2.5.2) neglects folds and other disordered regions. However, there is no evidence that this neglect is important, at least for $q < 0.4 \text{ Å}^{-1}$. The method has been compared with the use of the Debye equation (4 in

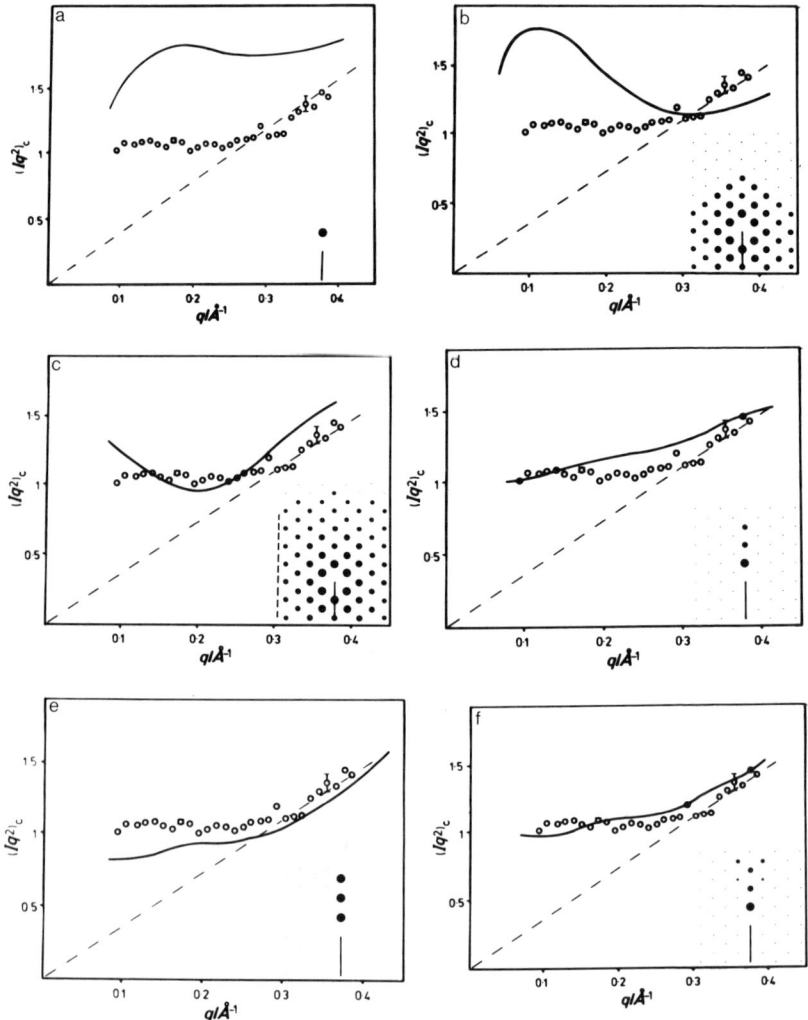

Fig. 10. Predicted scattering[97] for various subunit models with data from Reference 51. The points refer to experimental measurements and the lines to theoretical calculations. The insets show projected positions of the stems on the 001 plane; the top end of the vertical line corresponds to an existing (labelled) stem, and the areas of the larger circles indicate the probabilities of successive stems occupying each site. (a) Adjacent re-entry (interstem distance f is 4·9 Å). The size of the subunit is allowed to vary between 2 and 10 stems.

Section 2.5.2) for two stems separated by distances around 40 Å and a fold joining them, and gives similar results (Fig. 4 of Reference 97). Figure 10 also shows that two-dimensional random walks of stems ((b) and (c)) do not explain the observed scattering, even though the average distance between two stems along the chain may be similar to (d)–(f). The work in References 97–99, although it is based on matching intensities in different q ranges ($0\cdot01$–$0\cdot12$ Å^{-1} and $0\cdot17$–$0\cdot36$ Å^{-1}), and uses different methods of calculation, converges to structures which are very similar. The precise mix of stem–stem separations and their orientations in the lattice can be adjusted to some extent, but these do not fall outside the general description of the model as described at the beginning of this section. An important feature is the relative straightness of the subunits.

Although this section is devoted to polyethylene, it is worth mentioning at this point that for iPS crystallised under conditions equivalent to quenching PE, the conformation proposed is similar:[90] a 'garland' of short crystalline sequences corresponds to the sequence of subunits or clusters.

5.3 Relation to Conformations in the Melt

5.3.1 Comparison of Conformations

It is intriguing that the intensities before and after quench crystallisation are similar. Is this coincidence? If the Monte Carlo method is taken literally the answer is yes, since the method in effect chooses the conformation with maximum entropy for the crystalline state, independently of the melt state (energy is specified by the way the modelling is set up, so that free energy is being minimised). Equally, if space filling considerations are decisive for the fold type and hence the local structure, the fact that the correlation function (and hence the scattering) is the same as for the melt must surely be coincidental. On the other hand the solidification model[79,80] is based on the premise that

---

(b) Two-dimensional random walk of stems. Any neighbouring lattice position within 15 Å allowed, with a probability of $1/f$ so that different f values occur with approximately equal probability as shown in the inset. (c) As in (b) but with a cutoff for f of 25 Å. (d) Row model with decreasing probability as f increases: 55% of the folds are adjacent. (e) Row model with equal probabilities of f being $4\cdot9$, $9\cdot8$ or $14\cdot7$ Å. (f) Modified row model incorporating the occasional stagger: 46% of the folds are adjacent.

there is no coincidence, and that the melt structure is *cause* and the crystalline structure is *effect*. This idea would seem reasonable for large scale structure at least, so that (for example) the separations between subunits (clusters) is probably decided by the Gaussian coil of the melt. Does this 'freezing-in' extend to local conformations?

The idea of a particular kind of 'freezing in' of the molecule during crystallisation has been tested using analytical calculations.[97] The procedure is as follows:

1. A Gaussian coil is generated.
2. The three-dimensional probability distribution is projected on to the plane of a lamellar crystal.
3. The chain is divided up into lengths each sufficient to make one stem (including half a fold at each end of the stem).
4. A stem is created at the centre position of each of these short lengths of coil.

Figure 11 shows this procedure schematically. The scattering was then calculated using eqn. (17) with the result shown in Fig. 12. The lack of agreement is quite dramatic. The procedure in steps 1–4 is a way of specifying minimum diffusion during crystallisation, which, if not unique, is at least systematic.

The result is apparently paradoxical in that, in order to achieve a correlation function equal to that for the melt conformation, the average distance between stems needs to be reduced below that imposed by a freezing in of the melt. This is because the creation of a stem from a part of a coil changes the correlation function substantially and the clustering of the stems acts so as to compensate for this effect. A freezing in model does not hold which minimises *movements of chain* during crystallisation. It is possible (E. W. Fischer, private communication) that a different kind of 'freezing in' may minimise *disentanglement*, but it has not yet been shown quantitatively how this would lead to the correlation function being the same before and after crystallisation.

5.3.2 Restrictions Imposed by Relaxation Rates in the Melt
It might appear that the similarities there are between conformations before and after quench crystallisation (Section 5.3.1) must be explained in terms only of crystallisation rates compared with rates of coil relaxation. It may well be (e.g. for quench crystallisation and R_g values) that relaxation times are indeed too long for a result other than

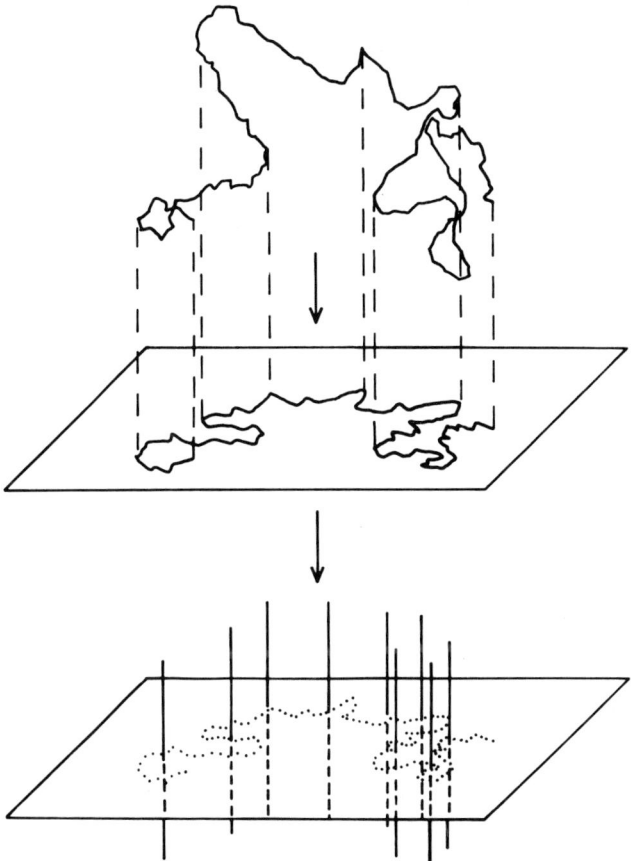

Fig. 11. Schematic representation of the 'freezing-in' model: top, a three-dimensional random coil; centre, the projection of this on to a plane; and below, the grouping of adjacent parts of the molecule into stems. For simplicity only one lamella is shown, but if the original coil occupied a space taken up during crystallisation by several lamellae, then parts of the molecule would be incorporated in each lamella.

the one observed. It does not follow however that relaxation rates are the single or even the dominant factor in determining conformations. Section 5.2 would indicate that, even for quench crystallisation, the structure on the scale of 30 Å is affected as much (if not more) by space filling at the (fold) surfaces than by purely rate effects. Experiments on low undercoolings (see Section 5.5 below) do not yet show

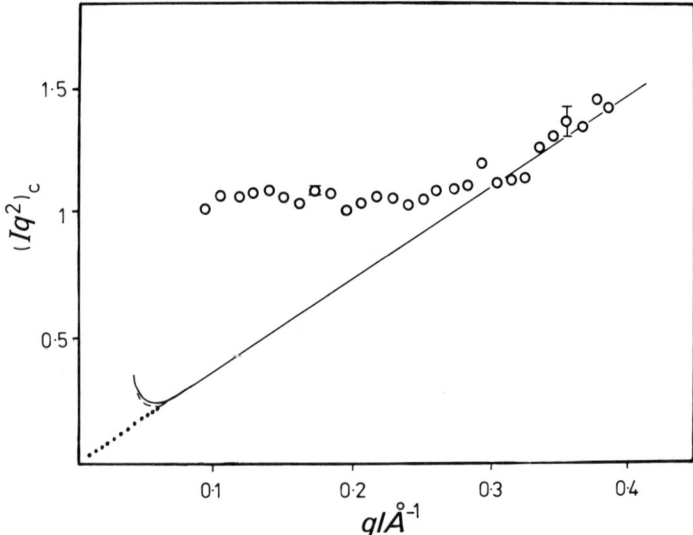

Fig. 12. Predicted scattering for a structure determined strictly on the principle of least movement of chains during crystallisation. The lower (broken) curve is for $n_s = 4$, the upper for $n_s = 12$. The dotted line shows the single-stem scattering.

that the *molecular conformation* (as opposed to the crystal morphology[68,69]) is radically different from that for quench crystallisation.

The topic of relating relaxation rates to crystallisation will not be tackled quantitatively: reference should be made to the 1979 Faraday Discussion.[104,105] Subsequent observation of rates of crystallisation orders of magnitude faster than for conventional quenching[106] may require revisions of some of the figures discussed in 1979. The observation of such very fast rates adds weight to the feeling that we may have exaggerated the importance of purely rate effects. This topic is continued in Section 5.6.

5.4 Melt Crystallisation at Low Supercoolings

Crystallising slowly (usually at low supercoolings) is clearly an important experiment for assessing the influence of the melt on crystal conformations. (Although crystallising at low supercoolings tends to give high crystallinities, high crystallinity is not in itself of so much interest in this context. Hence the effects of annealing will not be considered here.) R_g cannot be measured for PE other than in quenched melts because of fractionation, but lightly cross linking prior to slowly crystallising at 1 °C per day at least enables intensities at intermediate q to be measured. These are very similar to those for quench crystallisation (D. M. Sadler and S. J. Spells, to be published). For hydrogenated butadienes (which, for our purposes, can be considered as branched polyethylene) neither R_g values nor intensities at intermediate q are affected by crystallisation at 1 °C min^{-1} as compared with quenching. For iPS the intensities at intermediate q tend to be similar for all the modes of crystallisation, but R_g values become increasingly different from those in the amorphous state as the crystallisation rate is decreased. For isothermal and seeded crystallisation of PP (both low supercoolings) R_g values are somewhat different from those of melt and melt quenched material, though these effects tend to be confined to low molecular weights where one would expect[85] the R_g values to be influenced by the different lamellar spacings.

It is premature to generalise from these results, but it may well be that the conformation on the scale of a whole molecule (as assessed by R_g measurements) can be dependent on the crystallisation rates from the melt, but that the conformation on a local scale is less dependent.

5.5 Crystallisation of PE from Solution

The recent observation[55] that molecules are confined to one lamella simplifies considerably the range of models to consider. In order to study the local conformation the first approach[1,51] was to use eqn. (14) (for sheets) so as to test for adjacent folding (Figs 1 and 2). Although the existence of sheets is now agreed, two new features of the sheet model have been found which were not apparent before the application of neutron scattering. R_g values (Fig. 6) do not increase with M_w faster than with $M_w^{1/2}$ (as they would for simple sheets which were long compared with the lamellar thickness), but much less. Figure 13 shows the superfolding model (in a perspective sketch) which explains the results in a natural way.

The second new feature is the sharing of the folded sheet between two molecules, so that neither of them are always folding adjacently. This is apparent from the absolute values of intensity.[51,96] In Reference 51 the n_A values (eqn. (14)) were about half what was expected from the D_0 values for the sheets, though this observation was not interpreted in detail at the time. Within the 'diluted' folded sheets it is possible to have a range of degrees of adjacency: for example folding with a preference for a stem separation of two lattice units[96] would

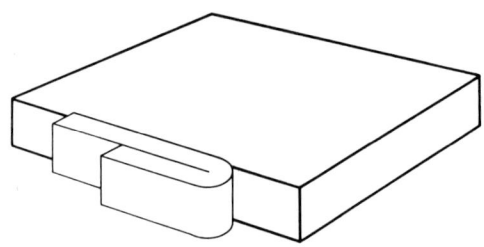

Fig. 13. Schematic view of a folded ribbon which 'superfolds' on the growth face of a polyethylene crystal.

have very low adjacency. Recent results[82,83] show that this cannot be the case since a maximum in the intensity at a Bragg spacing corresponding to a 'double fold' is not observed. Figure 14 includes measurements and predictions for an 80% preference for nonadjacency[96] and a 75% preference for adjacency[83] ($I_c(q) = I(q)/C_2(q)$). It can be seen that towards low q the two predictions come together, as expected since the 'average' dilution is by two in both cases and intensities at low q are independent of the details of the folding on a local scale. The existence of dilution is supported by infra-red spectroscopy.[107]

Difficulties in background subtractions and calculations at $q \simeq 1\,\text{Å}^{-1}$ were mentioned in Section 3. For a sample of 100% of perfect crystals $C_2(q)$ would be very small[83] at $q = 1 \cdot 4\,\text{Å}^{-1}$, whereas the measured intensities are not correspondingly small ($I(q) \simeq 0 \cdot 3$). (In order to judge this intensity value, it is useful to note that it would correspond to 30% of the hydrogen positions being completely random). Disorder both inside and outside the crystalline regions must be invoked.

The stacked sheet model of Stamm *et al.*[79,80] can be readily related to the model in Fig. 14. The 'bunches' which occur along the rows as a result of preferentially adjacent re-entry, when they coincide with bunches in adjacent rows, give small groups of stems which are very similar to the groups shown in Fig. 7 of Reference 79. Recent experiments[82] and calculations[108] have been performed to show that folding is not on (100) planes. The disagreements with Yoon and Flory[96] do not involve the basic sheet structure or the superfolding, but only the statistics of the folding within the rows.

The interpretation of the existence of rows of stems is straightforward in terms of a molecule attaching to a linear growth face. The interpretation of superfolding is not clear. It is probably associated

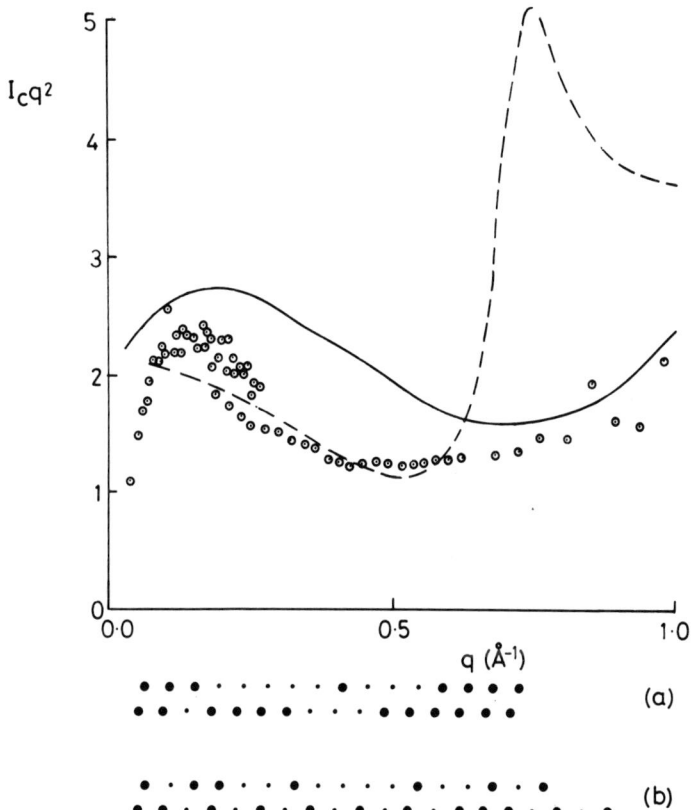

Fig. 14. Intensities at intermediate q which are compared with predictions for a statistical preference for 'next-but-one' folding[96] and for adjacent re-entry.[83] Insets illustrate stem positions (a) adjacent re-entry, ——— (b) next-but-one folding - - - -.

with the high supercoolings necessary for neutron scattering studies of PE, perhaps with intermolecular effects. The interpretation of the dilution is considered in the next section.

Results on solution grown crystals of iPS have also been interpreted in terms of adjacent re-entry,[89] though the *local* details of the conformation are still being investigated. Experiments are in progress to test this by searching for an intensity maximum at $q \simeq 0.5 \, \text{Å}^{-1}$ (Guenet, Keller, Sadler and Spells, to be published).

5.6 A New Interpretation

It may seem in the future that the recent debates on crystalline polymers have tended to take for granted the primary physical observation: the almost universal tendency for chains to fold. Polymer crystals remain unique in that they spontaneously incorporate an extremely high defect density by virtue of a very high surface to volume ratio (enough to depress the melting point by tens of degrees). The explanation usually accepted for this phenomenon is secondary nucleation,[4] together with the hypothesis that the secondary nucleus retains its dimension along the chain. (Many crystals grow by secondary nucleation which do not retain the nucleus size.) However, it now appears[6] that this explanation is not sufficiently general, since a requirement for secondary nucleation is a faceted (predominately smooth) growth face, whereas folding is observed both in the presence and in the absence of facets. It has been proposed[6] that a theory of crystal growth should incorporate the observation that growth faces may be faceted or not. It is not possible to describe in detail this new approach here. However, it enables some of the features of the molecular conformations described above to be rationalised in a fairly simple manner, so a brief summary of the arguments will be given.

The surface of a crystal may be disordered or ordered in an analogous way to a bulk phase being liquid or crystalline.[3] A 'crenellated' structure which has a higher surface energy than the molecularly smooth situations may be possible if there is a sufficiently large compensating surface entropy term. In recent years it has become increasingly clear that a transition to a surface disordered state can occur below the melting point of a crystal:[109,110] notably, experiments on ^{4}He crystals have shown a series of transitions for facets of different orientations.[111] There is no reason why surface disorder should not occur on polymer growth faces, though it is of course more difficult to specify what form it will take because of the connectivity along the chains. Figure 15 is the result of a preliminary simulation study, which does not allow for connectivity, and which differs from previous such calculations for non-polymers[109] only by the free surfaces at the top and bottom edges of the crystal. Its purpose is to convey the approximate degree of disorder necessary (Fig. 15(b)) for facets to be lost according to the currently accepted theories. It is now established that for an effectively infinite two-dimensional surface there is a temperature T_R at which faceting is no longer seen. For a surface of less

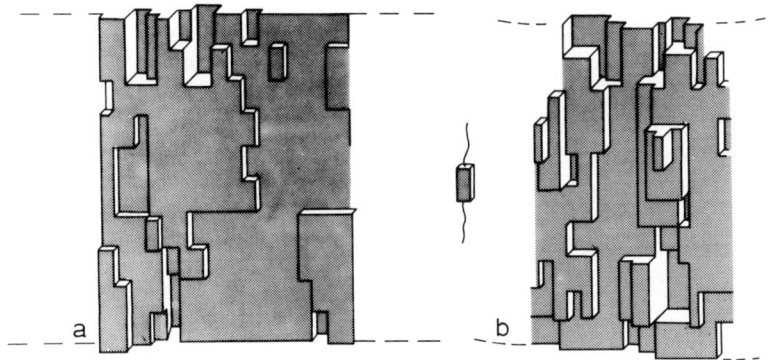

Fig. 15. Speculative view of growth surface roughness, where individual crystal units correspond to short sequences of chain (centre). The two versions correspond to different temperatures relative to band energies,[6] (a) $kT/\varepsilon_1 =$ 0·5, (b) $kT/\varepsilon_1 = 0·65$ (near what is termed the 'roughing' transition).

than 'infinite' extent there is still a fairly narrow temperature range over which the transition takes place. (This is discussed elsewhere.[6]) However for a wide temperature range, even below T_R, there is still a substantial degree of roughness (Fig. 15(a)).

For PE {110} facets are prominent at low absolute temperatures of crystallisation (from xylene solutions in practice). There is evidence,[6] from the morphology of twins, that at low enough temperatures there is secondary nucleation as envisaged previously. At high temperatures (crystallisation from other solvents than xylene, and probably from the melt) there are no facets and crystallisation cannot be by secondary nucleation in the usual sense. For PE both faceted and non-faceted growth is observed. For other polymers it is possible that growth may only be by one of these modes. Many crystals shown in micrographs are not definitely faceted; although their shapes may be closely related to simple ones such as hexagons, the outlines are often distinctly curved.

The possibility of rough surfaces can be seen as the consequence of one essential difference from previous models: for structures as in Fig. 15 to exist the energy for a stem to terminate (and for the chain to continue as a loop or fold) must be comparable with kT, whereas previously the stems were presumed to terminate with a fold of very high energy (around 8 kT). Such high energies suppress the disorder almost entirely.

Clearly, the implications for the way molecules will attach to the growth surface are fundamental. If the surfaces are smooth then there will be a tendency for the stems from one molecule to attach consecutively (adjacent re-entry), if diffusion and other factors allow this. On the other hand, a rough surface (Fig. 15) is essentially 'sticky' from the point of view of an approaching molecule. There may still be some preference for adjacent re-entry, but it will be less marked than for a smooth surface. Other factors will then become more important, notably mutual chain exclusion at the (fold) surfaces and an 'entropic' driving force in favour of attaching a stem or part of a stem from any chain in the vicinity of the growth face. The conformation for SX and MX are consistent with differences in growth surface roughness. For MX there is around 30–50% adjacency, with a high probability for the molecule to be attached independently at different locations, as would be expected for a rough surface. For SX the adjacency is much higher (around 75%) which agrees with a much less rough surface. The probability of non-adjacency (25%) is not at all negligible however, and leads to the dilution along the rows. This is consistent with the roughness still being finite.

In neither case (MX and SX) is the growth surface likely to be rough enough to prevent sequences of stems from the same molecule forming approximately straight rows, even though the degree of adjacency within the row will vary as discussed above.

The effects of supercooling, as opposed to simply the absolute temperature of crystallisation, have not been mentioned in this section. Previous studies have in fact proposed a growth surface roughening (of a simpler type than shown in Fig. 15) which is purely a kinetic effect.[4,5] The primary evidence, that of the existence of facets, does not at present show a large influence of supercooling.[6] It may be noted that the nearest approach to a nucleation controlled growth (as in 'Regime I')[4] is now proposed to be at the *lowest* temperatures of crystallisation, which is the reverse of what was previously thought.

At the beginning of this section the importance of explaining folding was emphasised. Although a detailed discussion is outside the scope of this review, a possibility will be mentioned. Computer simulations of crystal growth (Sadler and Gilmer, to be published) show that very simple kinetic restrictions on the addition of new units to the growth face may be sufficient to result in lamellar growth habits.

6 DEFORMATION

Only a small number of papers have appeared concerning deformation, though this is certain to be one of the most important applications of the technique. Extensive work has been done on changes of crystal morphology:[112] the spherulitic texture is changed to a fibrous one. Lamellae still exist, stacked in fibrils, and there are thought to be straight sequences of chains linking one crystallite to its neighbour, and hence increasing the modulus along the fibre. There are several technical difficulties associated specifically with deformation. Very often a stretched 'fibre' sample (with fibre-like texture even if not recognisably a *filamentous* one) has a very large number of voids. The subtraction of the coherent background then becomes problematic. Even if a subtraction should be possible in principle, the void formation is not usually fully reproducible. In addition R_z values can be expected to be high (Section 2.7).

Two systems have been found to be favourable experimentally: quenched PP[113] and PE.[114] In both cases the H matrix is polydisperse and the same for several D components. Even in PE the original isotope mixing was homogeneous; the success in attaining good mixing presumably depends on such details as the nucleation density.

The experimental methods differed: for PE data could be obtained to $q_{z'} = 0$ by making full use of the D11 area detector. Figure 16 shows a contour plot of difference intensity, with examples of rectangles over which the intensities were averaged to obtain plots of $I(q_{x'}, q_{z'})$ versus $q_{z'}$ (constant $q_{x'}$). For the $q_{x'}$ dependence there is of course little difficulty in taking data along the equatorial 'streak'. A technical point of interest was that the non-negligible width of the instrumental broadening (from beam *and* detector) 'smears' the intensities in the z' direction and reduces the maximum intensity by a factor of about two. (The shape of the instrumental function is approximately triangular.) The instrumental effects also depress somewhat the measured R_z values of course. The number of experimental data points which were used in deriving R_z was increased by using fits to a Debye function for a random coil.[115] This curious procedure for a crystalline polymer was justified by the idea that the structure of the molecule (in the undeformed state at least) follows the density distribution for the original molten coil. For PP the R_z values after deformation were not so high, and conventional one dimensional plots were made for the x' and z' dependence along a line passing through the main beam position.

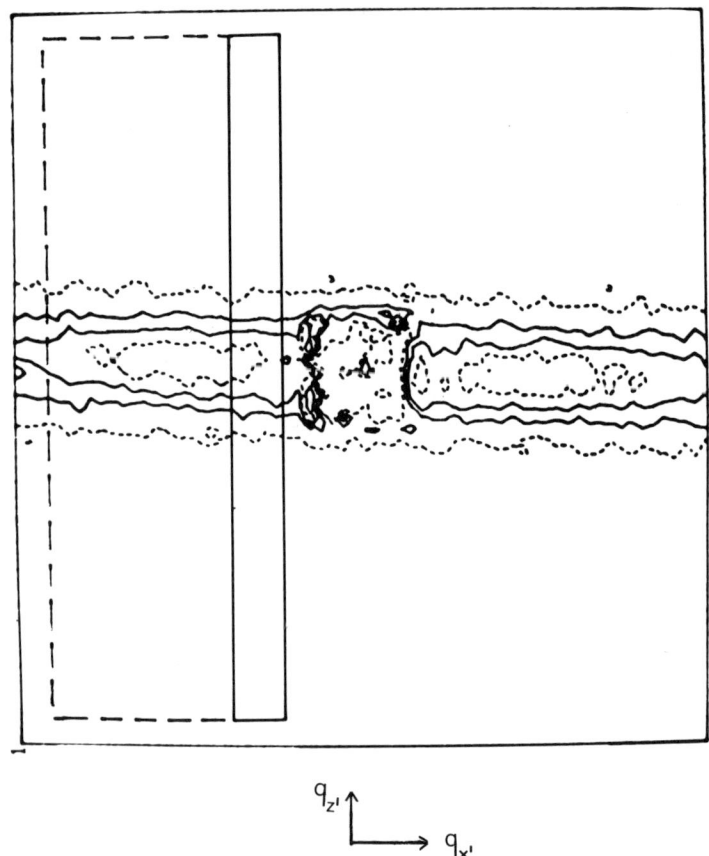

Fig. 16. Contour plot of difference intensity across the detector (beam position is central), for drawn quenched sheets of PE (DPE molecular weight is 62 000). Draw direction is vertical, specimen to detector distance is 9·14 m, wavelength 8 Å, and the dimensions of the detector array 64 × 64 cm.

For discussions of the extent to which molecular deformation is affine with the specimen deformation it is important to use only R values which refer to projections along single dimensions (e.g. R_x) and not R_g ($=(3)^{1/2}R_x$) for an isotropic molecule. This is what is done in the following.

For PE drawn near room temperature, with draw ratios 5–10, the R_{xy} dimensions decrease approximately affinely. The R_z dimensions increased by rather less than predicted by affine deformation for the lowest molecular weight DPE (6×10^4) but for the higher (10^5 and 2×10^5) the increase was nearer the affine value (instrumental effects are borne in mind). For SX (hot drawing) the results are quite different. The R values start smaller than for MX, and increase along z' by only about two in spite of a draw ratio of 20. The interpretation of the results turns on how many lamellae the molecules pass through prior to drawing. For MX several lamellae are involved and the molecule is well distributed even within one. As lamellae move and are broken up, the molecules on average follow the dimensions of the whole specimen. For SX however, each molecule is completely folded in *one* lamella. Movement and breakage of lamellae is then less effective in distorting the molecule. As concerns the fibre properties, it is perhaps surprising that the molecular extension for 'conventional' drawing (moduli c. 5 GPa) should be greater than for very high draw ratios (from SX, moduli c. 20 GPa). Hence there is no clear correlation between modulus and chain extension. A clear case of high modulus samples with virtually no extension[116,117] has already made this point, but in the present case the mode of sample production of both high and low modulus samples was basically the same (drawing).

For PP the changes in dimensions on drawing are less than affine even though the specimens were also melt grown. In this case however drawing was at 145 °C. Hence differences in drawing temperature should be borne in mind as possibly important in determining molecular deformation (see also the comparison of MX and SX of PE).

Scattering by fibre samples at intermediate q is clearly of great potential. Data for PP[113] and PE[118] have already been reported. They show clearly a dramatic anisotropy arising from the alignment of stems: just as for $hk0$ crystal diffraction peaks the intensity is strongly enhanced along the equator (x').

7 CONCLUSION

This chapter has been selective in order to give special attention to the interpretation of conformations in terms of crystallisation mechanisms: for example, annealing has only been mentioned in passing. I must also excuse myself for the brevity of the descriptions of work on polymers other than PE.

In the work on PE, only one significant disagreement persists concerning the actual geometry of the molecules (the nature of the dilution for solution grown crystals), and even here new data (Fig. 14) seem to settle the matter fairly clearly. Models for melt grown crystals do indeed differ in the randomness of the loops joining stems in a subunit, but this degree of detail cannot easily be tested against neutron scattering data.

In all cases in PE the stems form into rows parallel to the growth front of the crystal. The degree of adjacency within the rows is about 30–50% for melt growth and 75% for solution growth, and the lengths of the rows are about 20 Å and 100 Å respectively. The folds in the rows which are not adjacent are 'nearby' in that they rarely involve stem separations greater than about three nearest neighbour separations. For melt growth the separations between different rows in the same molecule are fairly random, giving an overall distribution in space comparable with the pre-existing molten coil (involving more than one lamella). For solution growth the different rows stack together forming a compact superfolded structure.

The interpretation of these conformations is of course less unambiguous, though a self consistent interpretation can be given as follows. There seems to be no need to invoke restrictions on rearrangements in the melt as a principal factor, but it probably controls the size of the subunits or clusters for melt growth. The separations of stems along the growth face is limited by packing in the fold surfaces, and is also controlled by the separations of niches which occur on the growth faces. These are more plentiful (the surface is rougher) for melt growth than for solution growth, and this could explain the corresponding differences in degrees of adjacency.

Experiments on oriented samples are becoming more important, where, for example, it has been possible to measure the size of molecules along the directions of the lamellar normal. Equally, studies on deformation will no doubt become increasingly important.

ACKNOWLEDGEMENTS

I wish to thank many colleagues for discussions, especially these at Bristol: Professor A. Keller and Drs P. Barham and S. J. Spells. I also wish to acknowledge the collaboration of Dr Spells in some of the work reported here and not yet published. I wish to thank Drs Guenet and Ballard for permission to reproduce figures.

REFERENCES

1. Sadler, D. M. and Keller, A. (1976). *Polymer*, **17**, 37.
2. Flory, P. J. (1962). *J. Am. Chem. Soc.*, **84**, 2857.
3. Burton, W. K., Cabrera, N. and Frank, F. C. (1951). *Phil. Trans. Roy. Soc.*, **243A**, 299.
4. Hoffman, J. D., Davis, G. T. and Lauritzen, J. I. (1976). In *Treatise on Solid State Chemistry*, Vol. 3, Chap. 7 (N. B. Hannay (Ed.)), Plenum Press, New York.
5. Hoffman, J. D. (1983). *Polymer*, **24**, 3.
6. Sadler, D. M. (1983). *Polymer*, **24**, 1401.
7. Keller, A. (1981). *Structural Order in Polymers*, Int. Union of Pure and Applied Chem. (F. Ciardelli and P. Giesti (Eds)), Pergamon Press, Oxford.
8. Khoury, F. and Passaglia, E. (1976). In *Treatise on Solid State Chemistry*, Vol. 3, Chap. 6 (N. B. Hannay (Ed.)), Plenum Press, New York.
9. Bassett, D. C. (1981) *Principles of Polymer Morphology*, Camb. Univ. Press.
10. Sadler, D. M. (1982). In *Static and Dynamic Properties of the Polymeric Solid State*, 81, (R. A. Pethrick and R. W. Richards) D. Reidel.
11. Wunderlich, B. (1973). *Macromolecular Physics*, Vol. 1, Academic Press, London.
12. Keller, A. (1957). *Phil. Mag.*, **2**, 1171.
13. Pennings, A. J. and Kiel, A. M. (1965). *Kolloid Z. u. Z. Polymere*, **205**, 160.
14. Bassett, D. C. (1976). *Polymer*, **17**, 461.
15. Kawai, T. and Keller, A. (1965). *Phil. Mag.*, **11**, 1165.
16. Jones, D. H., Latham, A. J., Keller, A. and Girolamo, M. (1973). *J. Pol. Sci. (Phys. Edn)*, **11**, 1759.
17. Atkins, E. D. T., Keller, A. and Sadler, D. M. (1972). *J. Pol. Sci. (Phys. Edn)*, **10**, 863.
18. Frank, F. C. (1963). *The Geometrical Thermodynamics of Surfaces*, Chap. 3, Conf. Am. Soc. Metals.
19. Keith, H. D. (1964). *J. Appl. Phys.*, **35**, 3115.
20. Bassett, D. C. Frank, F. C. and Keller, A. (1963). *Phil. Mag.*, **8**, 1739 and 1753.
21. Bassett, D. C. (1964). *Phil. Mag.*, **10**, 595.
22. Lindenmayer, P. H. (1963). *J. Pol. Sci. C*, **1**, 5.
23. Fischer, E. W. and Schmidt, G. F. (1962). *Angew. Chem.*, **74**, 551.
24. Hoffman, J. D. and Weeks, J. J. (1962). *J. Res. Natl. Bur. Stds*, **66A**, 13.
25. Fischer, E. W., Goddar, H. and Schmidt, G. F., (1969). *J. Pol. Sci. B*, **5**, 619.
26. Hoffman, J. D., Lauritzen, J. I., Passaglia, E., Ross, G. S., Frohlen, L. J. and Weeks, J. J. (1969). *Kolloid Z. u. Z. Polymere*, **231**, 564.
27. Sadler, D. M. and Keller, A. (1970). *Kolloid Z. u. Z. Polymere*, **239**, 641 and **242**, 1081.
28. Sadler, D. M., Williams, T., Keller, A. and Ward, I. M. (1969). *J. Pol. Sci. (Phys. Edn)*, **7**, 1819.

29. Patel, G. N. and Keller, A. (1975). *J. Pol. Sci. (Phys. Edn)*, **13,** 2259.
30. Peticolas, W. L., Hibler, G. W., Lippert, J. L., Peterlin, A. and Olf, H. G. (1971). *Appl. Phys. Lett.*, **18,** 87.
31. Folkes, M. J., Keller, A. and Pope, D. P. (1975). *J. Pol. Sci. (Lett.)*, **13,** 341.
32. Strobl, G. R. (1979). *Colloid and Pol. Sci.*, **256,** 584.
33. Windle, A. (1975). *J. Mat. Sci.*, **10,** 252.
34. Madelkern, L. (1979). *Disc. Far. Soc.*, **68,** 310.
35. Bacon, G. E. (1975). *Neutron Diffraction*, Oxford Univ. Press, London.
36. Kostorz, G. (Ed.) (1979). *Treatise on Materials Science*, Vol. 15 (Neutron Scattering), Academic Press, London.
37. Jacrot, B. (1976). *Rep. Prog. Phys.*, **39,** 911.
38. Allen, G. and Higgins, J. (1973). *Rep. Prog. Phys.*, **36,** 1073.
39. Guinier, A. and Fournet, G. (1955). *Small-Angle Scattering of X rays*, (trans. by C. B. Walker), Wiley, New York.
40. Kirste, R. G., Kruse, W. A. and Schelten, J. (1973). *Makromol. Chem.*, **162,** 299.
41. Cotton, J. P., Decker, P., Benoit, H., Farnoux, B., Higgins, G., Jannink, G., Ober, R., Picot, C. and des Cloizeaux, J. (1974). *Macromolecules*, **7,** 863.
42. Daoud, M., Cotton, J. P., Farnoux, B., Jannink, G., Sarma, G., Benoit, H., Duplessix, R., Picot, C. and de Gennes, P. G. (1975). *Macromolecules*, **8,** 804.
43. Higgins, J. and Stein, R. S. (1978). *J. Appl. Cryst.*, **11,** 346.
44. Ruland, W. (1976). *Colloid and Pol. Sci.*, **254,** 358.
45. Williams, C. E., Nierlich, M., Cotton, J. P., Jannink, G., Boue, F., Daoud, M., Farnoux, B., Picot, C., de Gennes, P. G., Ricarde, M., Moan, M. and Wolff, C. (1979). *J. Pol. Sci. (Lett.)*, **17,** 379.
46. Akcasu, A. Z., Summerfield, G. C., Jahshan, S. N., Han, C. C., Kim, C. Y. and Yu, H. (1980). *J. Pol. Sci. (Phys. Edn)*, **18,** 863.
47. Gawrisch, W., Fischer, E. W. and Brereton, M. (1981). *Pol. Bulletin*, **4,** 687.
48. Wignall, G. D., Hendricks, R. W., Koefler, W. C., Lin, J. S., Wai, M. P., Thomas, E. L. and Stein, R. S. (1981). *Polymer*, **22,** 886.
49. Zimm, B. H. (1948). *J. Chem. Phys.*, **16,** 1093.
50. Schelten, J., Wignall, G. D., Ballard, D. G. H. and Longman, G. W. (1977). *Polymer*, **18,** 1111.
51. Sadler, D. M. and Keller, A. (1978). *Macromolecules*, **10,** 1128.
52. Summerfield, G. C., King, J. S. and Ullman, R. (1978). *Macromolecules*, **11,** 218.
53. Debye, P., Anderson, H. R. and Brumberger, H. (1957). *J. Appl. Phys.*, **28,** 679.
54. Zimm, B. (1945). *J. Chem. Phys.*, **13,** 63.
55. Sadler, D. M. (1983). *J. Appl. Cryst.*, **16,** 519.
56. Summerfield, G. C. and Mildner, D. (1983). *J. Appl. Cryst.*, **16,** 384.
57. Schmatz, W., Springer, T., Schelten, J. and Ibel, K. (1974). *J. Appl. Cryst.*, **7,** 96.
58. Glatter, O. (1977). *J. Appl. Cryst.*, **10,** 415.

59. Ghosh, R. (1981). *A Computing Guide for Small Angle Scattering Experiments*, ILL report 81GH29T.
60. Tait, P. J. T., (1979). Zeigler–Natta and Related Catalysts, In *Developments in Polymerisation—2*, (R. N. Howard (Ed.)) Applied Science Publishers, London, p. 81.
61. Jacrot, B. and Zacci, G. (1981). *Biopolymers*, **20**, 2413.
62. Reinch, C. (1961). *Z. Phys.*, **163**, 424.
63. Beyster, J. R. (1968). *Nuclear Science and Engineering*, **31**, 254.
64. Sears, V. F. (1975). *Adv. Phys.*, **24**, 1.
65. Schelten, J., Ballard, D. G. H. and Wignall, G. D. (1974). *Polymer*, **15**, 682.
66. Sadler, D. M. (1971). *J. Pol. Sci. (Phys. Edn)*, **9**, 779.
67. Mehter, A. and Wunderlich, B. (1975). *Colloid and Pol. Sci.*, **253**, 193.
68. Bassett, D. C and Hodge, A. M. (1978). *Proc. Roy. Soc. (London)*, **A359**, 121.
69. Bassett, D. C. and Hodge, A. M. (1980). *Proc. Roy. Soc. (London)*, **A377**, 25 and 39.
70. Wu, W. (1983). *Polymer*, **24**, 43.
71. Wu, W. and Wignall, G. D., To be published.
72. Schelten, J., Wignall, G. D., Ballard, D. G. H. and Longman, G. W. (1977). *Polymer*, **18**, 1111.
73. Schelten, J., Zinken, A. and Ballard, D. G. H. (1981). *Colloid and Pol. Sci.*, **259**, 260.
74. Ballard, D. G. H., Longman, G. W., Crawley, T. L., Cunningham, A. and Schelten, J. (1979). *Polymer*, **20**, 399.
75. Lieser, G., Fischer, E. W. and Ibel, K. (1975). *J. Pol. Sci. (Lett.)*, **13**, 39.
76. Crist, B., Graessley, W. W. and Wignall, G. D. (1982). *Polymer*, **23**, 1561.
77. Schelten, J., Ballard, D. G. H., Wignall, G. D., Longman, G. and Schmatz, W. (1976). *Polymer*, **17**, 751.
78. Summerfield, G. C., King, J. C. and Ullman, R. (1978). *J. Appl. Cryst.*, **11**, 534.
79. Stamm, M., Fischer, E. W., Dettenmaier, M. and Convert, P. (1979). *Disc. Far. Soc.*, **68**, 263.
80. Dettenmaier, M., Fischer, E. W. and Stamm, M. (1980). *Colloid and Pol. Sci.*, **258**, 343.
81. Sadler, D. M. and Keller, A. (1979). *Science*, **203**, 263.
82. Wignall, G. D., Mandelkern, L., Edwards, C. and Glothin, M. (1982). *J. Pol. Sci. (Phys. Edn)*, **20**, 245.
83. Spells, S. J. and Sadler, D. M., *Polymer*, In press.
84. Ballard, D. G. H., Cheshire, P., Longman, G. W. and Schelten, J. (1978). *Polymer*, **19**, 379.
85. Ballard, D. G. H., Burgess, A. N., Crawley, T. L., Longman, G. W. and Schelten, J. (1979). *Disc. Far. Soc.*, **68**, 279.
86. Ballard, D. G. H., Burgess, A. N., Nevin, A., Cheshire, P. and Longman, G. W. (1980). *Macromolecules*, **13**, 677.
87. Stamm, M., Schelten, J. and Ballard, D. G. H. (1981). *Colloid and Pol. Sci.*, **259**, 286.

88. Guenet, J. M., Picot, C., and Benoit, H. (1979). *Macromolecules*, **12,** 86.
89. Guenet, J. M. (1980). *Macromolecules*, **13,** 387.
90. Guenet, J. M. (1981). *Polymer*, **22,** 313.
91. Guenet, J. M. and Picot, C. (1983). *Macromolecules*, **16,** 205.
92. Guenet, J. M. and Picot, C. (1979). *Polymer*, **20,** 1483.
93. Guenet, J. M., Picot, C. and Benoit, H. (1979). *Disc. Far. Soc.*, **68,** 251.
94. Fischer, E. W. and Herchenroder, P. (1980). *Reports to the ILL*, 394.
95. Yoon, D. Y. and Flory, P. J. (1977). *Polymer*, **18,** 509.
96. Yoon, D. Y. and Flory, P. J. (1979). *Disc. Far. Soc.*, **68,** 288.
97. Sadler, D. M. and Harris, R. (1982). *J. Pol. Sci. (Phys. Edn)*, **20,** 561.
98. Sadler, D. M. (1979). *Disc. Far. Soc.*, **68,** 429.
99. Guttman, C. M., DiMarzio, E. A. and Hoffman, J. D. (1981). *Polymer*, **22,** 587.
100. Frank, F. C. (1958). In *Growth and Perfection in Crystals* (Proc. Int. Conf. Crystal Growth) (Doremus, R. H., Roberts, B. W. and Turnbull, D. (Eds)), Wiley, New York, p. 529.
101. Frank, F. C. (1979). *Disc. Far. Soc.*, **68,** 7.
102. Sadler, D. M. (1979). *Disc. Far. Soc.*, **68,** 106.
103. Guttman, C. M., DiMarzio, A. and Hoffman, J. D. (1981). *Polymer*, **22,** 1466.
104. DiMarzio, E. A., Guttman, C. M. and Hoffman, J. D. (1979). *Disc. Far. Soc.*, **68,** 210.
105. Klein, J. and Ball, R. (1979). *Disc. Far. Soc.*, **68,** 198.
106. Barham, P. J., Jarvis, D. A. and Keller, A. (1982). *J. Pol. Sci. (Phys. Edn)*, **20,** 1717.
107. Spells, S. J., Sadler, D. M. and Keller, A. (1980). *Polymer*, **21,** 1121.
108. Stamm, M. (1982). *J. Pol. Sci. (Phys. Edn)*, **20,** 235.
109. Gilmer, G. H. (1980). *Science*, **208,** 355.
110. Muller-Krumbhaar, H. (1979). *Advances in Solid State Physics*, Vol. 19, J. Trevch (Ed.) Vieueg, Braunschweig, p. 1.
111. Avron, J. E., Balfour, L. S., Kuper, L. S., Landau, J., Lipson, S. A. and Schulman, L. S. (1980). *Phys. Rev. Lett.*, **45,** 814.
112. Peterlin, A. (1975). *Copolymers, Polyblends and Composites*, Am. Chem. Soc. Adv. Ser. 142, Am. Chem. Soc., Washington, DC.
113. Ballard, D. G. H., Cheshire, P., Janke, E., Nevim, A. and Schelten, J. (1982). *Polymer*, **123,** 1875.
114. Sadler, D. M. and Barham, P. J. (1983). *J. Pol. Sci. (Phys. Edn)*, **21,** 309.
115. Debye, P. (1945). Rubber Reserve Co. Technical Report CR637. Reprinted in *Light Scattering from Dilute Polymer Solutions* (D. McIntyre and F. Gornick (Eds)), Gordon and Breach, New York, (1964).
116. Odell, J. A. Grubb, D. T. and Keller, A. (1978). *Polymer*, **19,** 617.
117. Sadler, D. M. and Odell, J. A. (1980). *Polymer*, **21,** 479.
118. Sadler, D. M. and Odell, J. A. (1982). Report to the Institute Laue Langevin, p. 404.

Chapter 5

THE ABILITY OF SMALL-ANGLE X-RAY SCATTERING TO DISTINGUISH BETWEEN MORPHOLOGICAL MODELS OF CRYSTALLINE POLYMERS

I. H. HALL

Department of Pure and Applied Physics, UMIST, Manchester, UK

and

M. TOY

Department of Physics, Mathematics and Computing, Manchester Polytechnic, UK

1 INTRODUCTION

X-rays are scattered by crystalline polymers at angles of less than about 5° and Fig. 1 shows two typical patterns produced by oriented samples. In each case the meridian corresponds to the orientation axis. Patterns from isotropic samples are not shown; they can be visualised by imagining each photograph to be rotated about its centre.

A clear interpretation of this type of pattern is that electron density varies along the orientation axis (fibre axis) with a periodicity of the order of 100 Å and other structural evidence leads to the belief that there is a unit of structure comprising a stack of crystalline lamellae separated by less ordered ('amorphous') regions. The small number of broad diffraction peaks can have several possible causes. For example, there might be disordering in the periodicity within the stacks, caused by variation in either crystal or amorphous thickness. Periodicity might vary from stack to stack. The boundary between phases might not be sharp, there being a transition zone of intermediate density. The stacks themselves might be of short coherence length. Each of these factors and others, separately or in combination will affect the distribution of scattered radiation and the purpose of this chapter is to describe the

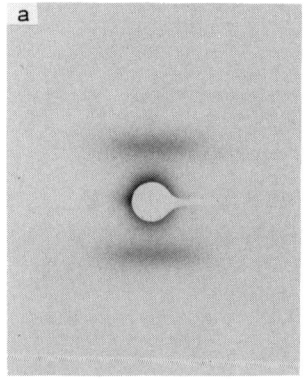
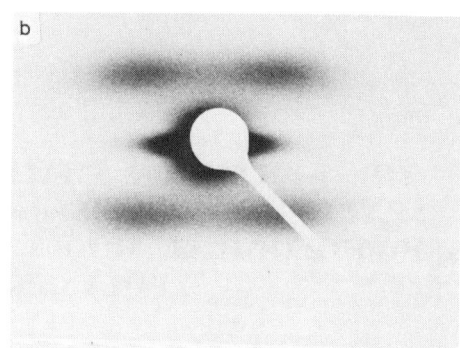

Fig. 1. Typical small-angle scattering pattern. (a) Well-annealed, highly oriented, melt-spun fibre of poly(tetramethylene terephthalate). (b) Rolled poly(ethylene terephthalate) sheet. (Fig. 1b was provided by Dr C. P. Buckley and is included with his permission.)

experimental studies of polymer morphology which have been made using small-angle X-ray scattering and to discuss whether any of the above details can be unambiguously deduced from them.

Morphological changes consequent upon various processing treatments have been followed by visual inspection of their diffraction patterns, and this has proved to be a valuable application. We shall not be concerned with these qualitative interpretations but instead with those experiments which have attempted to deduce quantitative information. Such studies can be subdivided into two distinct groups and these will be reviewed briefly before one of them is analysed in more detail.

In the first group, which will be called direct methods, structural parameters are deduced directly from the distribution of the intensity of the scattered radiation. In the second group, which will be called indirect methods, a model of the polymer structure is proposed and the distribution of the intensity of its scattered radiation calculated. This is then compared with the experimental distribution.

Although most studies using these techniques have been on isotropic polyethylene, crystallised either from solution or from melt, investigations of other polymers and drawn fibres will be included in this review. In intercomparing those results which have been derived using different methods of analysis of the diffraction pattern (or different techniques of investigation) it is important to remember that morphology will depend upon both the polymer and the details of its preparation, and that disagreements cannot be considered significant unless they have been derived from test-pieces prepared using identical methods.

It will be assumed that the reader is familiar with the ideas of Fourier transformation, convolution, reciprocal space and Ewald spheres. Whilst the non-specialist should be able to gain some idea of the power and limitations of small-angle scattering without a complete understanding of the sections dependent on these, anyone intending to do independent experimental work in this subject must become thoroughly conversant with these theoretical concepts. A brief and lucid review will be found in the article by K. C. Holmes and D. M. Blow in *Methods of Biochemical Analysis*, Vol. 13, (D. Glick (Ed.)) Wiley, New York (1966), pp. 113–239. The application of Fourier transform theory to X-ray diffraction is discussed by H. S. Lipson and C. A. Taylor in *Fourier Transforms and X-ray Diffraction*, Bell, London (1958), and by C. A. Taylor and H. S. Lipson in *Optical Transforms*, Bell, London (1964). A more mathematical treatment is given in Reference 24.

2 DIRECT METHODS

Even with direct methods it is necessary to propose a model for the polymer structure, but this is much less detailed than for indirect methods. Thus an isotropic, two phase structure, is assumed, in which there are sharp boundaries between crystalline and amorphous phases of density ρ_c and ρ_a respectively. The volume fraction of each phase is ω_c and ω_a. If, for an ideal point collimated beam, $I(\mathbf{s})$ is the fraction of the incident intensity scattered at an angle 2θ (the angle between the incident and diffracted beam), then a quantity Q may be defined such that:

$$Q = K_1 \int_0^\infty \mathbf{s}^2 I(\mathbf{s}) \, d\mathbf{s} = \overline{\Delta \eta^2} = K_2 (\rho_c - \rho_a)^2 \omega_c \omega_a \qquad (1)$$

where $s = (2 \sin \theta)\lambda$,† λ is the wavelength of the radiation used and $\Delta \eta$

† With X-ray scattering, the scattering vector $\mathbf{s}$ is usually defined as given here. The different definition used in neutron scattering (Chapter 4) should be noted; in that case the scattering vector $\mathbf{q}$ is defined as $\mathbf{q} = 2\pi \mathbf{s}$.

is the difference between the local density and the mean value. For a given material K_1 and K_2 are constants which depend upon the experimental conditions. (In this section simplified forms of equation will be given in which experimental factors are gathered into constants. These will be sufficient for the discussion to be presented. For more extensive treatments the reader is referred to the bibliography of Reference 1 and the other original references.) $Q = \overline{\Delta\eta^2}$ is true for any scattering system, but $\overline{\Delta\eta^2} \propto (\rho_c - \rho_a)^2 \omega_c \omega_a$ is only true for a two phase system with sharp boundaries.

It can be shown theoretically that for such a structure $s^4 I(\mathbf{s})$ approaches a limiting value at large $\mathbf{s}$ and if this is called T then:

$$\frac{2\pi T \omega_c \omega_a}{Q} = S \qquad (2)$$

where S is the area per unit volume of the boundary surface between the two phases. ($\mathbf{s}$ is large relative to small-angle scattering experiments. It is still smaller than the scattering vector from crystalline arrays of atoms).

Several experimental investigations have been based on eqn. (1). Fischer et al.[2] determined $\rho_c - \rho_a$ for sedimented crystalline mats of polyethylene and obtained a result in close agreement with the differences between the unit cell density (ρ_c) and a quantity obtained by extrapolating to room temperature measurements of the density of the melt at various temperatures (ρ_a). They therefore conclude that a two phase, crystalline–amorphous structure exists in these mats and contains few voids. (Voids would act as a third phase contributing to the scatter.) On the other hand, Hermans et al.[3] and Heikens,[4] using cellulose fibres, obtained a higher value of $\rho_c - \rho_a$ than expected and concluded from this and other experiments that voids contribute substantially to the scattering. With poly(ethylene terephthylate), Fischer and Fakirov[5] obtained a lower value than expected, and explained this by suggesting that defects in the crystalline phase lowered ρ_c.

Equation (2) appears to have found little application to solid polymers; indeed with many of these materials experimental measurements of $s^4 I(\mathbf{s})$ do not approach a constant limiting value at large $\mathbf{s}$. Ruland[6] has developed the simple two phase theory to show that this can be explained by assuming that a transition zone of intermediate density exists between the crystalline and amorphous phases. Vonk[7] has applied Ruland's theory and shown that if the density falls linearly from the crystalline to the amorphous value over a length E, then the intensity distribution which would exist if the phase boundaries were sharp must be multiplied by the factor $(\sin^2 \pi Es)/(\pi Es)^2$ to give that actually observed. After making certain approximations, it then follows that if $s^4 I(\mathbf{s})$ is plotted against s^2, this will tend to a straight line at large **s** instead of constant value, and from the slope of the line E may be determined. Vonk determines transition zone thickness of about 10 Å for a wide range of crystalline polymers.

Koberstein et al.[8] point out that the approximation Vonk uses is to truncate the series expansion of $(\sin^2 \pi Es)/(\pi Es)^2$ to two terms. This is only valid at small **s** whereas the linear region of the curve of $s^4 I(\mathbf{s})$ against s^2 occurs at large **s**. The two ranges will not necessarily overlap. However if electron density is assumed to change sigmoidally in the transition region an analysis which is exact over the whole range of **s** may be used. They extend these arguments to include slit collimated beams. With a sigmoidal electron density transition the approximations are less severe and the analysis simpler than with a linear one.

In the range of **s** in which the effects of transition zones are observed there is a significant contribution to the intensity from scattering caused by density fluctuations within phases, which must be subtracted. Wiegand and Ruland[9] have shown that this is distributed according to a power series in even powers of s, and of the methods of subtraction that have been used only two are justified by their theory. Vonk[7] assumed that:

$$I_b(\mathbf{s}) = I_b(0) + b_1 s^n \quad (3a)$$

where $I_b(\mathbf{s})$ is the background intensity, b_1 is a constant, n an even integer and $I_b(0)$ the background intensity at zero scattering angle. This is thus the first two terms of the power series. Ruland[10] used the expression:

$$I_b(\mathbf{s}) = I_b(0) \exp(b_2 s^2) \quad (3b)$$

where b_2 is a constant and this is another approximation to the series.

Intensity measurements are made in a range of **s** in which all other scattering has fallen to zero. The constants are determined from these measurements and values of $I_b(\mathbf{s})$ are then calculated and subtracted from the measured intensity over the whole range of **s**.

Koberstein et al.[8] have argued that intensity measurements must be continued into the wide-angle range to make this correction properly. Some investigators have assumed that I_b is constant; this is clearly inadequate.

According to Fischer and Fakirov[5] transition zones also reduce Q by an amount proportional to $E/\bar{x}$, where $\bar{x}$ is the average periodicity of electron density. The low value they obtained for Q could be explained by reasonable values of $E/\bar{x}$ as well as by defects in the crystalline phase.

Q can be determined directly from the scattering curve, but it is related by eqn. (1) to three parameters of the material, ρ_c, ρ_a and ω_c. The work described so far is limited to tests of the two phase model by ensuring that Q is consistent with other determinations of these parameters. Strobl[11] has developed the theory to enable further information to be extracted from the scattering curve. He makes the additional assumption that the thickness of all amorphous layers is constant and considers only the case of sharp phase boundaries.

The intensity distribution is proportional to the product of two factors $|F(\mathbf{s})|^2$ and $Z(\mathbf{s})$.† $F(\mathbf{s})$ is the Fourier transform of the electron density profile of the amorphous layer relative to the crystalline value (it is more usual to use the crystallite density profile) and is called the structure factor. $Z(\mathbf{s})$ is usually called the lattice factor and depends upon the periodicity and its variation. It is also known as the interference function, a name which might be preferable as a term also called the lattice factor is defined differently (see Chapter 6). A more precise definition will be given in Section 4.1.

Strobl showed, by considering two models in which the periodicity varied in different ways, that $\int Z(\mathbf{s})\,d\mathbf{s}$ over the range Δs was equal to unity provided $\Delta s = 1/\bar{x}$ and is centred on $s = m/\bar{x}$ where m is an integer. (This is, however, only true when the variation in periodicity is symmetrical about its average value.) Thus, if the scattering curve is integrated over one of these ranges the average value of the structure factor at the centre of this range can be obtained. These are determined at several values of $\mathbf{s}$ and extrapolated to give $(\bar{F})^2$ at $\mathbf{s}=0$ ($|\bar{F}(0)|^2$). Strobl and Müller[12] show that:

$$|\bar{F}(0)|^2 \propto \bar{x}^2 \omega_a^2 (\rho_c - \rho_a)^2 \qquad (4)$$

---

† This may be understood from eqn. (7a) of Section 4.1. The assumption of constant thickness of amorphous layers has two consequences: (a) the lattice variability is controlled by the crystalline phase and so the model belongs to the restricted paracrystalline class making eqn. (7a) applicable, and (b) $F(\mathbf{s})$ is the same for all long periods, whence $\bar{F}^2 = (\bar{F})^2$. It is further assumed that the stack of lamellae is long, and the result follows.

where the constant of proportionality may be determined from the chemical composition of the scattering material.

Thus, using eqns (1) and (4), ω_c, ω_a and $\rho_c - \rho_a$ may be determined. By measuring the density of the sample ρ_c and ρ_a may be found. Strobl and Müller applied this method to sedimented mats of polyethylene single crystals and to melt crystallised polyethylene. For the former material, several orders of reflection were visible and since the width of these increased linearly with order it was concluded that the periodicity was constant in one lamellar stack, but varied from stack to stack. For the other material, only the first order of reflection was resolved, but it was again concluded that the variation in periodicity between stacks was very much greater than within. This was done by dividing the scattered intensity by the deduced values of $|F(\mathbf{s})|^2$ to obtain $Z(\mathbf{s})$ and finding this to be incompatible with values calculated from models in which the variation was within stacks.

These methods have been further developed by Ruland,[13] who shows how to obtain the distribution functions of the distances between phase boundaries directly from the scattering curve. He gives methods for isotropic and oriented materials[14] and for both point and slit collimated beams. We will consider a point collimated beam and an isotropic scatterer. The scattered intensity $I(\mathbf{s})$ is measured and this is first corrected for density fluctuations within phases and then for finite phase boundary thickness. The first was made as described earlier (eqn. (3b)). the second used the same method as Vonk[7] to determine the boundary thickness and used this value to calculate the intensity distribution curve which would be observed with sharp boundaries. After making these changes $s^4 I(\mathbf{s})$ is asymptotic to a limiting value, T, at large $\mathbf{s}$. A function $G_1(\mathbf{s})$ is defined as $G_1(\mathbf{s}) \propto T - s^4 I(\mathbf{s})$ and the transform $g_1(\mathbf{r}) = 2 \int_0^\infty G_1(\mathbf{s}) \cos 2\pi \mathbf{r} \cdot \mathbf{s} \, d\mathbf{s}$ contains the distribution functions of the distances between interfaces.

$g_1(\mathbf{r})$ may be understood by reference to Fig. 2. Figure 2a represents a sequence of crystallite (clear) and amorphous (hatched) segments. The first peak of $g_1(\mathbf{r})$ (Fig. 2b) is the distribution function of the lengths d_1 and the second is of the lengths d_2, A and B being respectively their most probable lengths. In the case illustrated, the first relates to amorphous lengths and the second to crystallites, but the first is whichever of these two is shorter. The third peak is the sum of distributions of the lengths d_{12} and d_{21} and remaining ones may be identified similarly. Stribeck and Ruland[15] have applied the method to various polyethylenes. They found that although the crystallinity of

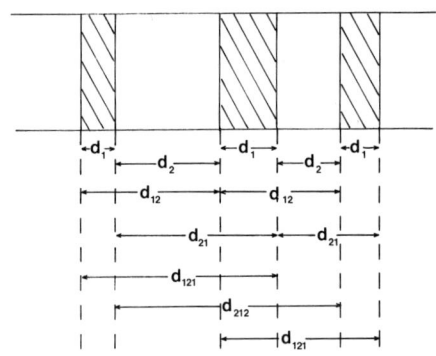

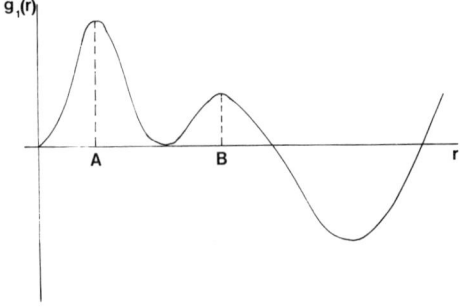

Fig. 2. (a) Sequence of crystalline (clear) and amorphous (hatched) segments. (b) The distribution function of distances between interfaces.

their samples varied widely, the thickness of the phase boundary, the amorphous layer, and its standard deviation were fairly constant for all samples, having values of about 10 Å, 35 Å and 15 Å respectively. Thus the differences between samples lay in the crystallite sizes. Values were obtained for their thickness, but overlaps of the various distribution functions prevented a more detailed statistical description of the structure being obtained. However $g_1(\mathbf{r})$ was calculated for several models and the best agreement with the experimental curve was obtained for one in which the stacks were perfectly periodic but in which x varied from one to another. Other models required ridiculously short stack lengths. The stack crystallinity was, in all cases, higher than the bulk measured by other means, indicating the presence of amorphous material which is not part of the lamellar structure. The volume fraction of this decreased sharply with increasing bulk crystallinity.

The experimental difficulties associated with all the investigations discussed so far are severe. There is, of course, the problem common to all scattering experiments in that the calculations refer to ideal point collimated beams, whereas slit collimation is used in many experiments and circular apertures of finite diameter in the remainder. In most of the experiments using slit collimation the theoretical reationships were convoluted with a narrow slit aperture of infinite length to give the 'slit-smeared' intensity before applying them. This procedure is less error prone than deconvoluting ('desmearing') the experimental observations. (See Section 5.1.1 for a fuller discussion of this point.)

The greatest experimental uncertainty concerns the range of integration of eqn. (1). In theory, it should be integrated from $s = 0$ to $s = \infty$ but there are difficulties at both ends of the range.

The region of very small s contains scattering due to voids, which cause the intensity to increase rapidly with decreasing s. Much of the region is experimentally inaccessible because of the main beam and so extrapolation is necessary. Fischer and Fakirov, and Strobl and Müller do this in a way which attempts to eliminate the void scatter, but at best this is an error prone procedure. Parasitic scattering on the edges of the main beam also contributes some intensity in this region and must be subtracted.

The situation at large s is even less satisfactory. Although the problem of background correction was discussed only with respect to measurement of transition zone thickness, it is important in any experiment where intensity measurement must be continued to large s. Where correction has been applied it has generally been assumed to be a constant and this is inadequate. Small errors will have a proportionately larger effect on Q since this is the integral of $s^2 I s$. If deconvolution is used this will also introduce errors, particularly into the extremes of the experimental range, and it has been recommended that where this is done data from the top 25% of the range of s be rejected.[16] In studies of polymer morphology, no attempts appear to have been made to assess the errors caused by these difficulties. However, in the related field of measurement of particle size distribution, Glatter[53,54] has studied the effects both of random errors in the experimental data and of finite experimental range on the accuracy with which the distribution is determined. Extension of this work to the problem discussed above could prove valuable.

With Strobl and Müller's work there is the additional uncertainty caused by extrapolation of the curve of $|\overline{F}(s)|^2$ against s to $s = 0$. They do this by fitting polynomials of various degree to their observed points, and judge an asymptotic value to which the extrapolation value is tending with increasing degree. However, it is well known that probable errors increase alarmingly with range of extrapolation even in linear cases. When a polynomial of uncertain degree is fitted to data of unknown accuracy and then extrapolated, the situation must be worse. It would be a valuable, though difficult, exercise to estimate the likely errors caused by the above uncertainties.

A further weakness of these methods is that assumptions have to be made about the nature of the scattering system (i.e. a model must be proposed) and the only check on the validity of this model is that the values which are deduced for various parameters seem reasonable in comparison with those deduced using other experimental techniques. Thus when Fischer and Fakirov obtained a value of Q which did not lead to the expected value of $\rho_c - \rho_a$, there was no way of deciding whether this was caused by transition zones or by crystallite imperfections. Similarly, the density values obtained by Strobl and Müller only have validity if the amorphous thicknesses are constant, the periodicity varies symmetrically above and below the mean, and the phase boundaries are sharp.

Although Koberstein *et al.* recommend the assumption of a sigmoidal electron density gradient in the transition zones, this is only because it leads to simpler and more accurate analysis than the assumption of a linear one. There is no evidence from experiments of this nature that either is correct. And there is no way of determining how such deviations from model assumptions will affect the results.

For these reasons, direct methods have only limited value in distinguishing between alternative morphological models and will not be discussed further in this review.

3 INDIRECT METHODS

Indirect methods may be subdivided into two types. In the first of these the distribution of scattered intensity from a proposed model is calculated and compared with that observed experimentally. We shall return to a more detailed consideration of this method after discussion of the second type which has been exploited principally by Vonk and Kortleve.[17]

This utilises a quantity known as the one-dimensional correlation function γ. If $\Delta\eta(x_1)$ is the difference between the electron density at a point x_1 and its average value, then $\Delta\eta(x_1 - x)$ is this same difference at a point distant x from x_1. The correlation function is then defined as:

$$\gamma(x) = \frac{\int_0^\infty \Delta\eta(x_1 - x) \Delta\eta(x_1) \, dx_1}{\int_0^\infty \Delta\eta^2(x_1) \, dx_1} \quad (5)$$

γ is thus the average of the product of the electron density differences at two points a distance x apart, and is a measure of the probability of finding the density the same at these two points, showing a peak at the most probable repeat distance. It may be calculated from the density distribution of a proposed model, but is also derivable from the observed intensity distribution. If this is from an isotropic sample:

$$\gamma(x) = \frac{4\pi \int_0^\infty s^2 I(s) \exp(2\pi i s x) \, ds}{\int_0^\infty s^2 I(s) \, ds} \quad (6)$$

The numerator of eqn. (6) is the Fourier transform of the intensity distribution, corrected for the fact that for an isotropic sample this is distributed over a sphere in reciprocal space. Except for an experimental constant, the denominator is Q of eqn. (1), which is equal to $\overline{\Delta\eta^2}$. The introduction of these quantities eliminates experimental constants, and enables relative rather than absolute intensities to be used.†

Thus the model may be tested by comparing $\gamma(x)$ predicted from eqn. (5) with that deduced from the experimental observations using eqn. (6).

Kortleve and Vonk[18] have determined the correlation functions from the scattering curves of samples of linear and branched polyethylene and have matched these with those calculated from a model structure. Blundell,[19] and Brown et al.[20] have determined the function for several different models to ascertain which would best

† From eqn. (6), $\gamma(x)$ is proportional to the Fourier transform of the intensity distribution, which is equivalent to the Patterson function of structural crystallography. Equation (5) is an expression equivalent to that for the interatomic vectors of a crystal. Thus, these two equations apply a technique familiar to crystallographers to scattering from one-dimensional disordered lattices.

reproduce that obtained experimentally, and it is clear from their results that variations both in the type of the model, and in the model parameters produce discernible differences in the correlation function. What is unclear in all this work is the uncertainty in the experimentally determined function arising from unavoidable errors in the intensity values.

The same experimental problems arise as in the direct methods, that is the range of integration must extend into experimentally inaccessible regions. At small angles (the backstop region), Vonk and Kortleve claim that any reasonable extrapolation is satisfactory and that neither the position nor the height of the first maximum is affected if quite absurd methods of extrapolation are used. However Blundell,[19] rather than determining the correlation function directly from the model, calculates its intensity distribution first and then evaluates the Fourier transform of this. He is then able to investigate the effect of extrapolation by replacing the back stop region of the theoretical intensity distribution with an extrapolated curve. He extrapolates his experimental curves with a curve proportional to s^2 and shows that if this is compared with the theoretical curve the position and height of the first positive peak are unaffected, as claimed by Vonk and Kortleve. However, the depth of the first negative peak is significantly changed, suggesting that parts of the correlation function are sensitive to this extrapolation. Vonk and Kortleve[17,18] admit that more care is necessary at the high angle tail. The intensity did not drop to zero in this region because of overlap with the wide angle scatter and so they shifted the base line up to the point of minimum intensity. 'Spurious detail' was still observed in the correlation function and so they constructed a tail, assuming constancy of $s^4I(s)$. They claim that this was satisfied over a wide range of s. Blundell used essentially the same method except that after subtracting the constant background he determined a transition zone thickness using Ruland's method[6] and multiplied his intensities by a factor including that thickness, which changed them to what they would have been if the system had had sharp phase boundaries. Vonk and Kortleve state that errors in the background correction cause uncertainty in the correlation function near the origin, but otherwise neither author comments on its sensitivity to this procedure. Brown et al.[20] make no mention at all of these difficulties.

Experimental uncertainty in the measured intensity is unavoidable. None of the authors give any indication of its magnitude. Since all use slit collimation, desmearing is necessary and this will increase the uncertainty particularly at the extremes of the experimental range.[16] These possible errors in the intensity data will be transmitted to the correlation function, and without knowing how large they are it is not possible to assess whether the differences between models, or between model and theory are experimentally significant. It has already been pointed out that the method is analogous to the use of the Patterson function in structural crystallography, and in this application the propensity of the errors mentioned above to generate spurious detail is well known. It would be surprising if the present application were less error prone. Because of these uncertainties, the remainder of this review will be concerned with the first type of indirect method; that is comparison of the theoretical intensity distribution for a given model with that determined experimentally. It is still necessary to know the accuracy with which intensity is measured if it is to be used to distinguish between models. This has been assessed in relatively few cases; however, because the comparison is with direct experimental data, judgement of the significance of differences between experiment and theory can be more readily made.

4 MORPHOLOGICAL MODELS

A basic unit of polymer structure is a stack along which electron density alternates between high values in regions of molecular order (crystallites) and low values in regions of disorder (amorphous regions). The simplest model incorporating this feature comprises stacks with no coherence between neighbours, the boundaries between the regions being perfectly sharp, flat and normal to the stack axis. The axial alternation of electron density is then the convolution of an electron density profile with a one dimensional lattice, whose spacings must be distorted from regularity because only a few, broad, diffraction spots are visible in the region of small-angle scattering; indeed with melt crystallised material it is only usual to observe one intensity peak.

These distortions may be classified as belonging to one of two types.[21,22] In the first, long range order is preserved, but at each lattice point there is either a random displacement or a random substitution of one particle with another. In the second, long range order is destroyed, the particle position being determined by placing one end of a lattice vector, which varies randomly about a mean value, on the position of the previous particle. It is this lattice, called paracrystalline, which describes the location of crystallite particles in a stack.

4.1 The Paracrystalline Models of Hosemann[23]

The lattice vector (**x**) at a point in a paracrystalline stack is equal to the sum of the lengths of the crystallite (**y**) and the adjacent amorphous region (**z**) at that point. It can be assumed either that its variation is solely due to the random variation in the length of the amorphous region from point to point, or that it is due to the combined, but independent, random variations in the lengths of both components.

The first case is known as the restricted paracrystalline model and it can be visualised by imagining a distorted lattice of the second type being laid down and then a crystallite being placed at each lattice point. The variations in the lengths of the crystallites constitute a substitutional disorder of the first kind. A model such as this is only physically realistic if the variations in both the lattice vector and the crystalline length are small compared with their mean values. Otherwise crystallites will overlap when long lengths coincide with short vectors. The intensity distribution from such a model is given by:[22,24]

$$I(\mathbf{s}) = \Delta \eta^2 [N\{\overline{F^2} - (\bar{F})^2\} + \{(\bar{F})^2/\bar{x}\}Z(\mathbf{s}) * |\Sigma(\mathbf{s})|^2] \tag{7a}$$

$I(\mathbf{s})$ is the intensity of scattering of one stack relative to that of an isolated single electron placed at its centre, and now relates to an oriented, not an isotropic sample. N is the number of crystallites the stack contains. $F(\mathbf{s})$ is the structure factor which has already been defined (Section 2), but now relates to a crystallite, not the amorphous layer. $Z(\mathbf{s})$, the lattice factor or interference function (also introduced in Section 2), is the Fourier transform of a function $z'(\mathbf{u})$, where $z'(\mathbf{u})\,d\mathbf{u}$ is the probability of finding a lattice point in a length $d\mathbf{u}$ at any point distant **u** from the origin.

If one lattice point is taken as origin, and $dP(\mathbf{u})$ is the probability of finding a second lattice point in the short length $d\mathbf{u}$ situated at the end of the vector **u**, then a quantity $P(\mathbf{u})$ may be introduced such that:

$$dP(\mathbf{u}) = \frac{P(\mathbf{u})\,d\mathbf{u}}{\bar{x}}$$

It follows that:

$$z'(\mathbf{u}) = \delta(\mathbf{u}) + P(\mathbf{u})/\bar{x}$$

where $\delta(\mathbf{u})$ is the delta function. (This is necessary because of the certainty of finding a lattice point at the origin).

The symbol ∗ denotes the operation of convolution and $\Sigma(\mathbf{s})$ is the Fourier transform of a function σ which has the value unity within the stack and zero outside. It is thus related to the stack length L and as this approaches infinity $Z(\mathbf{s})\ast|\Sigma(\mathbf{s})|^2$ approaches $LZ(\mathbf{s})$; thus eqn. (7a) becomes:

$$I(\mathbf{s}) = \Delta\eta^2 N[\{\overline{F^2} - (\bar{F})^2\} + (\bar{F})^2 Z(\mathbf{s})] \tag{7b}$$

Hosemann[22] has shown that $Z(\mathbf{s})$ may be determined from the Fourier transform G_x of the normalised distribution function of $\mathbf{x}$ ($g(\mathbf{x})$) and is given by:

$$Z(\mathbf{s}) = (1 - |G_x|^2)/(|1 - G_x|^2) \tag{7c}$$

In these and subsequent equations, $\Delta\eta$ is the electron density difference between amorphous and crystalline regions, not the difference from the mean value (as it was in previous sections). In order to determine $\bar{F}$ and $\overline{F^2}$ it is necessary to know the distribution function of crystallite lengths ($h(\mathbf{y})$).

Note that the model can equally well apply to the situation where the lattice variation is defined by the variation in crystallite lengths and the amorphous lengths constitute a substitutional disorder. The essential feature is that the lattice variability is controlled by only one of the phases. Thus the model used by Strobl and Müller[12] is of this type.

The second case is known as the general paracrystalline model, and it may be visualised by imagining each lattice vector being constructed by choosing at random a crystallite length and an amorphous region length and adding them together. According to Hosemann[22] the intensity scattered by a finite stack of N crystallites is now $I(\mathbf{s}) = I_B(\mathbf{s}) + I_C(\mathbf{s})$ where:

$$I_B(\mathbf{s}) = \frac{\Delta\eta^2 N}{(2\pi s)^2}\left[\frac{|1-H_y|^2(1-|J_z|^2)+|1-J_z|^2(1-|H_y|^2)}{|1-H_y J_z|^2}\right] \tag{8a}$$

and:

$$I_C(\mathbf{s}) = \frac{\Delta\eta^2}{2(\pi s)^2} Re\left[\frac{J_z(1-H_y)^2\{1-(H_y J_z)^N\}}{(1-H_y J_z)^2}\right] \tag{8b}$$

H_y and J_z are the Fourier transforms of the normalised distribution functions of crystallite lengths and disordered region lengths ($h(\mathbf{y}), j(\mathbf{z})$). The units of I are the same as for eqn. (7). If N is large (i.e. the stack approaches infinite length), I_c may be neglected over the observable range; the effect of small N will be discussed later.

4.2 Distribution Functions

In order to use eqns (7) and (8), $\bar{F}$, G_x, H_y and J_z must be evaluated which requires knowledge of the distribution functions $g(\mathbf{x})$, $h(\mathbf{y})$ and $j(\mathbf{z})$. Whilst it is unlikely that any simple mathematical function can be fitted to the distribution which actually exists within a fibril (for example, crystallite lengths are likely to be distributed discontinuously since they probably change by integral numbers of unit cells), it is our concern to investigate whether any such function will lead to an intensity profile compatible with experimental observation, or indeed if the profiles produced by different distributions are experimentally distinguishable. Thus we shall evaluate the power and limitations of small-angle X-ray scattering to elucidate this aspect of polymer structure. For this purpose distribution functions will be classified according to whether they are symmetric or asymmetric about the mean value. For convenience they will be expressed as functions $j(\mathbf{z})$, i.e. as distributions of amorphous lengths. As will be described in Section 4.3, they can be applied to either phase.

4.2.1 Symmetric Distribution Functions

Two symmetric distribution functions have found application. The first of these is the well known Gaussian function:

$$j(\mathbf{z}) = \sigma_z^{-1}(2\pi)^{-1/2} \exp\left\{\frac{-(z-\bar{z})^2}{2\sigma_z^2}\right\} \qquad (9)$$

where σ_z is the standard deviation which gives a measure of the dispersion of lengths about the mean value $\bar{z}$.† The other is a top-hat function whose application in this context was first suggested by Tsvankin.[25,26] In this it is assumed that all lengths between an upper and lower limit are equally probable.

Unless the standard deviation is much smaller than the average length, the Gaussian distribution will lead to negative segment lengths, which is physically unrealistic. Also, where second order reflections are visible, in the diffraction pattern they do not usually occur at twice the value of **s** for the first order. While this does not necessarily imply that the length distribution is asymmetric (it could be caused by distortions of the lattice), the possibility cannot be ignored. For these reasons, asymmetric distributions have been investigated.

† The symbols σ, γ and m which are used in describing distribution functions, also occur in Sections 2, 3 and 4.1 with different meanings. Since there is no danger of confusion, the original author's usage has been retained.

4.2.2 Asymmetric Distribution Functions

Tsvankin[25,26] has suggested application of the exponential function:

$$j(\mathbf{z}) = \frac{1}{\bar{z}} \exp(-z/\bar{z}) \qquad (10)$$

but this is very restrictive because the mean length and standard deviation are both $\bar{z}$ and cannot be varied independently. Blundell uses instead:[19]

$$j(\mathbf{z}) = \frac{1}{\sigma_z} \exp\left\{\frac{-(z-z_0)}{\sigma_z}\right\} \quad \text{for } z > z_0$$
$$= 0 \quad \text{for } z < z_0 \qquad (11)$$

σ_z is now the standard deviation of the distribution and $z_0 + \sigma_z$ its mean length.

The function:

$$j(\mathbf{z}) = \frac{(z-\varepsilon)}{(\gamma\bar{z})^2} \exp\left\{\frac{-(z-\varepsilon)}{\gamma\bar{z}}\right\} \qquad (12)$$

has been proposed by Reinhold et al.[27] In this $\varepsilon = \bar{z}(1 - 2\gamma)$ and γ is a parameter controlling both the width and the skew of the distribution. Skew is positive (i.e. values very much smaller than average are less probable than those very much larger) if γ is positive and vice versa. Asherov and Ginsburg[28] have proposed the function:

$$j(\mathbf{z}) = \frac{(\nu z)^m}{z(m-1)!} \exp(-\nu z) \qquad (13)$$

where m and ν are parameters controlling the width and skew of the distribution and are related by $m = \nu\bar{z}$. If $m = 1$ this reduces to eqn. (10) and if $m = 2$ it becomes eqn. (12) with $\gamma = \frac{1}{2}$. As m becomes larger, the distribution becomes more symmetric and with values greater than about 20 resembles the Gaussian.

Three quantities are important in an asymmetric distribution, the mean length, the standard deviation and the degree of asymmetry. Ideally, it should be possible to vary all three independently. This is not possible with any of the above distributions. In eqns (12) and (13) the asymmetry and width are interrelated, in (10) and (11) the asymmetry is that of the exponential function, and in eqn. (10) the mean length and the standard deviation are equal. Blundell[19] has attempted to introduce more freedom by proposing a distribution in which eqn.

(11) (the exponential) is convoluted with eqn. (9) (the Gaussian). The resulting expression does not have a simple analytic form, but it is the Fourier transform (which may be expressed as the product of the transforms of the individual functions) which appears in the expression for intensity and this may be readily evaluated. By varying the relative magnitudes and variances of each function, distributions of varying width and skew can be generated.

4.3 Application of Distribution Function to Models

Various authors[25-35] have calculated the distribution of intensity from models containing the above distribution functions but we shall consider only the few investigations in which comparisons have been presented between calculated curves and those derived experimentally.

The only models to be studied in which both phases have the same type of symmetric length distribution are those with Gaussian distributions. Blundell[19] and Toy[32] have allowed the standard deviation of each phase to be independent of the other, whereas Crist and Morosoff[33,34] have restricted themselves to cases where the ratio of standard deviation to mean length is the same for both.

Models in which both phases have the same type of asymmetric distribution have been studied by Toy,[32] who has used both Reinhold and Asherov distributions, by Crist and Morosoff[33,34] who have used the Reinhold distribution keeping γ the same on both phases; and by Blundell[19] who used both the modified exponential (eqn. 11), and the exponential convoluted with Gaussian.

There is, of course, no reason why both phases should have the same type of distribution, and models in which the crystalline phase has a symmetric length distribution and the amorphous an asymmetric one could be realistic. Tsvankin[25,26] suggested a model in which the crystallites have a 'top-hat' distribution and the amorphous phase an exponential (eqn. 10), and this has been investigated by both Toy[32] and Crist and Morosoff.[33,34] Blundell[19] has studied models in which the crystallite distribution is Gaussian and the amorphous is modified exponential.

Asherov and Ginsburg[28] propose that the distribution of crystallite lengths should be Gaussian, and that of amorphous lengths according to eqn. (13) (the Asherov distribution). They report having calculated intensity distributions for various combinations of m, σ_y (the standard deviation of the crystallite lengths), and the ratio of mean crystallite to mean amorphous length (this quantity is usually expressed as the crystallinity ϕ; $\phi = \bar{y}/\bar{x}$), but since no actual curves are presented, their work is of limited value in the present context.

Kilian and Wenig[35] assumed a model different in principle from all these in that the distribution of amorphous lengths is symmetric (Gaussian) and that of the crystallites asymmetric (exponential). They assumed values for the statistical parameters and investigated only the effect of varying the coherence length of the lamellar stack.

Other combinations of distributions are possible, but do not appear to have received attention, and are probably not worth studying unless it can be shown that the intensity distributions from the models already proposed differ significantly from experiment.

4.4 Other Morphological Possibilities

The model described above is one of the simplest which will account for the distribution of intensity along the meridian of the diffraction pattern, but there are many ways in which the morphology may differ from this and yet produce patterns which are qualitatively similar.

If the stacks are of finite length, then eqn. (7b) is invalid and the function $\Sigma(\mathbf{s})$ must be included in the expression for the intensity scattered by the restricted paracrystalline model. I_C (eqn. (8b)) must be included with the general model. The effect of these changes is to broaden the diffraction peak and to introduce a second scattering component around the main beam which, since it lies in the region obscured by the backstop, is difficult to observe experimentally.

In most of the investigations mentioned above it has been assumed that all the lamellae stacks belong to the same statistical population of crystallite and amorphous lengths. This is not necessarily so and Blundell[19] has investigated the consequences of allowing the crystallinity to vary from stack to stack whilst maintaining the average amorphous length the same for all. Thus those stacks with higher crystallinity would have greater crystallite length and so a greater long period. The effect would therefore be to broaden the diffraction peak. This is only one of many ways of introducing variability between stacks; other ways would produce a similar effect.

Tsvankin has suggested models which include a transition zone between phases in which the electron density changes linearly over a finite distance, and Blundell,[30,31] from calculations of intensity distributions for models with and without such zones, concludes that they cause the intensity to fall off gradually with increasing scattering angle, the effects only being noticeable at angles greater than two or three times that of peak intensity. In later work[19] he corrects his experimental curves for their presence but gives no indication of the magnitude of this correction. Crist[33,34] and Toy[32] also include them in some of their models.

Vonk[36] has studied the effect of allowing the boundary surface to deviate from flatness. In an isotropic material its effect is again to broaden the diffraction peak and reduce its intensity, which is qualitatively similar to that of increasing the width of the distribution function, of reducing the length of the lamellar stack, or of allowing inter-stack variability. However, whereas these increase scattering at very small angles, distortion of the boundary surface reduces it, and so measurements as close to the main beam as possible would be necessary to distinguish it from the other possibilities. The intensity distribution along lines normal to the meridian is also affected by these distortions and so experiments with oriented material might have more prospect of success.

This two-dimensional distribution of scattering intensity is related to the three-dimensional nature of the morphology. In Section 7 we shall briefly discuss the experimental and theoretical investigations that have been performed to elucidate this, but the one-dimensional studies will be the main concern of this review.

5 EXPERIMENTAL PROCEDURES

If it is to be possible to distinguish between the various morphological models described above by comparing their calculated scattering curves with those determined experimentally, it is essential that these experimental curves be determined with sufficient resolution and accuracy, and that these factors be known. In this section we shall discuss those features causing loss of resolution and inaccuracy which are common to all small-angle scattering apparatuses. We shall not be concerned with the designs of particular cameras.

5.1 Factors Causing Loss of Resolution

In deriving the equations for scattered intensity for various models it is assumed that the specimen is irradiated by a plane, monochromatic, wave front of an extent which is infinite compared with the structural detail, and the intensity of the scattered wave is determined at an unambiguously defined angle. In any practical experimental equipment the specimen will be irradiated by a non-parallel beam of finite cross-section and range of wavelength, and the radiation will be scattered by a finite volume of specimen into the entrance slit of a detector which accepts a range of scattering angles. All of these features will cause loss of resolution (line broadening) and further loss might be caused by the method of recording the detector signal.

If the radiation was incident on a perfectly periodic structure of infinite extent all scattering would occur at one angle (i.e. the detector signal should be a delta-function with respect to angle). Each feature mentioned above taken separately will broaden this delta-function into a profile of finite width; the total broadening will be the convolution of all these profiles. Any recorded signal is the scattering distribution of the specimen convoluted with the total broadening profile, which therefore represents the loss of resolution due to instrumental features.

Typically, the intensity distribution is a broad peak with an angular width at half peak intensity between 0·5 and 1 of the angle at which the peak occurs.[19,32] To distinguish detail in the peak the width of the broadening function must be at least an order of magnitude less than this half-width, i.e. it should be no more than 0·05 of the angle of peak intensity (or typically, not greater than about 2'). Only Toy[32] provides information enabling this criterion to be checked. Crist and Morosoff[34] indicate beam width but take no account of other broadening features.

We will now discuss in some detail the determination of the broadening function of each instrumental feature.

5.1.1 Broadening Due to Beam Collimation

A typical collimating system is illustrated in Fig. 3. If the specimen were a perfectly periodic structure, a diffraction image would be

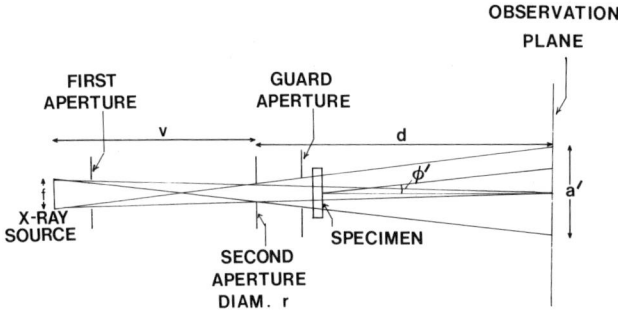

Fig. 3. Schematic pin-hole collimation system.

formed which (for small diffraction angles) would have the same intensity profile as that of the main beam at the observation plane, but whose centre is displaced through the appropriate diffraction angle. The broadening function is therefore this intensity profile.

With aperture collimation, any adjustment to reduce a' (Fig. 3) will also reduce intensity. The simultaneous provision of adequate intensity and resolution is one of the major difficulties in small-angle scattering experiments and has caused slit collimation to be widely used. The narrow width gives high resolution in a direction perpendicular to the slit length; the length increases the intensity. The diffraction image is, however, rectangular and the observed scattering is the convolution of that due to the specimen with this rectangular object and is called the slit-smeared distribution. For a specimen with fibre orientation this is unimportant if only the intensity distribution parallel to the fibre axis is required, and the slit length is perpendicular to this axis. In practice it is difficult to achieve this setting with sufficient accuracy. The method has found widest application with isotropic scatters and the observed scattering distribution is then very different from that of the specimen.

When using direct methods (Section 2), the usual method has been to convolute the theoretical distribution with a narrow slit of infinite length and to apply the experimental data to the resulting equation to determine the various required parameters.

'Desmearing' procedures have been developed whereby the intensity distribution due to the specimen is recovered from the observed distribution (see, for example, References 37–40). These are essentially deconvolution procedures, which, as is well known, are prone to magnify random errors, convert them to systematic ones, and to introduce spurious detail.[41] This is particularly true near the ends of the experimental range.[16] Critical comparison of calculated and recorded curves then becomes dangerous. Unfortunately, the otherwise excellent work of Blundell[19] suffers from this defect. Although the remarks have been directed to the use of rectangular slits, they also apply when circular apertures are used. It is always dangerous to attempt to recover information by deconvolution, and more reliable to convolute the calculated distribution with the instrumental broadening function. But it then becomes imperative that this function is not so broad (i.e. resolution is so poor) that it obscures the detail which is necessary to distinguish between models.

It is also worth noting at this stage that precisely the same smearing occurs if a rectangular entrance slit is used for the detector, whether it be a microdensitometer or photon counter. This appears to have been overlooked in some work. With oriented specimens, it is sometimes advantageous to use a rectangular entrance slit and in this case slit collimation may be used. This will be discussed in Section 5.3.

The general problem of maximising intensity for a given resolution has been considered by Bolduan and Bear,[42] Guinier and Fournet[37] and Huxley.[43] Features from all of these will be included in the following, which is a simplification adequate for most experimental situations.

The X-ray tube is assumed to have a focal spot of fixed size, and is viewed at such a take-off angle that it appears to be a square of side f. For maximum intensity Huxley[43] shows that the first aperture must not obscure any of the focal spot, and both he and Bolduan and Bear[42] show that extreme rays from the focus through the second aperture must intersect at the observation plane. The guard aperture obscures parasitic scatter from the second aperture and should be just larger than the extreme beam diameter at its location. Provided the above conditions are met, the size and location of the first aperture are unimportant and it will be omitted from the present treatment. The only effect of the guard aperture on present considerations is to displace the specimen from the second aperture. Provided this displacement is small compared with v, it can be neglected.

For extreme rays to intersect at the observation plane:

$$\left(\frac{d+v}{d}\right)r = f \tag{14}$$

Suppose the maximum acceptable angular width of the collimation broadening function is w_c, then:

$$dw_c > a' = \frac{rd + fd + rv}{v}$$

For maximum intensity, d must be as small as possible, and so:

$$d = \frac{rv}{vw_c - f - r} \tag{15}$$

Eliminating d between eqns (14) and (15) gives:

$$v = \frac{2f}{w_c} \tag{16}$$

In using eqn. (16) it must be remembered that w_c is less than the total loss of resolution which can be tolerated, as broadening due to other features must be included. On the other hand, it is the extreme width of the collimation broadening function, and so will be somewhat greater than the loss of resolution caused. A reasonable value would be of the order of half the total acceptable broadening, which using the criteria given earlier would be about 1'. For a microfocus X-ray tube, with a focal spot of 0·1 mm square, v would be 688 mm, which is large but not unreasonable. For a fine focus tube, with $f = 0·4$ mm, this resolution would be impracticable, since $v = 2750$ mm.

Since the broadening function (B) is the intensity profile of the main beam at the observation plane, it can be measured directly. If this is done, the angular range accepted by the entrance slit of the detector should be the same as that used in the scattering measurements (see Section 5.1.4). Alternatively, it can be calculated by assuming that the brightness of the focal spot of the X-ray source is uniform. In that case, with collimators set as in Fig. 3 and using eqn. (16), it is given by:

$$B(\phi') \propto \cos^{-1}\frac{2\phi'}{w_c} - \frac{2\phi'}{w_c}\left[1 - \left\{\frac{2\phi'}{w_c}\right\}^2\right]^{1/2} \qquad (17)$$

where ϕ' is defined in Fig. 3.

5.1.2 Broadening Due to Spectral Distribution of X-rays

At small angles, X-rays are scattered through an angle proportional to their wave length. Thus the intensity distribution of the specimen is broadened by the spectral distribution of the X-rays used. Normally, this is a doublet (the $K\alpha$ and $K\beta$ lines) superimposed on a weaker continuous spectrum. The characteristic lines are so narrow that the broadening each causes individually is negligible compared with other sources and can be neglected. On the other hand, using copper radiation as an example, the $K\beta$ line would produce an intensity distribution in which the peak is displaced from that of the $K\alpha$ line, by about 0·1 of its scattering angle. This would cause a broadening comparable with the desired resolution. The continuous spectrum is sufficiently weak compared with the characteristic that it is unlikely to produce significant distortion of the scattering pattern. However, since the total energy in the continuous spectrum cannot be ignored, its contribution is important if absolute intensities are being measured, and monochromatisation is then essential.

Thus, provided only relative intensities are being measured, broadening due to the spectral distribution of the X-ray source can be ignored if β-filtration is used. If the detector is a proportional counter, this can be adjusted to accept only a narrow wavelength range ($\pm 5\%$ of the mean value is easily achieved) which will cut out the $K\beta$ line and provide more than adequate monochromatisation.

5.1.3 Broadening Due to Specimen Size

To achieve maximum intensity, all of the X-ray beam must be intercepted by the specimen, and the line broadening due to this lateral extent is included in the collimation broadening already discussed. There is also an optimum thickness, at which scattered intensity is a maximum, and for a polymer this is approximately 1 mm. The broadening profile caused by such a specimen would typically have a maximum angular width of about 0·01 of the angle at which peak intensity occurs, which is small enough to be neglected, when compared with the other broadening functions.

5.1.4 Broadening Due to the Detector

Since the intensity of scattered radiation at the observation plane will fall as d increases, it would appear that this distance should be kept as small as possible. However, other factors modify the situation. If photographic detection is used, it is true that an adequate exposure will be obtained more quickly when d is small. But to obtain a quantitative intensity distribution, the film must be scanned by a microdensitometer and to avoid excessive noise in the output trace its entrance slit must be appreciably wider than the grain size of the film. A lower limit is thus placed on width of this slit. If this is t_m and if w_D is the maximum acceptable angular width of the broadening function due to the detector, then d is given by $d = t_m/w_D$. The function is, of course, a 'top-hat' of width w_D. Taking 0·1 mm as a reasonable estimate of t_m and using the earlier criterion that 2' represents the maximum acceptable total instrumental broadening, then detector slit broadening must be of the order of 1' giving $d = 333$ mm. The diameter r of the second aperture can be calculated from eqn. (15) as 0·03 mm for a microfocus tube.

When ratemeter measurements are used, the broadening function is again a 'top-hat' of the same angular width as the entrance slit. If its linear width is t, then $t = dw_D$. The number of photons entering the slit per unit time will be proportional to t but inversely proportional to its total distance from the source. Thus the count rate is proportional to

$dw_D/(d+v)^2$. This is a maximum when $d = v$. Using the same resolution criterion as above, then for a microfocus tube $d = 688$ mm, $r = 0·05$ mm and $t = 0·2$ mm. The total length of the camera, and the diameter of the aperture are such as to make this resolution barely practicable, and the best that can be achieved will be less than is experimentally desirable.

If the beam intensity profile is recorded experimentally, the true profile will be convoluted with the width of the detector entrance slit. Thus provided d and t are the same for this measurement as for the scattering curves, separate correction for the entrance slit width will not be necessary; the measured profile will be the convolution of both broadening functions.

So far, only the effects of entrance slit width have been considered; it also has a length. Thus, as with slit collimation of the incident beam, the detector output is the convolution of the scattering pattern with a rectangular aperture. With fibre oriented scatterers, the main interest is in the intensity distribution along a line parallel to the fibre axis, and provided care is taken that the slit length is accurately perpendicular to this direction no distortion should occur. However, with isotropic scatterers great care must be taken to ensure that the length of the slit is sufficiently short that the region of the distribution being investigated is not distorted.

The same considerations apply when a diffraction photograph is scanned with a microdensitometer. Crist and Morosoff[34] obtained intensity distributions from isotropic materials in this way, and were apparently aware of this difficulty for they state that the slit dimensions were chosen so as not to distort the pattern. However, since they do not give any dimensions, the small amount of distortion which must have occurred cannot be estimated.

Raster scanning microdensitometers using pin-hole apertures are now available, but we are not aware of these being applied to small-angle scattering.

5.1.5 Broadening Due to the Recording System

The signal from either the ratemeter or microdensitometer must be recorded, and this might introduce further loss of resolution. Toy[32] allowed the detector entrance slit to scan the scattered radiation at a fixed angular speed, and plotted count-rate against time (which can be converted to angle) on a chart recorder. The recording system has a response time, during which a finite angle will be scanned and the total instrumental response further broadened.

Since the arrival of a photon in the counter is a random event, there is a considerable statistical fluctuation in count-rate, which is reduced by lengthening the time over which it is averaged at a given angle. This is achieved by using a slow scanning speed, and causing the signal to decay exponentially with a controllable decay time. Thus if photons were diffracted at only one angle, and a narrow slit were used, the chart recorder would show a sharp leading edge when this angle is reached followed by an exponential decay. For a given decay time and scan-rate the tail could be given an angular scale, this profile would be the response–time broadening function.

Similar considerations apply to microdensitometers, except that there is no reason for a very slow response time, and this can be chosen to produce negligible broadening. Step-scanning (in which the photon detector is held at a fixed angle while counting proceeds) avoids this function, as does the use of a linear position sensitive detector. With this counting proceeds at all angles simultaneously, though the angular width of each channel imposes a limit on resolution.

Complications arise because the response–time broadening function is asymmetric. It is good experimental practice to record the intensity distribution by scanning continuously in one direction from high positive angles, through zero to negative ones, and to average the two peaks. However, if I'_L and I'_R are the two recorded distributions, B the broadening function and B_R its reflection in its leading edge, then $I'_L = I * B$ and $I'_R = I * B_R$ where I is the intensity distribution seen by the counter. Since convolution is a linear operation, $I'_L + I'_R = I * (B + B_R)$. Thus if this practice is followed, the correct broadening function is the average of B and B_R, and this will correct for errors introduced both by the response time, and by the averaging. Toy[32] followed this procedure.

5.2 Accuracy of Measurement of Intensity Distribution

Of the two methods available for the measurement of the intensity of the scattered radiation, only ratemeter measurements enable a quantitative estimate of the uncertainty of the measured value to be made. If N is the total number of counts, then the uncertainty δN in N has a 95% probability of being less than $0.51(N^{1/2})$. With continuous scanning, N would be the total number of counts at a given angle in the time it takes the width of the slit to traverse that angle. Thus having chosen the various instrumental parameters discussed above to maximise intensity at the required resolution, the scanning speed must be chosen to be the maximum at which the error will be acceptably small.

A disadvantage of continuous scanning is that the proportional error in the measurement of the weaker intensities will be greater than in the stronger ones. Though step-scanning is less convenient, longer counting times can be used for the weak intensities, and accuracy maintained.

5.3 Correction for Specimen Orientation

Theoretical calculations of the scattering curve assume that the lateral dimensions of the lamellae stacks are infinite. If this was so, the diffracted intensity from a specimen comprising perfectly oriented stacks would be confined to a line parallel to the stack axes. In practice, for oriented material it is either broadened to a band (Fig. 1a) or distributed in a four-point diagram (Fig. 1b). For isotropic material, it is spread into rings. In each of these cases, corrections must be applied to the measured intensity before comparing it with that calculated for a particular model.

Considering oriented material, there are three possible reasons for the lateral spread of intensity from the line which was predicted.

1. The stack axes will not all be exactly parallel. The intensity at a point on the line distant **s** from the origin of reciprocal space will be spread over a spherical cap of area proportional to $\mathbf{s}^2$. In this situation the best experimental procedure is to use a pinhole entrance to the detector, and multiply the measured intensity at **s** by $\mathbf{s}^2$ to correct it to the value for a perfectly oriented specimen. If a slit is used, with its length normal to the orientation direction and greater than the lateral spread of intensity at the detector plane, two difficulties will arise. Firstly, the disorientation must be small enough for the curvature of the spherical cap to be negligible compared with the thickness of the slit, and secondly the recorded intensity will be that contained in the surface in which the spherical cap intersects the Ewald sphere (which can be regarded as a plane in the region of small-angle scattering), from which the total intensity in the cap must be determined. This can only be done if its lateral distribution is known. However for any distribution met in practice, multiplication by s is likely to be a reasonable approximation.
2. The lateral dimensions of the stack are small, causing the line of intensity in reciprocal space to be spread into a cylinder. As with

the previous case, the total intensity in a disc normal to the stack axes is required, whilst (using a detector slit of sufficient length) that contained in the intersection of the disc with the Ewald sphere is measured. However, the radius of the disc is now independent of **s**, and so provided the relative lateral intensity distribution is also independent of it, the correction factor will be a constant and may be ignored.

3. The normals to the lamellae faces are inclined to the stack axes. In this case a four-point diagram will result (the 'points' are rings in reciprocal space—see Section 7 and Fig. 8). Variation in the inclination of the normals (either between or within stacks) without changing the periodicity along the stack axis and, whilst keeping the stacks perfectly oriented, will cause the ring to spread into a circular band normal to the meridian, and of area proportional to s^2. If a detector slit is used which is longer than the lateral spread of intensity at the observation plane, the situation is equivalent to case 1 and multiplication by **s** will provide a correction of reasonable accuracy.

It is not possible to determine the cause of the lateral spread if the only information available is a one-dimensional distribution of intensity, and in this case the proper choice of correction factor cannot be made. Uncertainty will therefore exist in any results obtained, even if these are only the one-dimensional lattice statistics. The correct procedure is to measure the intensity distribution in reciprocal space in the number of dimensions appropriate to the test-piece symmetry, and to consider the diffracting entity as a three-dimensional scatterer using the methods outlined in Section 7.

More often, isotropic material has been used, which is a special example of case 1 and the measured intensity multiplied by s^2. However, Crist and Morosoff[34] have pointed out that where other causes of lateral broadening are also present, then the disc of infinitesimal thickness produced by an oriented specimen is spread into a spherical shell of finite thickness for an isotropic one. Multiplication by s^2 will not then recover the intensity distribution. With oriented material, it is essential to have information on the nature of the morphology from other experiments. For example, electron microscopy of polyethylene[45] has shown that this comprises inclined lamellae stacks, and that the variation in inclination of the lamellae surfaces is the major cause of lateral broadening. In other experiments, comparison of small- and wide-angle scattering patterns has been used. In this way Point[56] arrives at a conclusion similar to that above for both polyethylene and polyamides. Toy[32] was concerned with the interpretation of the pattern shown in Fig. 1a and showed that the observed lateral spread was an order of magnitude greater than would be expected were it caused by the disorientation observed in the wide-angle pattern. He also estimated the crystallite width from both the wide-angle equatorial reflections, and the narrow-angle lateral broadening, obtaining a value of about 200 Å in each case. He thus concluded that broadening was caused by small crystallite width, and did not apply any correction factor. This conclusion can be criticised because paracrystalline effects would cause the crystallite width to be underestimated from the wide-angle measurements, and so variation in lamellae inclination cannot be ruled out. However, it should be noted that if this was the cause, then to give the pattern observed in Fig. 1a either lamellae must

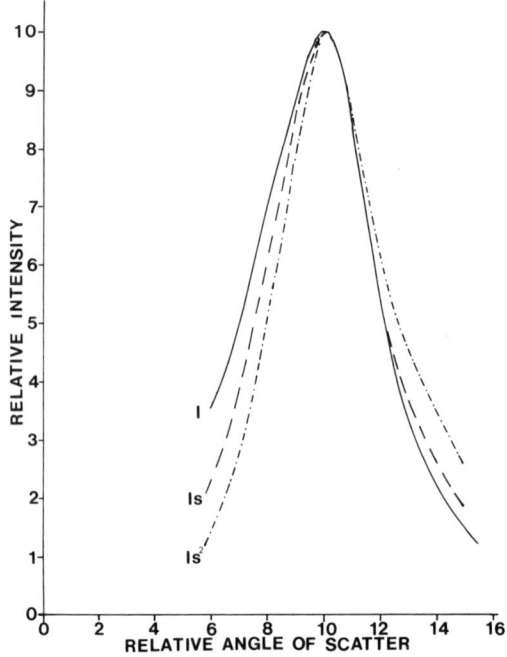

Fig. 4. Effect of applying orientation correction factors s and s^2 to Toy's experimental results.[32] Both axes are scaled so that, in each case, the peak occurs at (10, 10).

be inclined on both sides of the stack axis, or there is an appreciable contribution from the crystallite width. Otherwise intensity would decrease near the meridian as in Fig. 1b. In Fig. 4 both correction factors (s and s^2) have been applied to Toy's results and it is seen that with the experimental conditions he used (relative intensities, and scattering angles expressed relative to that of peak intensity) their effect is significant, though not large. Thus when oriented samples are used, uncertainty over this factor will cause an additional uncertainty in the line profile.

In situations where it is advantageous to use a long slit for the detector, then slit (instead of pin-hole) collimation may also be used thereby increasing the intensity. The slit length (or its effective length if this is controlled by the specimen width) must be short enough to cause negligible increase in the lateral spread of intensity, and all slits must be accurately perpendicular to the orientation direction.

5.4 Correction for Background Scatter

If ratemeter measurements are used, removal of background scatter originating outside the specimen is straightforward; it is only necessary to measure the intensity distribution without a specimen in the beam, reduce this to allow for specimen absorption, and subtract it from that measured with the specimen in place. In doing this, however, it must be remembered that in the tails of the distribution, the intensity is the difference of two nearly equal numbers. Hence unless they are determined with high precision, the result could be grossly inaccurate.

Correction for liquid-like scatter originating within the specimen is more difficult. From the discussion of Koberstein et al.[8] the methods that have been used where intensity profiles are compared are inadequate, and application of the methods discussed in Section 2 should be considered.

5.5 Assessment of Reported Experiments

Crist and Morostoff[34] used pin-hole collimation of nickel filtered radiation and photographic recording. They mark the width of the primary beam at half maximum intensity on their experimental scattering curves, but give no other details which enable the degree of distortion to be assessed. They use isotropic samples and correct for disorientation as described above.

The beam widths they show are much larger than 0·1 of the peak width at half maximum intensity, the value we have suggested to be the largest which will give adequate resolution. However, they are primarily concerned with the locations of the peak intensities. Only small errors, estimated to be about 2%, are likely to be introduced into these measurements. They also report values for the peak width, which are likely to be considerably broadened by the beam widths shown. No correction is given for this broadening nor are estimates of its magnitude made.

Blundell[19] uses slit collimation, and a linear position sensitive detector. Desmearing procedures are applied, but it is not stated whether these correct for an infinite slit of infinitesimal width, or for the beam profile actually used. No corrections appear to have been made for line broadening by the detector, nor is information given which enables its magnitude to be estimated. Thus the effect of instrumental broadening on his experimental intensity distributions is somewhat ambiguous. Isotropic material was used and correction made for disorientation as described above. He also subtracted a constant background intensity and corrected for the finite thickness of transition zones following Vonk.[7] In view of the discussion of Koberstein et al.[8] both of these steps are suspect.

Toy[32] used pin-hole collimation, with a fine focus X-ray source and a take-off angle such that the focal spot appeared to be 0·4 mm square. The detector was a proportional counter accepting a wavelength range of 5% of that of the Cu $K\alpha$ line and centred on that value. Continuous scanning was used. All broadening functions were determined but, as shown above with a source of the size used it was impracticable to achieve the suggested resolution of 0·1 of peak width at half maximum intensity (the first aperture could have been used to reduce the size of the source, but this would have been accompanied by an unacceptable loss of intensity); in fact, the width of the total instrumental broadening functions was about five times this value. It probably represents the best compromise between resolution and intensity with the apparatus available but about half the total broadening occurred because it was necessary to use a very slow response time and average the two slightly different peaks obtained in a scan (see Section 5.1.5).

Background scatter originating outside the specimen was removed by subtracting an intensity scan made without a specimen present, but no correction was made for that originating within. It was estimated that when errors from all sources were taken into account the uncertainty in the measured intensity varied from about ±20% at 50% of maximum intensity on the low angle side, through ±3% at peak intensity to ±20% at 10% of maximum intensity on the high angle side. These uncertainties were in the mean of 10 scans, and the statistical uncertainty in the count-rate made only a small contribution to their magnitude.

Oriented samples were used and it was assumed that lateral spread of the diffraction was caused by narrow stack width. As already discussed (see Section 5.3) the neglect of disorientation effects was justified, but not the variation of lamellae inclination. It is possible, therefore , that the correction for orientation effects is wrong, and that the models should have been fitted to the curve for I_S in Fig. 4, not that for I.

The work of Harrison et al.,[55] though limited in scope, is important in implication. Its scope is limited from the point of view of this review because it is concerned only with sedimented crystalline mats of polyethylene displaying several orders of diffraction and utilises only the peak width at half height, not the full profile. It is important because, using slit collimation, it specifically investigates the effect of collimation line broadening on the experimental results and conclusions drawn from them. They show that slit widths typically used broaden the width of the first order diffraction peak by about 50%. The implications of this will be discussed in Section 8.

6 COMPARISON OF THEORETICAL AND EXPERIMENTAL INTENSITY DISTRIBUTIONS

It will be clear from the foregoing discussion that none of the reported experimental investigations satisfies all the requirements of a critical study of the morphology of crystalline polymers. Nevertheless, those of Crist and Morosoff,[34] Blundell[19] and Toy[32] come close to meeting them, and their comparisons of calculated and measured intensity distributions will now be considered.

Crist and Morosoff used two materials for their comparisons, polyoxymethylene (POM) and linear polyethylene (PE), both in all three of the conditions quenched, annealed and slow cooled. They also included a solution crystallised polyethylene. All of these samples, except the quenched PE showed two peaks in the intensity distribution curve and it was assumed that these were first and second orders of diffraction from the same population of lamellae. In partially disordered systems such as these, the ratio of the angles at which these peaks occur will differ from 2·0, the differences depending upon the model chosen, and this was used to discriminate between possible models. However there is a controversy concerning the origin of the second peak where two are observed; some ascribe it to a separate population by lamellae. Peak width and the general shape of the curve were also used, but critical comparisons involving these parameters were not made.

Blundell used two grades of low-intensity polyethylene (LDPE) and included quenched specimens as well as those slow cooled at different rates of cooling. Initial sorting of models was done using parameters from the correlation function as well as the width of the intensity distribution at half intensity and then detailed shape comparisons were made. For reasons given in Section 3, only the comparisons using the intensity distribution will be discussed here; those involving the correlation function will be omitted. The experimental curves he shows have only a first order peak.

Toy used only one sample, a highly oriented fibre of poly(tetramethylene terephthalate) (4GT) annealed to a high degree of crystallinity. This again displayed only a first order diffraction peak. Models were sorted initially using the peak width at half intensity and two parameters which are functions of the peak asymmetry, then detailed shape comparisons were made for promising cases.

Both Toy and Blundell calculated the intensity distribution using non-dimensional coordinates, then chose a value of mean long spacing to bring the peaks into co-incidence along the abscissae, and finally scaled the intensities to make the peak heights agree.

In assessing Toy's results, the uncertainties concerning his correction for specimen orientation must be remembered. However, even if it is assumed that lateral spread of the reflection was entirely due to variation in lamellae tilt, because of the use of non-dimensional coordinates, significant effects only occur in the tails of the intensity distribution. Thus whilst the model statistics which give the best fit to the observed curve would be changed a little, conclusions concerning the type of model and insensitivity of results to choice of model parameters are unlikely to be seriously affected.

6.1 Symmetric Distribution Functions on Both Phases

The investigations of this class of model use Gaussian distributions (eqn. 9). Crist[33] showed that the ratio θ_2/θ_1 depends only on $\sigma_z/\bar{z}$ (i.e. is independent of ϕ) for ϕ in the range 0·15 to 0·85 and is less than 2, but only the case in which $\sigma_y/\bar{y} = \sigma_z/\bar{z}$, was considered. Of the materials he studied only for the PEs was $\theta_2/\theta_1 < 2$. With an apparently arbitrary choice of $\phi = 0\cdot 85$, the general shape of the experimental and calculated curves matched except that the experimentally measured peak width was broader than that calculated. Various explanations were suggested, but since instrumental broadening corrections were not made, this discussion is suspect.

For an infinite lamellae stack, and for a range of parameters in which $\sigma_y/\bar{y} \geqslant \sigma_z/\bar{z}$ (i.e. the variation of crystallite lengths is greater than that of amorphous lengths) Blundell was unable to satisfy the initial sorting criteria; the closest fit is shown in Fig. 5 (the values of σ_y and σ_z in the captions of these and other figures are for a lattice with a long period of unity). Considering only $\sigma_y/\bar{y} = \sigma_z/\bar{z}$, and $\phi = 0\cdot 5$, a good fit could be obtained if the stack was made finite, but only if its length was that of about two lamellae, which is physically unreasonable. Allowing crystallinity to vary between stacks also produced a good fit as is shown in Fig. 5, although if the accuracy of intensity measurement is comparable

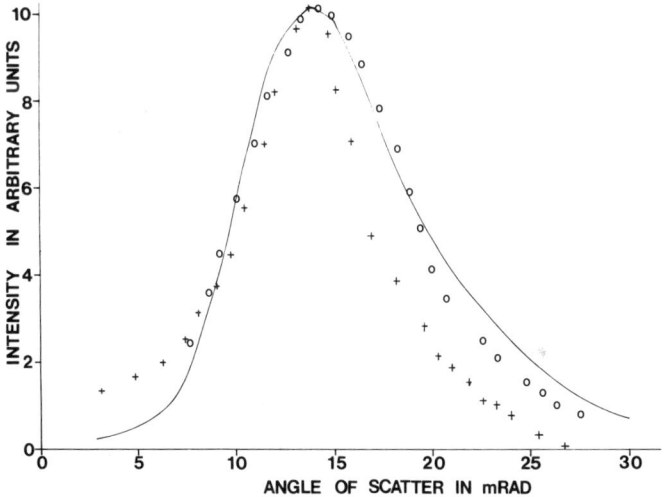

Fig. 5. Experimental intensity plot for quenched LDPE (MFI $0\cdot 3$) with superimposed, scaled, theoretical points for Gaussian distribution. ○ variable stack model, $\phi = 0\cdot 5$, $\sigma_x = 0\cdot 175$, $\sigma_c = 0\cdot 15$; + infinite stack model, $\phi = 0\cdot 5$, $\sigma_y = 0\cdot 299$, $\sigma_z = 0\cdot 025$. ($\sigma_x = (\sigma_y^2 + \sigma_z^2)^{1/2}$, σ_c = standard deviation of inter-stack variability in crystallinity. (Reproduced from Blundell, D. J., (1978). *Polymer*, **19**, 1258, by permission of the publishers, IPC Business Press Ltd.)

with that of Toy, the discrepancies would be significant. This could be because of the restrictions on $\sigma_y/\bar{y}$, and ϕ; greater freedom in the adjustment of these parameters might have improved the fit.

The ambiguity concerning the effect of instrumental broadening remains; if its effect is as great as Harrison et al. indicate,[55] this would account for the discrepancies between the infinite stack model and experimental curve in Fig. 5.

Toy investigated a wide range of parameters with $\sigma_y/\bar{y} \leq \sigma_z/\bar{z}$ (different from Blundell) and $0.48 \leq \phi \leq 0.85$. (The lower limit was the bulk crystallinity by density of his test piece). The best fit is shown in Fig. 6, and it is seen that the model curve lies within the error bars over the entire experimental range. By varying one parameter whilst keeping the other two constant a curve within error bars could be obtained with

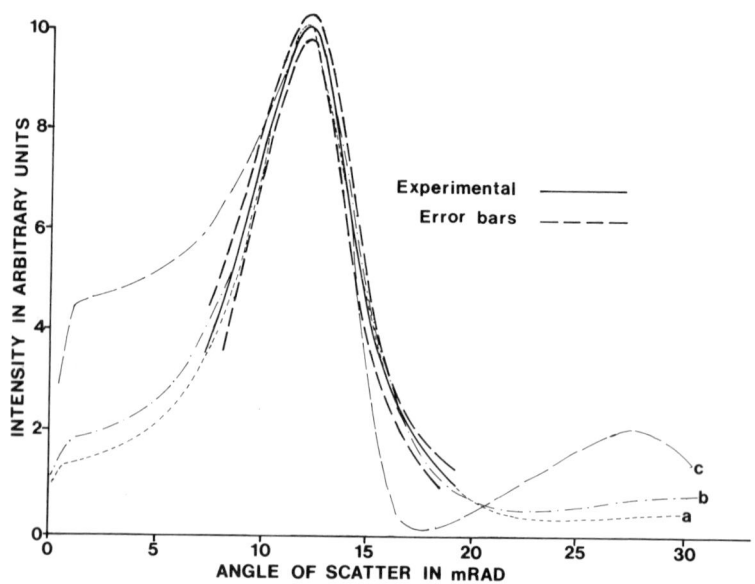

Fig. 6. Experimental intensity distribution for 4GT with superimposed calculated curves. (Dashed lines indicate uncertainty in experimental measurements). (a) Gaussian distribution with $\phi = 0.6$, $\sigma_y = 0.117$, $\sigma_z = 0.26$. (b) Reinhold distribution, with $\phi = 0.6$, $\gamma_y = 0.35$, $\gamma_z = 0.45$. (c) 'Top-hat' distribution of crystallite lengths in range $\mathbf{y} = 0.7 \pm 0.015$, and exponential distribution of amorphous lengths with $\sigma_z = 0.3$. $\phi = 0.7$. A transition zone of thickness 0.007 is assumed.

$\phi = 0.6 \pm 0.02$, $\sigma_y = 0.117 \pm 0.016$, $\sigma_z = 0.262 \pm 0.006$. By simultaneously varying two or more parameters the range of fit could be widened. The tails were the most sensitive parts of the curve to variations in these parameters.

The case with $\sigma_y/\bar{y} \geq \sigma_z/\bar{z}$ was not studied in detail, but it was found that if two scattering curves were calculated, one for $(\sigma_y/\bar{y})/(\sigma_z/\bar{z}) = A$, the other for the same values of $\bar{y}$ and $\bar{z}$ but with $(\sigma_y/\bar{y})/(\sigma_z/\bar{z}) = 1/A$ (i.e. the crystallinity and mean long period were the same in each but the relative standard deviations were interchanged), they differed by less than the experimental uncertainty in the measured curve. Thus with this model it would not be possible from the scattering curve alone to determine whether it is the crystalline lengths or the amorphous lengths which are most variable.

Although good fits with experimental data were obtained, Toy rejected these models because there was a substantial proportion of negative lengths, and this was considered physically unreasonable.

6.2 Asymmetric Distribution Functions on Both Phases

Crist[33] studied the application of the Reinhold distribution to this class of model using the same value of γ for both phases, and showed that when disorder was large enough for there to be only two orders of diffraction visible θ_2/θ_1 was less than 2 for negative γ and greater for positive. Thus, in this criterion alone, the PE samples could equally well be described by this model as the one with symmetric distributions. However, when the general shape of the curves were evaluated (again, only those with $\phi = 0.85$ are reported) they differed from the experimental. Hence it was concluded that this distribution was unsatisfactory. The POMs (except the quenched sample) had $\theta_2/\theta_1 = 2$, and neither Reinhold nor Gaussian distributions would lead to this value with only two orders of diffraction visible. With quenched POM the ratio was 2·5 which was, again, outside the range of any model studied.

Toy found that to match his experimental curve with this distribution it was necessary to allow γ to assume different values on each phase. A fit could be obtained with negative γ, but this led to negative lengths and was rejected as being physically unreasonable. Fits within error bars were achieved over a range of parameters with positive γ and Fig. 6 shows the experimental and calculated intensity distributions at one set of values within this range. The range over which a fit within error bars could be obtained was found by allowing one parameter to vary,

Table 1. Parameters of Reinhold distribution giving adequate fit with experimental intensity distribution of 4GT

ϕ	γ_y	γ_z
0·515±0·045	0·13±0·03	0·41±0·03
0·61±0·03	0·13±0·03	0·45±0·01
0·655±0·015	0·17±0·02	0·56±0·02
0·69±0·03	0·13±0·03	0·49±0·03

whilst keeping the other two fixed. This was done at four sets of values; the ranges are given in Table 1, and Fig. 7 illustrates the variations in the calculated scattering curves at some of the extremes of the range. The crystallinity in particular cannot be closely defined, but greater discrimination could be achieved if absolute intensities were measured, since calculation shows the absolute intensity of the peak to drop sharply with crystallinity at values greater than about 0·6.

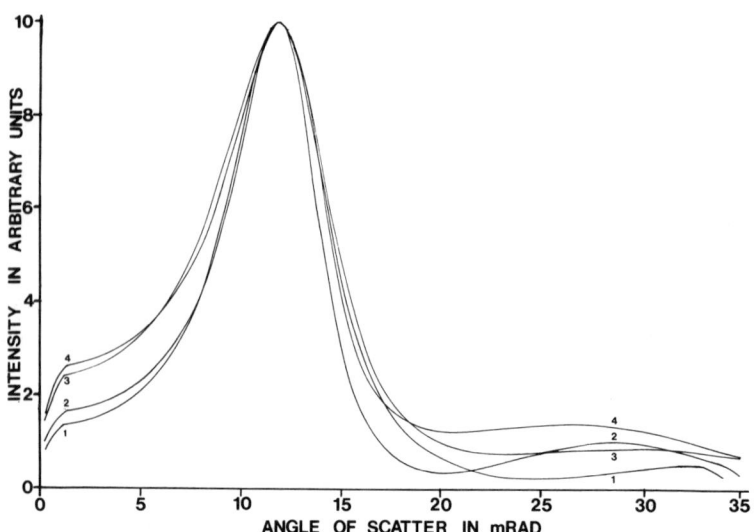

Fig. 7. Calculated intensity distributions for Reinhold distribution with; (1) $\phi = 0·47$, $\gamma_y = 0·10$, $\gamma_z = 0·38$. (2) $\phi = 0·60$, $\gamma_y = 0·10$, $\gamma_z = 0·44$. (3) $\phi = 0·60$, $\gamma_y = 0·16$, $\gamma_z = 0·46$. (4) $\phi = 0·72$, $\gamma_y = 0·16$, $\gamma_z = 0·52$.

It is, again, the low and high angle tails of the scattering curve which are most sensitive to parameter fluctuations. Small γ_y coupled with high crystallinity tends to produce an increase in intensity at high angles with a second order peak beginning to appear. At angles between this and the first order peak, intensity is increased by increasing either γ_y or γ_z or both. The same variation will also increase the intensity in the low angle tail.

Toy also investigated the application of the Asherov distribution (eqn. (13)) to both phases, but encountered difficulties because mean length, asymmetry and dispersion are all interrelated. Thus as m increases, the distribution approaches the Gaussian, but with a narrow dispersion that cannot be independently varied. With these restrictions it was not possible to find a combination of parameters which matched the experimental curve.

With parameters giving a good fit, interchanging the ratio of γ to the mean length between amorphous and crystalline regions produced changes in the calculated scattering curve which would be experimentally undetectable. Thus, it would again be impossible to determine whether maximum variability occurred in the lengths of the crystalline or amorphous regions.

Blundell[19] considers cases in which both distributions are exponential (eqn. 11) and in which they are exponential convoluted with Gaussian. Calculations with only a limited range of parameters are reported, and in all cases the standard deviation of the lengths of both phases are the same. None of the models satisfied the initial sorting criteria.

6.3 Symmetric Distribution Functions on One Phase, Asymmetric on the Other

Both Crist and Toy investigate a model in which the crystallite lengths have a 'top-hat' distribution and the amorphous lengths an exponential one (eqn. 10). A characteristic of this model is a much higher intensity in the low angle tail (see Fig. 6) than was found in any of the experimental curves. This model was therefore rejected. (Toy included a narrow transition zone in his model; the effects of such zones will be discussed in the next section).

Blundell considered models in which the crystallite lengths have a Gaussian distribution, and the amorphous lengths a modified exponential (eqn. 11). With the very restricted range of parameters he reports, the initial sorting criteria are not satisfied.

6.4 The Effects of Transition Zones

None of the investigations in which experimental and calculated intensity curves are directly compared provide much indication of whether transition zones, if they exist, could be detected. Toy, using the model of Section 6.3, recalculates some curves increasing the transition width tenfold, but his only comment is that they do not remove the objectionable features of this particular model. Crist and Morosoff do not consider the possibility of transition zones in the interpretation of their experimental results. Presumably they considered this pointless since Crist[33] had shown that (with Gaussian lattice statistics) such zones would affect neither the angle at which the peak intensity of first order scattering was detected, nor the width of this peak at half maximum intensity and that θ_2/θ_1 would only be changed by about 3%. These were the main quantities they were measuring. The only quantity to be changed appreciably would be the absolute intensity. These conclusions agree with those of earlier calculations by Blundell.[31]

Thus if transition zones are to be detected by comparing experimental intensity distributions with those calculated from model structures, the measurement of absolute intensity is essential. The only evidence for their existence from small-angle scattering experiments is that discussed in Section 2.

7 STUDIES OF THREE-DIMENSIONAL MORPHOLOGY

Small-angle diffraction patterns of the type shown in Fig. 1b are commonly observed, and these would be produced by structures of the type illustrated in Fig. 8a. The lattice, which is a series of points parallel to the axis,[44,45] has a Fourier transform comprising planes normal to the axis with spacing a'' corresponding to the lattice vector. The transform of the shape of the crystallites is cigar shaped with its axis inclined at an angle ψ (equal to the inclination of the lamellae normals) to the meridian. Diffraction spots are found at the intersections of these two transforms as illustrated in Fig. 8b. Lateral broadening of the spots can be caused by either the finite diameter of the lamellae stacks, or fluctuations in the inclination of the lamellae surfaces throughout the stack. Gerasimov and Tsvankin[46] have calculated the two-dimensional intensity distribution to be expected from such a model, assuming identical, parallel fibres, randomly oriented about their axes, and with neighbours incoherent. All lamellae surfaces

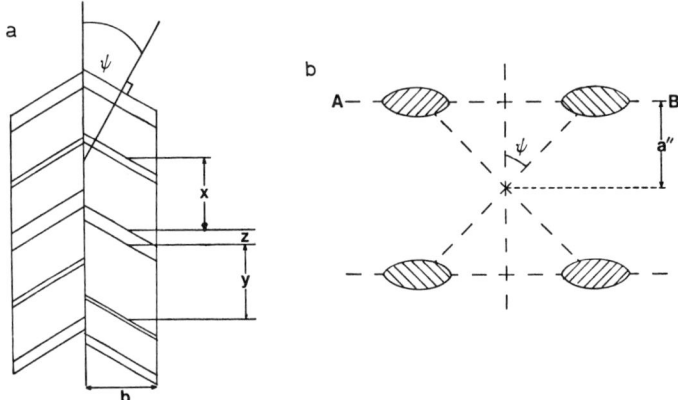

Fig. 8. (a) An inclined lamellae stack. (b) Its four-point diffraction pattern.

were assumed parallel. They showed that the important parameters were $(b \tan \psi)/y$ and b/y. If $(b \tan \psi)/y$ is $\sim 0.3-0.7$ a single meridional reflection was observed, becoming four-point for values $\sim 0.7-1.0$. For still larger values the lobes of the pattern are no longer parallel to the equator, and at about 1·7 they become radial. Bar-type reflections correspond to $b/y \sim 1.0$; as this value increases to about 4 they become globular.

Gezalov, Kuksenko and Slutsker[47] have developed two expressions for the fibril width for a model in which the lamellae surfaces are normal to the fibril axis but is otherwise the same as that described above. The first is the Scherrer equation $b = \lambda \Delta(2\theta)_{0.5}$ where $\Delta(2\theta)_{0.5}$ is the width at half intensity of the reflection parallel to the equator. The second involves an experiment in which the fibre is tilted through an angle α towards the X-ray beam, the peak intensity measured, and $[-\log I(\alpha)/I(0)]^{1/2}$ plotted against $\tan \alpha$. The slope of this line is $1·21b/x$ if the lamellae stack is assumed to have a square cross section of $1·03b/x$ if it is assumed to be circular. They obtained reasonable agreement between the two methods for a poly(caproamide) sample.

Crist[48] has extended those ideas by considering a model in which fibril diameters are distributed. In this case the Scherrer equation becomes:

$$\frac{\overline{b^4}}{\overline{b^3}} = \frac{1·04\lambda}{\Delta(2\theta)_I}$$

$\Delta(2\theta)_I$ is the integral breadth of the reflection, $\overline{b^3}$ is the average value of b^3 and similarly for $\overline{b^4}$. Thus a quantity $k\bar{b}$ is determined where k is of the order unity, its exact value depending on the nature of the distribution. If point collimation is used it is not necessary to tilt the fibre to use the second method, it is sufficient to measure the intensity profile $I(\mathbf{s}_2)$ along a line parallel to the equator, where $\mathbf{s}_2$ is zero at the meridian. Then for small $\mathbf{s}_2$:

$$I(\mathbf{s}_2) = K \exp\left[\frac{-\pi s_2^2}{4}\left(\frac{\overline{b^6}}{\overline{b^4}}s_2^2\right)\right]$$

where K is a constant depending upon the experimental conditions. Thus $\overline{b^6}/\overline{b^4}$ may be determined, and a quantity $k_1\overline{b^2}$ will be obtained where k_1 is defined similarly to k above. Clearly, the two methods will only give identical results if all lamellae stacks have the same width. More generally $(k_1\overline{b^2})^{1/2} > k\bar{b}$.

These methods were applied to four different fibres of nylon 6, and in all cases the above inequality was observed. Other line broadening mechanisms have been neglected in this treatment, they would reduce $k\bar{b}$ below the true value, but their effect upon $k_1\overline{b^2}$ is less obvious.

Kaji et al.[49] and Hosemann et al.[50] object to the explanation of the four-point diagram given above, on the grounds that it presupposes all lamellae have identical shapes, which is known to be untrue from electron microscopy. They propose instead that a lamellar stack consists of a number of fibrils with lateral coherence. The inclination of the phase boundary to the chain direction in an individual fibril is not then important. What is important is the inclination of the mean lamellar surface which is controlled by the width of the fibril and relative longitudinal displacement of neighbours. They call the crystallites microparacrystallites and the observed four-point diagram comprises the 001 reflections of this paracrystalline lattice.

Gerasimov et al.[51] and Vonk[52] have both calculated intensity distributions for this type of model. Gerasimov et al. perform two sets of calculations. In the first, various values of b, y and $\cos\psi$ were considered both with and without lateral coherence. For a given geometry, coherence causes a change in the intensity distribution, particularly if $(b\cos\psi)/y \sim 1$, but from the range of conditions covered in the diagrams presented it is not clear whether it is possible to produce a given distribution by choice of an appropriate fibril geometry whatever assumption is made about coherence. In the second set they consider the effect of dividing lamellae into different numbers of microparacrystallites, and do this for different values of their width, thickness and tilt. In all cases considered the intensity distribution is little affected by the division.

Vonk calculated two-dimensional correlation functions for inclined lamellae and microparacrystalline models, comparing these with the experimentally determined functions from oriented fibres of low-density polyethylene, after various annealing treatments (these gave four-point patterns with diffraction lobes parallel to the equator). Different ranges of geometrical parameters were selected for the calculated functions from the two models. For the microparacrystalline model the range was that for which the calculations of Gerasimov and Tsvankin predict a four-point pattern with lobes parallel to the equator (the observed type), whereas for the inclined lamellar model it was the one for which a radial four-point pattern would be expected. It is not clear why these different ranges were selected; had the same been used for each model the conclusion in favour of the microparacrystalline model would have been more convincing. Experimental determination of the two dimensional correlation function will be prone to the difficulties regarding accuracy that have been mentioned earlier for the one-dimensional function.

Thus the key question which must be answered before further progress can be made in interpreting the three-dimensional structures of polymers from small-angle scattering is whether or not lateral coherence exists between fibrils. The limited work so far done favours coherence, but the evidence is by no means clear, and critical comparisons of measured intensity distribution from different materials with those calculated from various models, taking account of all the experimental inaccuracies discussed earlier, will be needed to clarify this issue.

8 CONCLUSIONS

Much of the experimental work which has been discussed can be criticised in that inadequate attention has been paid to the errors present in the raw data and the way these propagate through the frequently complicated mathematical operations performed on these data. This is especially true in those investigations where structural parameters have been deduced directly from experimental data (direct methods), where deconvolution procedures have been used to recover an intensity distribution from a signal smeared by a rectangular beam profile, and where the correlation function has been computed.

Where the directly measured scattering curve has been compared with that calculated from a proposed model, it is unclear whether proper allowance has been made for all instrumental broadening effects, and in the one study where it is clear that this has been done, the resolution is much lower than is desirable for the discrimination required.

Three studies directly compare experimental and calculated intensity distributions and are in agreement that models in which the distribution functions of crystallite and amorphous lengths are highly asymmetric lead to intensity distributions differing significantly from those observed. Thus even with the limitations of resolution and ambiguities concerning proper correction for instrumental broadening which have been discussed, this degree of discrimination can be achieved. They are in disagreement over whether a model in which all lamellae stacks are of infinite length and of the same crystallinity will give a scattering distribution which matches that observed experimentally. Toy found moderately skewed distribution of amorphous and crystallite lengths which gave an intensity distribution differing insignificantly from that he observed. Those obtained by the other two investigators were narrower than the ones they observed. This disagreement might arise because there are genuine differences between the materials studied. On the other hand, it could be because the resolution and correction for line broadening is inadequate, or because the range of parameter values they explored was too narrow.

The model proposed by Blundell having an intensity distribution close to that he observed, that is a model in which crystallinity varies between lamellae stacks, is in agreement with the results of Strobl and Müller[12] and Stribeck and Ruland[15] who used direct methods of analysis with similar materials. However in their studies of the effect of collimation line broadening Harrison et al.[55] show that this type of conclusion is sensitive both to the amount of broadening present and to the method used to correct for it. Using what seemed to be the most satisfactory method, they could explain their results equally well by a model in which lamellae thickness varied between stacks, and by one in which it varies within them.

Toy is unable to discriminate between moderately skewed (Reinhold) length distributions and symmetrical (Gaussian) ones. Some of the crystallite and amorphous length distributions which gave an

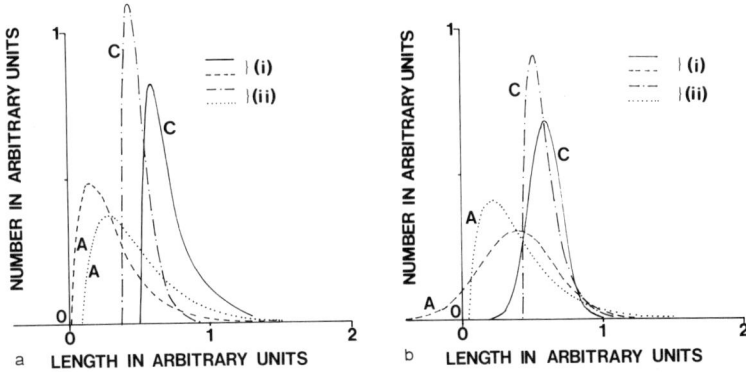

Fig. 9. Distribution of amorphous (A) and crystallite (C) lengths. (a)(i) Reinhold distribution, $\gamma_y = 0\cdot13$, $\gamma_z = 0\cdot49$, $\phi = 0\cdot69$; (ii) Reinhold distribution, $\gamma_y = 0\cdot13$, $\gamma_z = 0\cdot41$, $\phi = 0\cdot515$. (b)(i) Gaussian distribution; $\sigma_y = 0\cdot117$, $\sigma_z = 0\cdot26$, $\phi = 0\cdot60$; (ii) Reinhold distribution; $\gamma_y = 0\cdot135$, $\gamma_z = 0\cdot45$, $\phi = 0\cdot60$.

adequate match with his experimental curves are illustrated in Fig. 9. In Fig. 9a two distributions are shown from opposite extremes of the range of Reinhold distributions which matched the experimental scattering curve, and in Fig. 9b the Gaussian is compared with the Reinhold from the centre of the range. All the distributions are of similar width indicating that the scattering curve is more sensitive to this than to the exact shape. The differences of shape are quite severe, showing that the experiments with the resolution achieved by Toy are unable to distinguish between quite dissimilar lengths distributions. It is possible to argue on physical grounds that negative length segments or ones of only one or two ångströms are unreasonable, and the Gaussian and Reinhold distribution with $\phi = 0\cdot69$ can then be rejected, thus narrowing the range of possibilities. This is, however, bringing in other criteria, and not discrimination purely from the scattering distribution. Added uncertainty is introduced into the description of the structure because, provided the widths of the thickness distribution functions are expressed as a fraction of their mean value, they can be assigned to either phase and give experimentally indistinguishable intensity distributions.

The magnitude of the experimental task is indicated in Fig. 10, where the calculated scattering curves (prior to convolution with instrumental broadening functions) are compared. These are shown

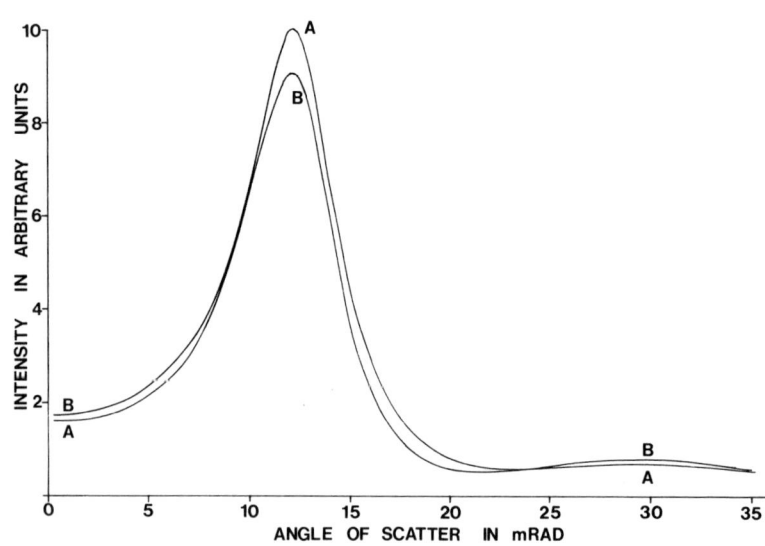

Fig. 10. Theoretical scattering curves (before instrumental broadening) of length distributions giving the best match to experimental data. (A) Gaussian distribution; $\sigma_y = 0.117$, $\sigma_z = 0.26$, $\phi = 0.60$. (B) Reinhold distribution; $\gamma_y = 0.135$, $\gamma_z = 0.45$, $\phi = 0.60$.

with the correct relative intensities. Whilst measurement of absolute intensities would help to differentiate between models, an accuracy better than ±10% would be necessary. A comparative study conducted between various laboratories suggests this would be difficult to achieve,[16] particularly in the tails of the curve. It is also clear from Fig. 10, that where relative intensity measurement is used, discrimination between models will be obtained in the tail region of the curve rather than near the peak, and that particular attention must be paid to collecting accurate data in this experimentally difficult region.

It is unlikely that conclusive evidence for the existence of transition zones will ever be obtained from small-angle scattering and methods for obtaining quantitative information about morphology transverse to the axes of the lamellae stacks are only beginning to be exploited. Until these have been subjected to further critical investigation, their precision is uncertain. If the stacks are treated as three-dimensional scatterers, many parameters are required to characterise the model. If they are treated as one-dimensional lattices uncertainties arise concerning the proper way to correct the experimental data for specimen orientation. Hence small-angle scattering is likely to be of most use when used in conjunction with other techniques of morphological examination. If, for example, it could be established by electron microscopy that the correct model was one of inclined lamellae stacks with lateral broadening caused by variations in the inclination of normals to the lamellae surfaces, then it might be possible to establish the statistics of stacking from the small-angle scattering pattern. Even to do this would require considerable improvement in the resolution of experimental data over what has been achieved to date.

REFERENCES

1. Alexander, L. E. (1969). *X-ray Diffraction Methods in Polymer Science*, Wiley, New York.
2. Fischer, E. W., Goddar, H. and Schmidt, G. F. (1967). *Polymer Letters*, **5**, 619.
3. Hermans, P. H., Heikens, D. and Weidinger, A. (1959). *J. Pol. Sci.*, **35**, 145.
4. Heikens, D. (1959). *J. Pol. Sci.*, **35**, 139.
5. Fischer, E. W. and Fakirov, S. (1976). *J. Mat. Sci.*, **11**, 1041.
6. Ruland, W. (1971). *J. Appl. Cryst.*, **4**, 70.
7. Vonk, C. G. (1973). *J. Appl. Cryst.*, **6**, 81.
8. Koberstein, J. T., Morra, B. and Stein, R. S. (1980). *J. Appl. Cryst.*, **13**, 34.
9. Wiegand, W. and Ruland, W. (1979). *Prog. Colloid Pol. Sci.*, **6**, 355.
10. Ruland, W. (1977). *Colloid and Pol. Sci.*, **255**, 417.
11. Strobl, G. R. (1973). *J. Appl. Cryst.*, **6**, 365.
12. Strobl, G. R. and Müller, N. (1973). *J. Pol. Sci. (Phys. Edn.)*, **11**, 1219.
13. Ruland, W. (1977). *Kolloid Z.u.Z. Polymere.*, **255**, 417.
14. Ruland, W. (1978). *Kolloid Z.u.Z. Polymere.*, **256**, 932.
15. Stribeck, N. and Ruland, W. (1978). *J. Appl. Cryst.*, **11**, 535.
16. Hendricks, R. W. and Shafer, L. B. (1978). *J. Appl. Cryst.*, **11**, 196.
17. Vonk, C. G. and Kortleve, G. (1967). *Kolloid Z.u.Z. Polymere.*, **220**, 19.
18. Kortleve, G. and Vonk, C. G. (1968). *Kolloid Z.u.Z. Polymere.*, **225**, 124.
19. Blundell, D. J. (1978). *Polymer*, **19**, 1258.
20. Brown, D. S., Fulcher, K. U. and Wetton, R. E. (1973). *Polymer*, **14**, 379.
21. Zernicke, Z. and Prins, J. A. (1927). *Z. Phys.*, **41**, 184.
22. Hosemann, R. and Bagchi, S. N. (1962). *Direct Analysis of Diffraction by Matter*, North Holland, Amsterdam.
23. Hosemann, R. (1950). *Z. Phys.*, **128**, 464.
24. Guinier, A. (1963). *X-ray Diffraction in Crystals, Imperfect Crystals and Amorphous Bodies*, Freeman, San Francisco.
25. Tsvankin, D. Ya. (1964). *Pol. Sci. USSR*, **6**, 2304.
26. Tsvankin, D. Ya. (1964). *Pol. Sci. USSR*, **6**, 2309.

27. Reinhold, C., Fischer, E. W. and Peterlin, A. (1964). *J. Appl. Phys.*, **35**, 71.
28. Asherov, B. A. and Ginsburg, B. M. (1978). *Pol. Sci. USSR*, **20**, 1009.
29. Slutsker, L. I. (1975). *Pol. Sci. USSR*, **17**, 262.
30. Blundell, D. J. (1970). *Acta Cryst.*, **A26**, 472.
31. Blundell, D. J. (1970). *Acta Cryst.*, **A26**, 476.
32. Toy, M. (1980). Ph.D. Thesis, University of Manchester.
33. Crist, B. (1973). *J. Pol. Sci. (Phys. Edn.)*, **11**, 635.
34. Crist, B. and Morosoff, N. (1973). *J. Pol. Sci. (Phys. Edn.)*, **11**, 1023.
35. Kilian, H. G. and Wenig, W. (1974). *J. Macromol. Sci. (Phys.)*, **B9(3)**, 463.
36. Vonk, C. G. (1978). *J. Appl. Cryst.*, **11**, 541.
37. Guinier, A. and Fournet, G. (1955). *Small-Angle Scattering of X-rays*, Wiley, New York.
38. Dijkstra, A., Kortleve, G. and Vonk, C. G. (1966). *Kolloid Z.u.Z. Polymere.*, **210**, 121.
39. Hendricks, R. W. and Schmidt, P. W. (1967). *Acta Phys. Austriaca*, **26**, 97.
40. Vonk, C. G. (1971). *J. Appl. Cryst.*, **4**, 340.
41. Cooper, M. J. (1977). *Phys. Bull.*, **28**, 463.
42. Bolduan, O. E. A. and Bear, R. S. (1949). *J. Appl. Phys.*, **20**, 983.
43. Huxley, H. E. (1953). *Acta Cryst.*, **6**, 457.
44. Hay, I. L. and Keller, A. (1967). *J. Mat. Sci.*, **2**, 538.
45. Grubb, D. T., Dlugosz, J. and Keller, A. (1975). *J. Mat. Sci.*, **10**, 1826.
46. Gerasimov, V. I. and Tsvankin, D. Ya. (1969). *Pol. Sci. USSR*, **11**, 3013.
47. Gezalov, M. A., Kuksenko, V. S. and Slutsker, A. I. (1970). *Pol. Sci. USSR*, **12**, 2027.
48. Crist, B. (1979). *J. Appl. Cryst.*, **12**, 27.
49. Kaji, K., Mochizuki, T., Akiyama, A. and Hosemann, R. (1978). *J. Mat. Sci.*, **13**, 972.
50. Hosemann, R., Loboda-Čačkovič, J. and Kaji, K. (1978). *J. Appl. Cryst.*, **11**, 540.
51. Gerasimov, V. I., Zanegin, V. D., and Tsvankin, D. Ya. (1978). *Pol. Sci. USSR*, **20**, 954.
52. Vonk, C. G. (1979). *Colloid and Pol. Sci.*, **257**, 1021.
53. Glatter, O. (1977). *Acta Phys. Austriaca.* **47**, 83.
54. Glatter, O. (1980). *J. Appl. Cryst.*, **13**, 7.
55. Harrison, I. R., Kozmiski, S. J., Varnell, W. D. and Wang, J.-I. (1981). *J. Pol. Sci. (Phys. Edn.)*, **19**, 487.
56. Point, J. J. (1977). *J. Pol. Sci. (Pol. Symp.)*, **59**, 87.

Chapter 6

LONG-WAVELENGTH X-RAY SCATTERING TO STUDY CRYSTAL MORPHOLOGY

H. K. HERGLOTZ

Engineering Research and Development Division, E. I. du Pont de Nemours & Co., Wilmington, Delaware, USA

1 INTRODUCTION

The existence of large-scale organisation in a solid composed of long chain molecules has been described in other chapters of this book and requires no further discussion. The Engineering Physics Laboratory of E. I. du Pont de Nemours was interested in introducing studies of the phenomenon by techniques other than the generally practised methods, because of the observation that these classical tools failed in some cases to account for macroscopic properties. For example, in the development of reverse osmosis membranes for water desalination,[1] no correlation could be established between manufacturing variables, operating performance and structural morphological data (wide-angle/small-angle X-ray scattering of 0·154 nm radiation). More about this case can be found later in Section 6.

The limitations of classical X-ray wavelengths in unravelling features of the polymeric solid greater than the size scale of the crystallographic unit cell are made evident in Fig. 1. While 0·154 nm copper radiation is a good match for the interatomic distances of the polyethylene structure and produces interference maxima, the same cannot be said about repetitive morphological features. Figure 1b is an example of the lamellar arrangement of chain-folds frequently displayed in the literature (Reference 2, p. 446, also see Chapter 7 of this book). On the scale of Fig. 1b copper radiation cannot be included because of its mismatch in relation to the morphological features. Aluminium radia-

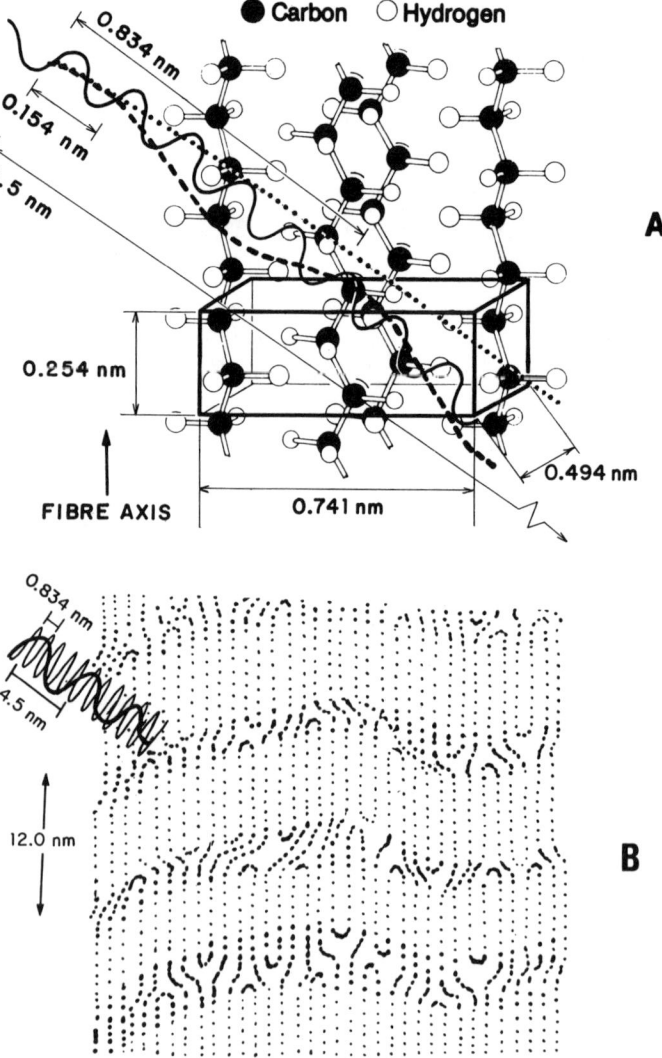

Fig. 1. (a) Crystallographic, orthorhombic unit cell of polyethylene and symbolic waves of 0·154 nm copper radiation, 0·834 nm aluminium radiation and 4·47 nm carbon radiation. The match between unit cell dimensions and copper radiation is obvious, while aluminium and carbon radiation produce no interference maxima. ($\sin\theta = n\lambda/2d > 1$) Cu ——; Al - - - -; C . . . . .
(b) Morphological scale chainfold stacks with same radiations ($\lambda = 0·154$ nm too small for this scale). The better match of long wavelengths is obvious.

tion of 0·834 nm and carbon radiation of 4·47 nm fare much better in this respect. Since the existence of morphological features with various size scales (physicists often prefer the expression 'spatial frequencies') and degrees of periodicity has been established beyond doubt in the polymeric solid, it was worthwhile to take a closer look at the reasons for and difficulties with making use of long-wavelength X-rays, because their lower than conventional frequencies definitely provide a better match for the spatial frequencies found in polymers.

2 REASONS FOR AND DIFFICULTIES WITH LONG-WAVELENGTH (LOW FREQUENCY, LOW ENERGY) X-RAYS FOR INVESTIGATING THE SOLID POLYMER

The drawings of Fig. 1 appear so convincing that one has to ask: 'Why didn't everybody interested in polymer morphology recognise and utilise them?' One reason is that structural X-ray analysis of polymers derived much of its procedure from the older metallurgy which, however, shares as many features with polymer physics as there are profound differences. In the latter category belong the existence of the strong polymeric chain organisation with its covalently bonded backbone versus the individualistic metal atom. Both metallic and polymeric solids are polycrystalline aggregates, but the metallurgical crystallite is no counterpart for the polymeric 'domain', 'lamella' and 'fibril', even if the word 'crystallite' is sometimes used informally for all these entities. Polymeric chains cross boundaries between morphological features, provide coherence during deformation and play all sorts of tricks the metallurgical polycrystal cannot achieve. In summary, 'morphology' in metals is primarily the arrangement of crystallites, observable by the light microscope, while in polymers the extended X-ray range is primarily competent for its investigation.

Furthermore, polymers are almost exclusively composed of elements in the second row of the periodic table, while all metals popular in technology belong to higher rows of the table. This composition has implications on scattering versus photoelectric absorption coefficients; essentials of X-ray structure analysis.

There have been statements about the usefulness of long-wavelength X-rays for small-angle scattering, most prominently suggested in Reference 3 (p. 654). There, the justification is based primarily on increased intensity due to larger slits at a permissible residual divergence of the beam. Our justification was based on three salient features which set long wavelengths apart from conventional X-rays.

1. Larger scattering and Bragg angles at the same 'spatial frequencies' in the sample (Figs 2 and 3).
2. Larger amplitude differences of waves scattered from domains of unequal electron density (Fig. 4).
3. Small penetration depth (large photoelectric absorption coefficient) making surface and gradient studies possible (Fig. 4, Table 1).

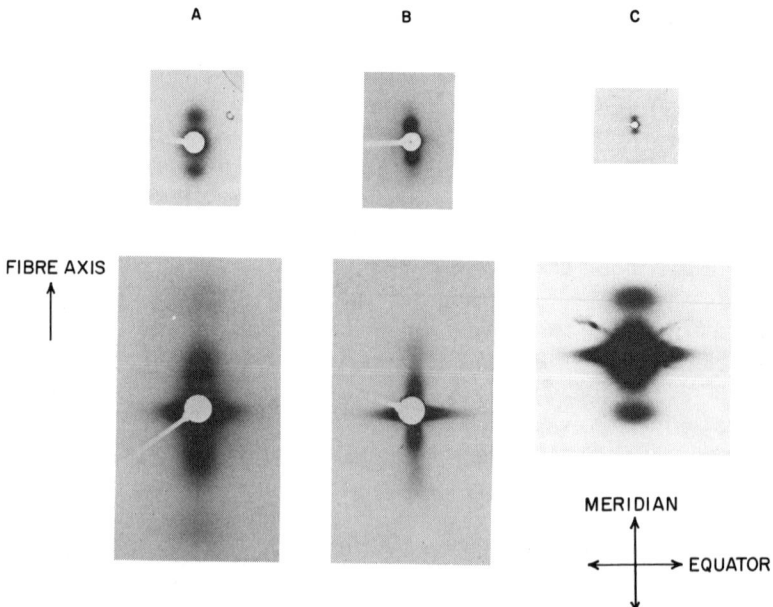

Fig. 2. Comparison of patterns from three samples, prepared with 0·154 nm and 0·843 nm radiations: (a) Hollow nylon fibre (described in more detail in Section 6). (b) and (c) Polypropylene fibres spun under different conditions. Sample to film distance $STFD = 175$ mm except in 0·834 nm-pattern of sample (c), where it was 320 mm. Cu-radiation: 50 kV, 7·5 mA, 2 h; Al-radiation: 14 kV, 12 mA, 15 h, except sample (c), which was prepared with a newer instrument at 20 kV, 4 mA, 17 h.

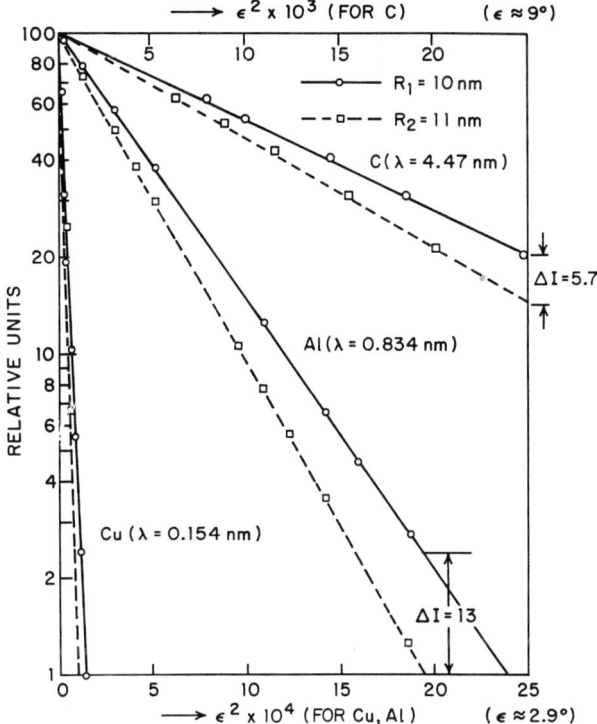

Fig. 3. Comparison of copper, aluminium and carbon radiations in their faculty to discern between 'scatterers' of radii of gyration R of 10 and 11 nm. See also Reference 7. Note the larger scattering angles, higher intensities and intensity differences ΔI obtained with long wavelengths.

Considering each of these benefits individually: first, it is easy to demonstrate the larger Bragg angle. Figure 2 displays the diffraction patterns of nylon and polypropylene samples with copper and aluminium radiation at the same or similar sample to film distances. No carbon radiation patterns are included, since at the large Bragg angles for carbon radiation ($\theta > 10°$), maxima from these samples are too broad to be discernible. Later examples will show C-radiation interference maxima. Scattering by aperiodically arranged scatterers, as characterised by their 'Guinier radius of gyration' R and ordinate

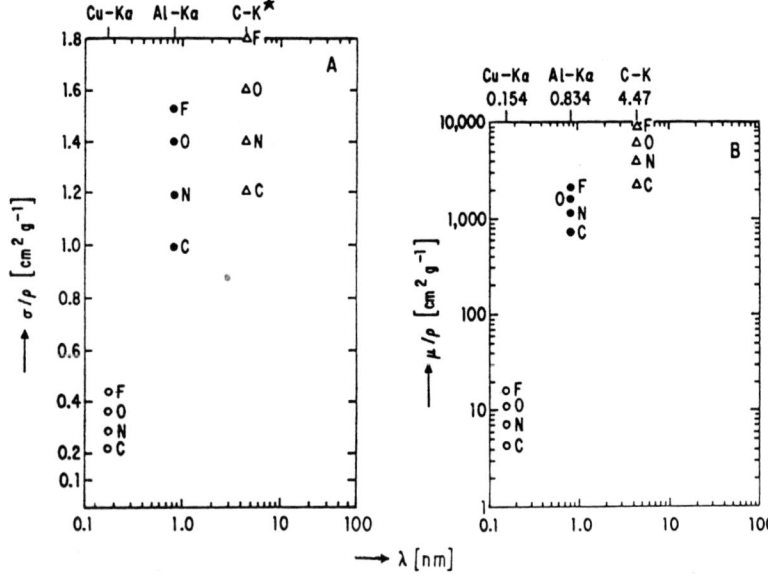

Fig. 4. (a) Classical (= Thomson) scattering cross-sections for light elements and characteristic radiations of Cu, Al, C; (b) total absorption cross sections for the same elements and radiations.[10]

Table 1. Mass absorption coefficients μ/ρ [cm^2 g^{-1}], pathlength $I_0/100$ [cm] and optimal thicknesses $1/\mu$ [cm] of poly(ethylene) (PE), poly(ethylene terephthalate) (PET), and poly(para-phenylene terephthalamid) (PPD-T), for Cu-, Al- and C-radiations

		PE	PET	PPD-T
	μ/ρ	3·72	6·43	5·38
Cu	$I_0/100$	1·24 = 12·4 mm	0·72 = 7·2 mm	0·86 = 8·6 mm
	$1/\mu$	0·27 = 2·7 mm	0·16 = 1·6 mm	0·19 = 1·9 mm
	μ/ρ	623	870	851
Al	$I_0/100$	7·4×10^{-3} = 74 μm	3·8×10^{-3} = 38 μm	3·7×10^{-3} = 37 μm
	$1/\mu$	1·6×10^{-3} = 16 μm	8×10^{-4} = 8 μm	8×10^{-4} = 8 μm
	μ/ρ	2034	3350	2476
C	$I_0/100$	2·3×10^{-3} = 23 μm	1·0×10^{-3} = 10 μm	13×10^{-4} = 13 μm
	$1/\mu$	5×10^{-4} = 5 μm	2·0×10^{-4} = 2 μm	4×10^{-4} = 4 μm

intercept n^2,[4] differs for the three wavelengths in the manner evident from Fig. 3 and visible on the equator and in the halo of the pattern of Fig. 2a. Increasing the sample to film distance (at severe restriction of beam divergence) has made copper radiation competitive in some, but not every respect, with aluminium radiation. There will be more about this later in Section 6.

The second of the above arguments, i.e. larger amplitude differences, is based on the classical or Thomson scattering coefficients of Fig. 4. A larger wavelength entails larger Thomson coefficients for the light elements from which polymers are made. Practical scattering coefficients grow even faster with increasing wavelength than the theoretical values displayed here (see Reference 5, Fig. III-3, p. 122). This is particularly true for heavy elements in the 1 Å ($= 0 \cdot 1$ nm) range of Fig. III-3, but also becomes promenent for light elements in the long-wavelength region. The scattering amplitudes A_1 and A_2 from two neighbouring 'domains' which differ in density, i.e. containing different numbers of atoms, whether in an aperiodic domain array or in the periodic arrangement of Fig. 1b, will be proportional to the Thomson coefficients (σ_T) of the sum of atoms in the domains. Superposition of the two or multiple waves at some distance produces an interference pattern with maxima proportional to the difference of σ_T. See Reference 6, p. 124, equation (14.6):

$$I = I_e |F|^2 |G|^2$$

where (in von Laue's nomenclature) $I =$ the intensity of the scattered wave, $I_e =$ Thomson's scattering factor (usually called σ_T), $F =$ the structure amplitude and $G =$ the lattice factor.† A demonstration of the effect will be presented later. More convincing yet is the effect of larger λ when one looks at the equatorial scattering mentioned previously, which is often attributed to 'voids', an expression that might not be too objectionable in view of the general naïvity of our models of polymeric solids. It is a fact that huge equatorial spikes appear often on small-angle patterns of highly crystalline polymers (Fig. 2), which can be considered as coming from aperiodic density differences in the radial direction (perpendicular to the fibre axis) and can be treated and evaluated by Guinier approximation.[4]

† Note that this nomenclature is different from that used in Chapter 5. This equation is analogous to eqn. (7b) applied to a perfect lattice. $Z(s)$ of that equation, which is sometimes called the lattice factor, is obviously defined differently from G here, though both factors play the same role [Ed.].

3 INSTRUMENTATION

The previously mentioned lack of observable X-ray differences in chemically identical polymers with different physical properties convinced us that some morphological features escape detection by classical X-ray methods. Encouraged even more by the considerations of Section 2, we decided to develop the instrumentation shown in Fig. 5 for the routine characterisation of polymers by the characteristic radiation of aluminium ($\lambda = 0.834$ nm)[7,8] and of carbon ($\lambda = 4.47$ nm).[10] The instrument shown in Fig. 5, which is newer than those described in References 7 and 8, is suitable for either aluminium and carbon radiation or any other radiation that is needed and which originates from a stable target material with sufficient heat dissipation

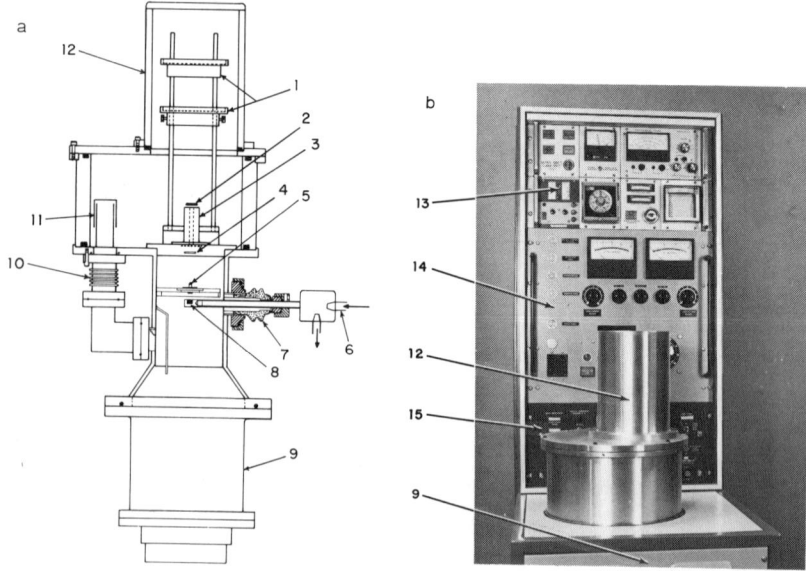

Fig. 5. (a) Design of an instrument for long-wavelength X-ray work and (b) its control panel. 1. Variable distance film holders, 2. Sample, 3. Collimator, 4. Optical filter (no vacuum seal), 5. Cathode assembly, 6. Target cooling, 7. High voltage insulator, 8. Target with exchangeable buttons, 9. Turbo-pump, 10. Vacuum communicator between X-ray source and camera compartment, 11. Visible light trap, 12. Vacuum housing, 13. Control panel for vacuum, cathode heating, target current, 14. High voltage power supply, 15. Turbo-pump controls.

Table 2. Transmittance I/I_0 of gases at various pressures for the characteristic radiations of aluminium (0·834 nm) and carbon (4·47 nm) for 10 cm pathlength

	$\lambda = 0·834$ nm			$\lambda = 4·47$ nm		
	760 Torr^a	1 Torr^b	10^{-3} Torr^c	760 Torr	1 Torr	10^{-3} Torr
Air	$4·88 \times 10^{-7}$	0·981	0·999	$3·4 \times 10^{-23}$	0·934	0·999
H$_2$O-Vapour	$1·94 \times 10^{-5}$	0·986	0·999	$1·4 \times 10^{-18}$	0·947	0·999
He	0·97	1·000	1·000	$9·0 \times 10^{-3}$	0·994	0·999

$^a = 1·01 \times 10^5$ Pa, $^b = 1·33 \times 10^2$ Pa, $^c = 1 \, \mu$m Hg.

capability. D. R. Lynch of this laboratory had prime responsibility for design and construction of this instrument. Because of the very high absorption by any window material, particularly with carbon radiation, no permanently sealed X-ray tubes can be visualised for these radiations. The windows through which the radiation must exit would be much too fragile for this type of operation. These long-wavelength X-rays do not propagate even in air at atmospheric pressure, as Table 2 shows, so that at least a forevacuum is needed for their propagation in the camera. The choice of a 10 cm path length for the values recorded in Table 2 is realistic, since most information retrieved in our operation was obtained with similar sample to film distances. For aluminium radiation, a 3 μm window seals the high vacuum of the X-ray source from the forevacuum of the camera. The window also acts as a filter (Fig. 6) that reduces bremsspectrum thus assuring a nearly monochromatic beam. For carbon radiation we used a 'window' of $\sim 1 \, \mu$m thick collodion film (prepared by casting on a glass plate) which contained dispersed soot for optical opacity. It absorbed all visible light from the hot cathode while passing 47% of the carbon radiation. The bremsspectrum from a carbon target interferes even less with the results. The integral intensity of the bremsspectrum is proportional to the atomic number Z (Reference 5, p. 89, equation (2.47)). Therefore for a carbon target, it is less than half of that from aluminium.

The wavelength of the bremsstrahlung-maximum can be approximated by:

$$\lambda_{max} = 1·5 \times \frac{hc}{eV}$$

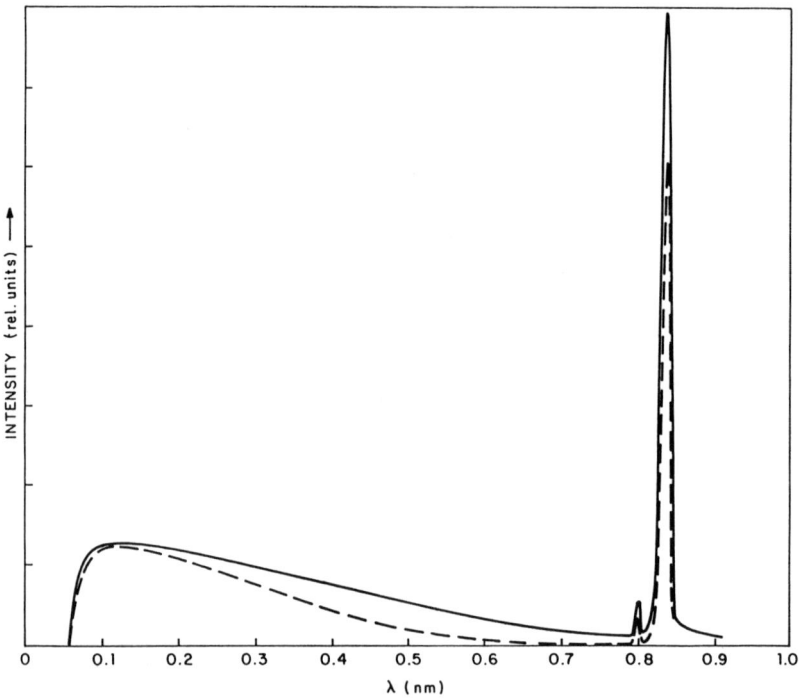

Fig. 6. Schematic wavelength-distribution of radiation from an aluminium target excited by 20 keV electrons (———), and after passage through a 3 μm aluminium filter (– – – –).

where h = Planck's quantum, c = velocity of light, e = electron charge, V = operating voltage of X-ray source. If the source was operated at 20 kV, as ours was most of the time, then the shortest possible wavelength (at zero intensity) was 0·06 nm and the maximum occurred at ~0·09 nm (Fig. 6). The morphological features of 12·0 nm periodicity of Fig. 1b diffract the bremsspectrum at a (weak, broad) Bragg maximum at $\theta = 0\cdot21° = 13\cdot0'$, which is barely visible next to the much more conspicuous maximum from $\lambda = 0\cdot834$ nm at 1·99° Bragg angle. To discriminate further against the bremsspectrum, we have taken advantage of differences in film sensitivities as explained and demonstrated in Fig. 7 and its caption. The figure shows the drastic sensitivity difference between single layered X-ray film and optical, panchromatic Tri-X film. The lead stearate crystal reflector used for this figure was

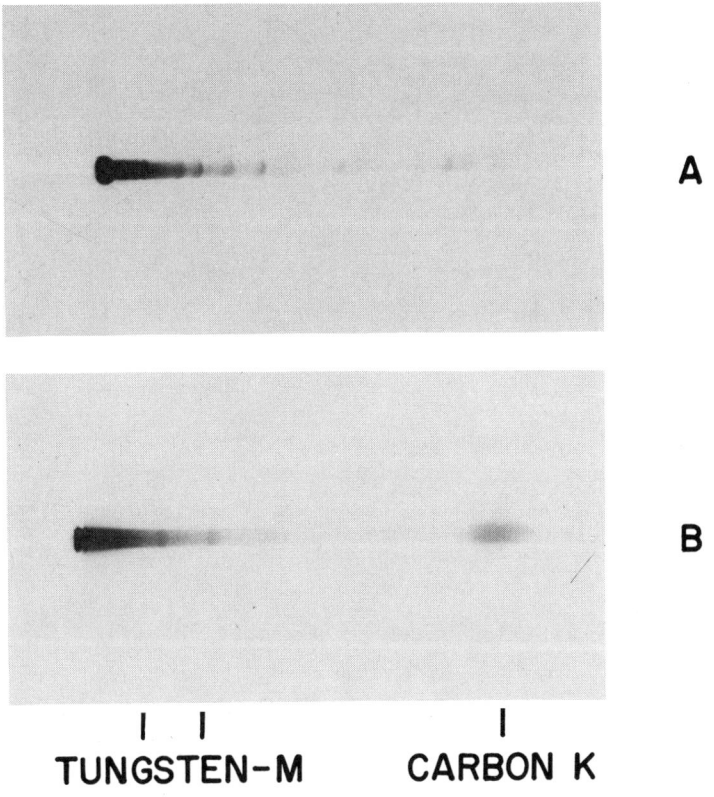

Fig. 7. (a) Lead stearate ($d = 5\cdot04$ nm $= 50\cdot4$ Å) on single coated X-ray film, 3 h exposure; (b) on optical TRI-X film, 1 h exposure. Sample to film distance: 29·89 mm. Rocking angle of the lead stearate pseudocrystal in both cases: 0–30°. Note the higher sensitivity of the TRI-X film for C-radiation, while the spurious tungsten-M-lines at shorter wavelength (~0·7 nm) are much weaker than on the X-ray film.

later also used as an intensity monitor, to ensure at regular intervals that the beam intensity I_0 has not changed, e.g. by contamination of the window or geometrical shift of the cathode. The use of film recording and a pin-hole, rather than electronic recording and slit-geometry had good reason. These features assure easy recognition of unexpected patterns, which will become apparent in examples presented later.

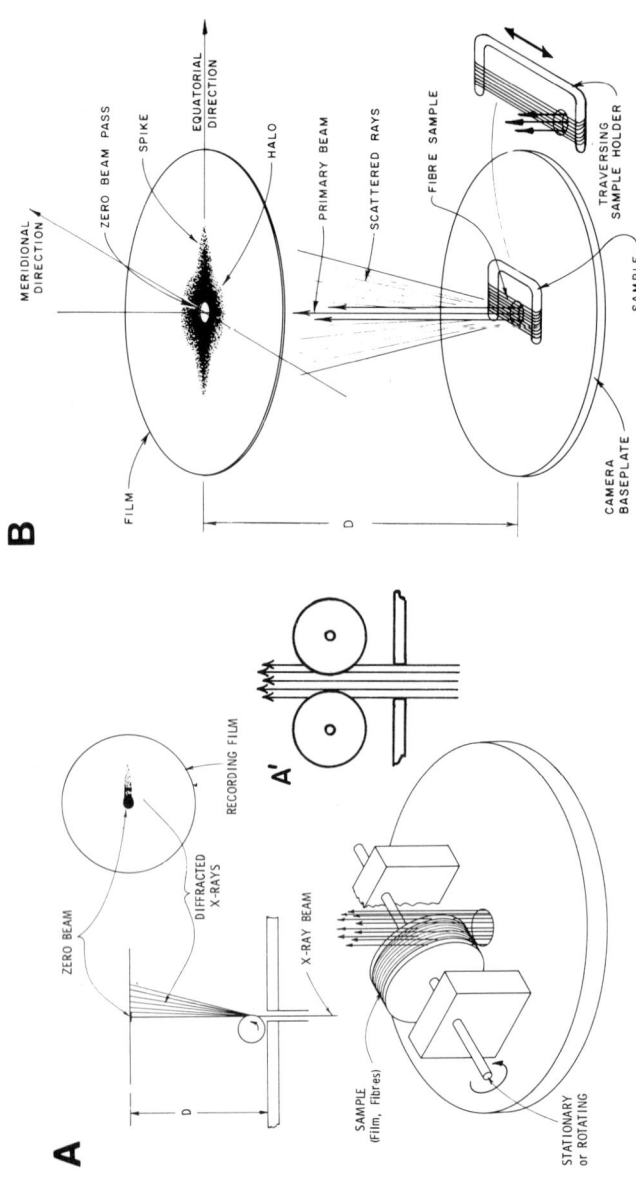

Fig. 8. (a) Grazing incidence arrangement for films or fibres which are wound on a single drum; (or twin, a'). Drum diameter and rotation of the drums during exposure to increase sampled volume are options. (b) Sample arrangement for 'transmission' characterisation of fibres. Motion of fibre during exposure to increase tested volume is again optional.

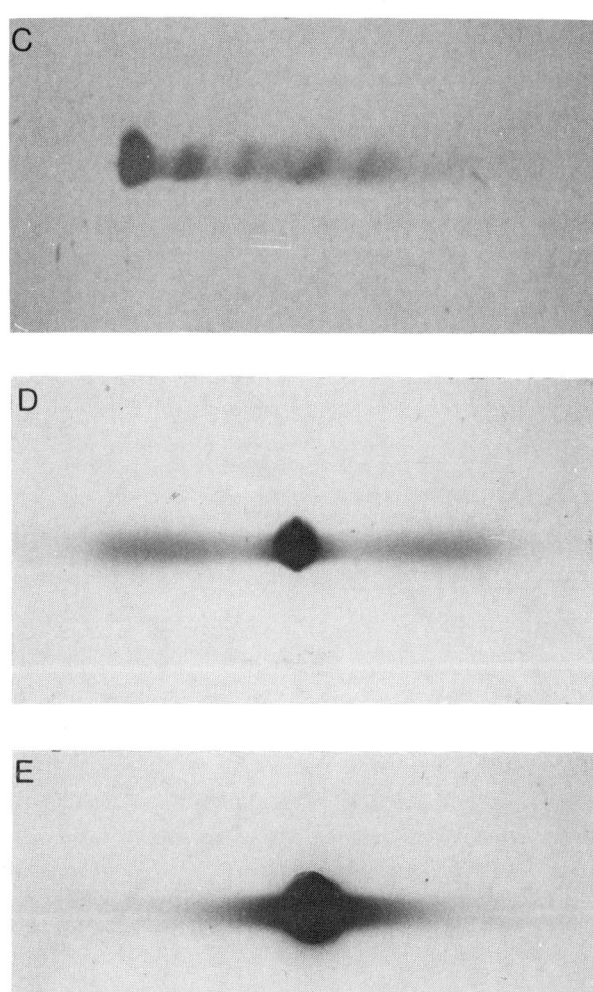

Fig. 8. (c) Pattern of heat treated (225 °C, 300 s), PET film prepared with arrangement (b) (20 kV, 8 mA, 30 min, 10 cm sample to film distance). (d) Pattern from a sample on twin drums; Sample: silver halide film, 20 kV, 8 mA, 1 h, 10 cm sample to film distance. (e) Pattern of PPD-T fibres prepared with arrangement A traversing sample arrangement; carbon radiation, (20 kV, 8 mA, 6 h; 15 cm sample to film distance).

4 SAMPLE ARRANGEMENT AND PATTERN RECORDING

The high absorption of aluminium radiation and even higher of carbon radiation, as mentioned previously, is evident from the mass absorption coefficients μ/ρ, $I_0/100$ pathlengths and optimal thicknesses $1/\mu$ of some common polymers shown in Table 1. Therefore, millimetre thick

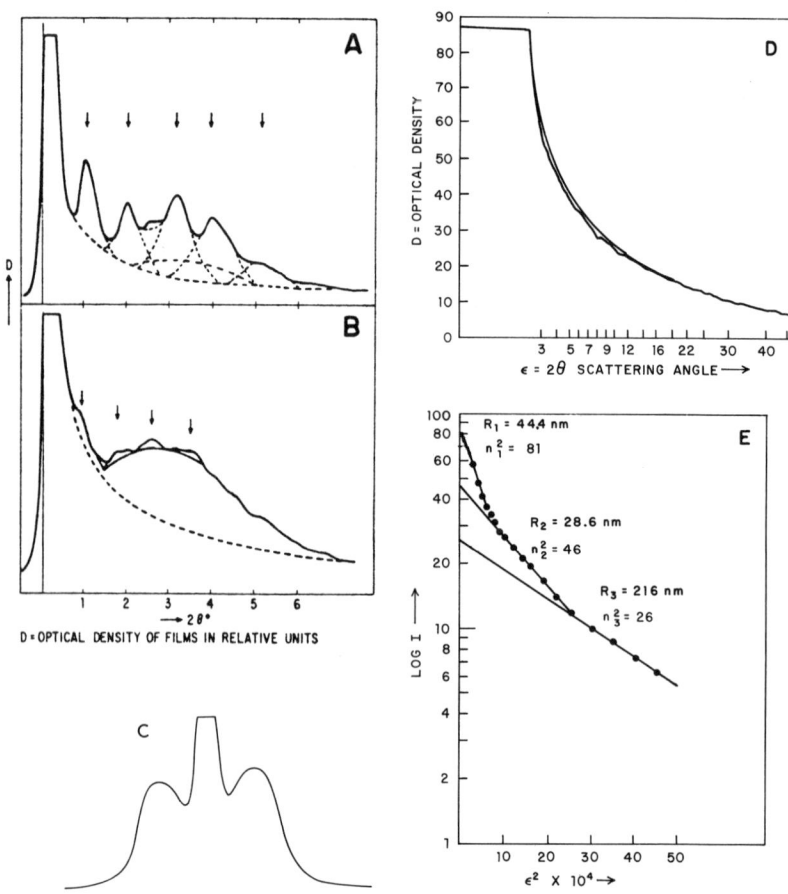

Fig. 9. Densitometer tracks of the patterns of Fig. 8. (a) of 8c (b) of pattern taken from same sample as 8c but after 5 months storage of sample (c) of 8d (d) and (e) of 8e with Guinier-plot.

multilayer fibre samples or stacks of film layers used with copper radiation are not practicable for Al- or C-radiation where a single layer of fibres a few μm in diameter, or single films of the same thickness-range are appropriate. Figure 8 shows sample arrangements for fibres and films, including examples of patterns obtained with these arrangements. Figure 9 completes the information by presenting densitometer curves and a Guinier evaluation.[4] Figure 10 describes the contributions to the pattern that come from various depths of the fibre. The high absorption of aluminium radiation, topped yet by that of carbon radiation was mentioned as a curse with regard to tube window

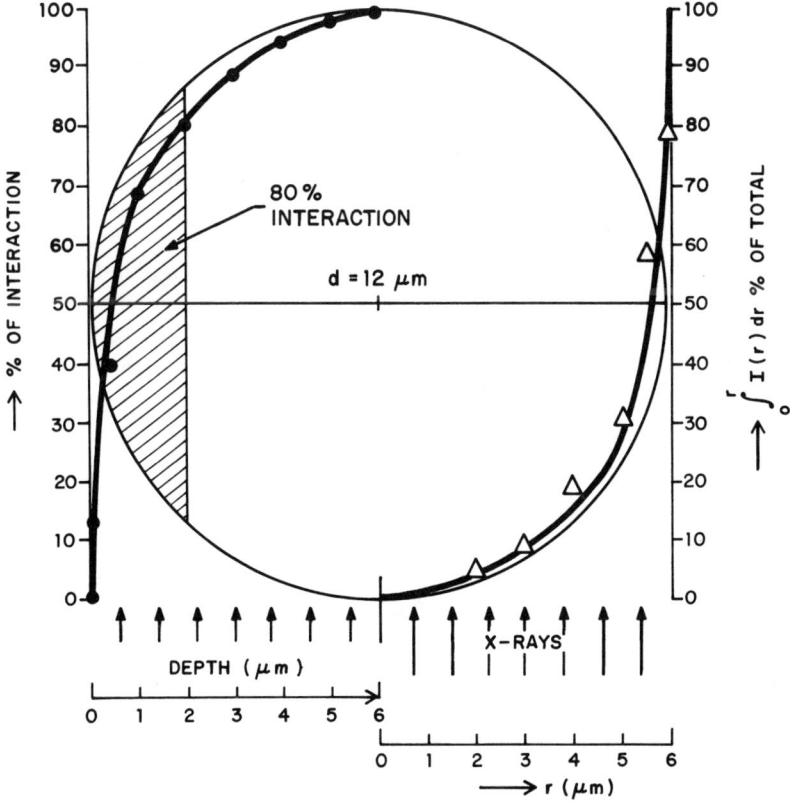

Fig. 10. Sample cross-section of a 12 μm PPD-T fibre and the fraction of sample contributing to the pattern of Fig. 8e. The Al-pattern is formed by the entire fibre, since $(I_0/100) = 37$ m for Al-radiation (Table 2).

and sampled volume; but it is a blessing insofar as the choice of wavelength and sample arrangement allow us to probe radial morphology gradients in fibres. The caption of Fig. 9 gives details about sample geometry and exposure conditions.

5 DUAL WAVELENGTH SCATTERING: $\lambda_1; \lambda_2 = 0\cdot154; 0\cdot834$ nm

Sometimes it was found advantageous to observe the wide-angle 'crystallographic' pattern with copper radiation and the small-angle 'morphological' pattern with aluminium radiation simultaneously from the same sample. This is particularly true when the information, both crystallographic and morphological, has to come from the same section of the sample, since treatment might not have been uniform throughout the sample. Figure 11 presents two examples taken from Reference 9. To identify which parts of the dual wavelength pattern come from aluminium and which from copper radiation, the pattern can be recorded on a stack of two X-ray films, the first very thin and single layered, the second double layered. Cu-radiation will pass through the first film leaving hardly any exposure, while being completely stopped in the second, thick film. Al-radiation will expose only the first film, never reaching the second. More about this method is found in References 8 and 9, and information about its usefulness in Section 6.

6 EXAMPLES OF APPLICATION

6.1 Aluminium Radiation: $\lambda = 0\cdot834$ nm

Figure 2 eloquently shows the difference between small-angle patterns for copper radiation and aluminium radiation. Again, the captions provide information about samples and exposure conditions. The example of Fig. 2a is taken from a material of considerable commercial significance, namely from acid treated, hollow nylon fibres developed for reverse osmosis desalination of sea water. The particular sample represented in Fig. 2a was chosen from a large series of fibres which differed in spinning conditions and subsequent acid treatment. The wide-angle section of the dual wavelength pattern of Fig. 11 (see Section 5) revealed a highly crystalline nylon with 'crystallinity indices'[1] of approximately 100 after acid treatment, compared with the much lower starting value of 80 for the as-spun hollow fibre.

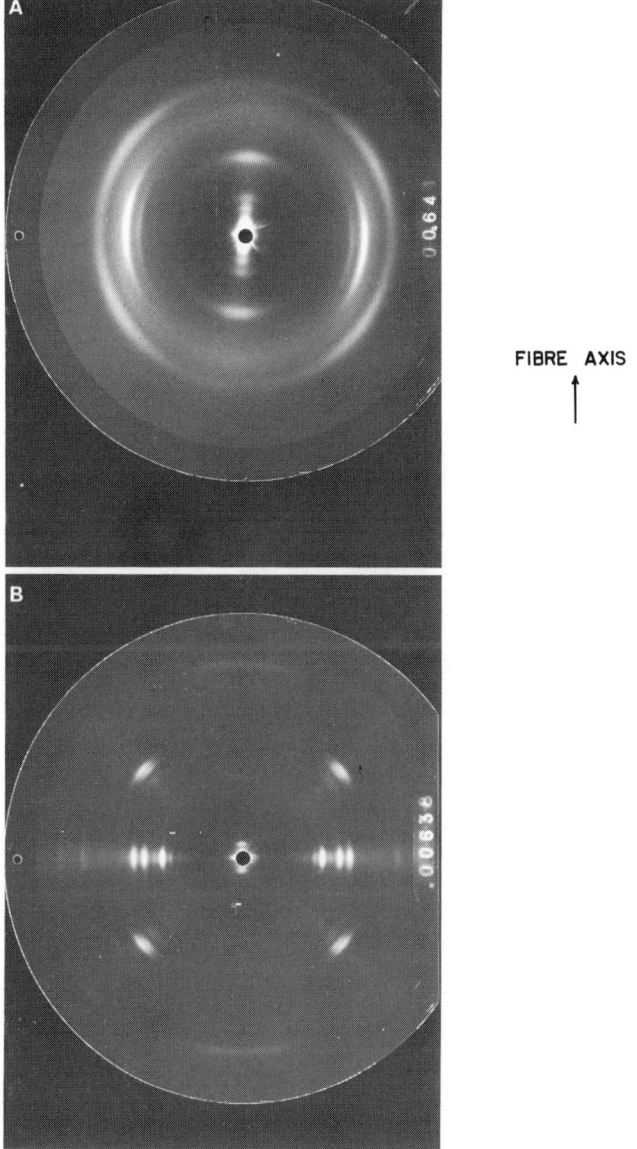

Fig. 11. (a) Dual wavelength (0·154, 0·834 nm) pattern of acid treated hollow nylon fibre (20 kV, 2 mA, 3 h, 5 cm sample to film distance). (b) The same of polypropylene fibre.

The small-angle pattern of Figure 2a, which is also visible on the dual wavelength pattern of Fig. 11, can be considered as a composite of three components:

(a) a meridional interference maximum derived from stacks; s-vectors nearly coincident with fibre axis (see Chapter 5 of this book, Fig. 1)
(b) equatorial aperiodic scattering spike, often casually called 'void scattering'; from laterally (radially) aperiodic density fluctuations
(c) a circular 'halo' from randomly, aperiodically arranged 'scatterers'.

All three features varied with spinning conditions and subsequent acid treatment; but only the 'halo', when characterised by Guinier approximation[4,8] connected the salient physical property, i.e. water permeability, with the ordinate intercept n^2, a feature of the Al-radiation small-angle patterns which itself varied systematically with acid treatment (Fig. 12). The ordinate intercept n^2, which played a key role in this investigation, is defined by:

$$n^2 = [(\rho - \rho_0)V]^2$$

where $(\rho - \rho_0)$ stands for the density difference between scatterers and matrix and $V =$ total volume of scatterers. That larger volume of scatterers (= hollows) entails better water permeation is plausible.

This case is a classical example for the triangular relationship of Fig. 13,[10] which motivates much of the industrial research in polymer morphology. The reverse osmosis case also makes two other points relevant to the topic of this book:

(a) that even highly crystalline polymers like that of Fig. 12 consist of numerous, sizeable non-crystalline components exhibited in the halo and spike of Fig. 2a,
(b) that these non-crystalline morphological features can determine some macroscopic properties, as in this case those of water permeability.

6.2 Carbon Radiation: $\lambda = 4 \cdot 47$ nm

There have been numerous examples where morphological features, revealed only by long-wavelength X-rays, were the key to transport properties (permeability, dyeability) and to mechanical properties (tenacity, elongation). In the choice of examples suitable for demonst-

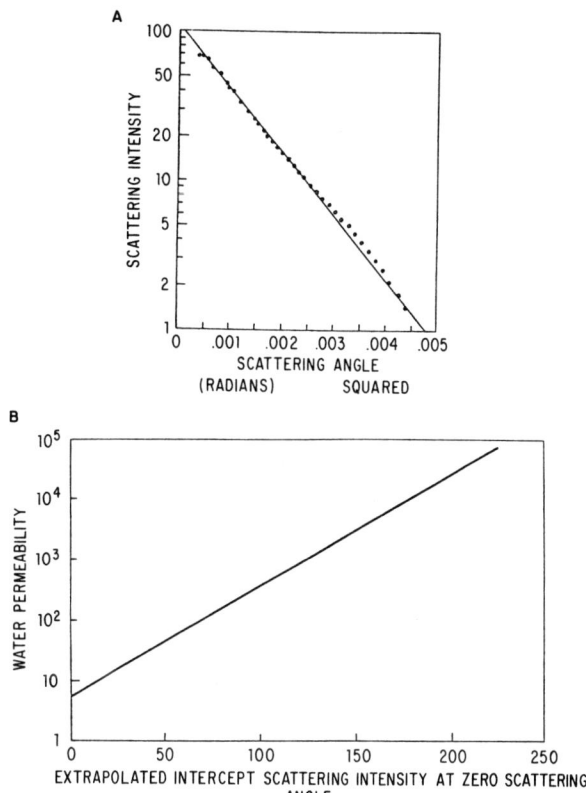

Fig. 12. (a) Guinier plot of halo in pattern of acid treated hollow nylon fibre. (b) Water permeability versus ordinate intercept n^2 of various acid treated hollow nylon fibres. From US Patent 3,551,331.[1]

ration of the usefulness of this novel, but so far little-used method, an application to films was chosen. Polyethylene terephthalate film (containing some plasticisers), that was heat treated briefly near its melting point in a frame, i.e. under dimensional constraint, showed with carbon radiation the pattern of Fig. 8c when studied by the method of Fig. 8a. The five interference maxima reveal a very high degree of order in the thin surface layer interacting with carbon radiation.[10] Repetition of the pattern under identical conditions 5 months later revealed a pattern with a degree of order which had decreased considerably during the 5 months while the sample had rested at room temperature. Crystallinity

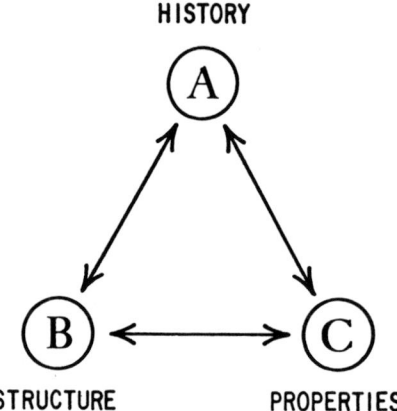

Fig. 13. Symbolic relationship between: history of polymer (thermal, mechanical treatment); structure (of all spatial frequencies); and properties (as marketable product).

and orientation were not affected. The quintessence from this experiment is that rearrangement in the polymeric solid can occur at room temperature which only carbon radiation can 'see'.

What features give rise to the interference maxima of Fig. 8c? Treating them like Bragg-maxima results in the figures given in Table 3. But can a beam of C-radiation with a divergence of $\sim 1\cdot 0°$ enter into a smooth film on a 2 cm diameter drum? The total reflection curve of

Table 3. Evaluation of pattern of Fig. 9c by Bragg equation

Order of diffraction	r (mm)	$2\theta° =$ arctan r/D	$n \times d$ (nm)
1	3·15	1·80	142·3
2	6·30	3·60	142·3
3	9·80	5·60	137·3
4	12·40	7·07	145·0
5	15·80	8·98	142·8

r = distance of interference maximum on film from zero beam, D = sample to film distance, here 100 mm.

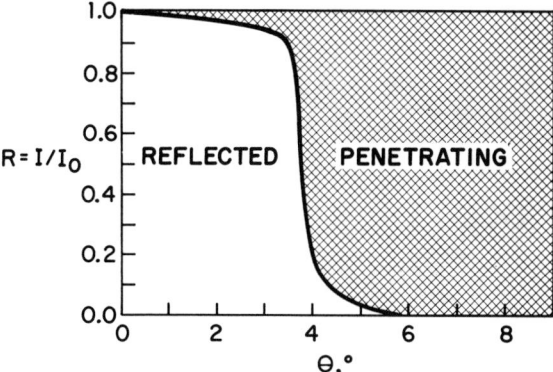

Fig. 14. Reflection curve R vs. incidence angle θ of carbon radiation on paraffin (or polyethylene, polypropylene). Information from Reference 11.

carbon radiation on paraffin has been calculated (Fig. 14),[11] and that of poly(ethylene terephthalate) will be different only to a degree insignificant for this consideration. The maximum angle of incidence of the beam of 0·5 mm diameter on a 2 cm drum (Fig. 8a) is 12·8°. Divergence of the beam adds roughly another degree, so that only a small part, 0·025 mm or 1/20 of the beam diameter of 0·5 nm, is totally reflected and the rest can enter the sample.

An experiment with a 'synthetic' morphology model has provided some clues to the origin of the fringes of Fig. 8c. A 'phase grating' was prepared by B. L. Booth and E. T. Kurtzner of this laboratory for an experiment in the fashion described in Fig. 15. Two coherent wavefronts from a laser interfered in a photopolymer to generate density differences with a spatial frequency of 6000 mm^{-1}, i.e. a lattice parameter of 166·67 nm. Details of the method of preparation of such periodic structures are given in References 12 and 13. When carbon radiation was impinging on this laser made lattice in the way described in Fig. 8a, the pattern of Fig. 16 was obtained. On the original photograph 15 fringes can be discerned. The similarity with Fig. 8c is striking. The only differences are the higher repetitiveness of the laser produced lattice, compared with the 'grown' lattice of Fig. 8c; and the lower surface perfection of the laser made lattice, which is responsible for the warped maxima. In other words, the relative fluctuation $\Delta d/d$ (in which d = lattice parameter) is larger in the grown than in the

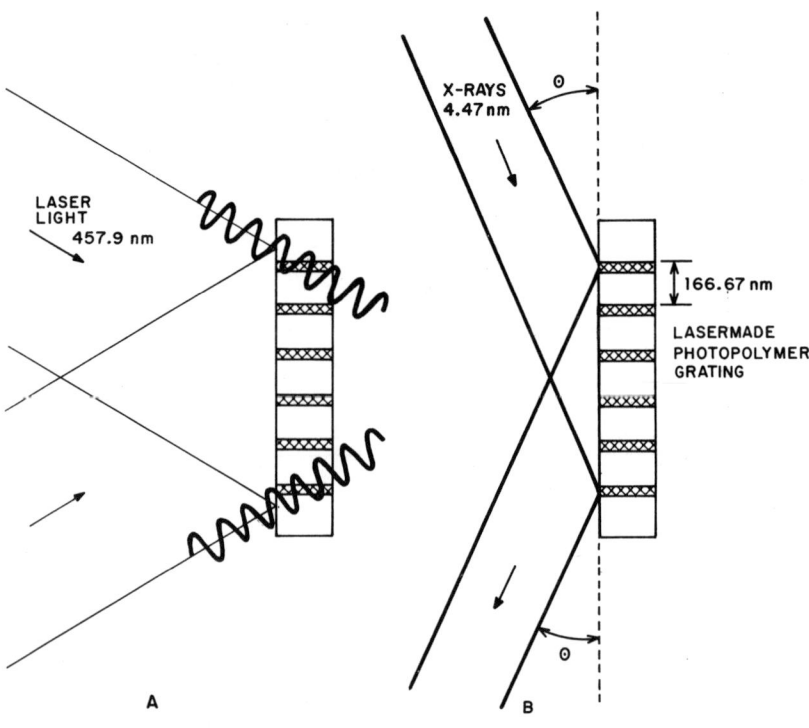

Fig. 15. (a) Preparation of 'phase-grating' (= periodic density fluctuations) by two coherent laser beams, impinging on a photopolymer.[12,13] (b) Bragg-diffraction of carbon radiation on this grating.

fabricated polymer lattice. Evaluation by a microdensitometer and calculation of d by the Bragg equation (not corrected for refraction), yields the figures in Table 4, which are satisfactory in view of the simple drum arrangement of Fig. 8a, the asymmetric nature of the pattern which makes it difficult to establish the exact point ($\theta = 0$) and the irregular shape of the interference maxima.

This model experiment has given confidence to the interpretation of the maxima in Fig. 8c. They stem from highly periodic density fluctuations in the heat treated PET with a spatial frequency $10^6/142 \approx 7042 \text{ mm}^{-1}$ (Table 3).

LONG-WAVELENGTH X-RAY SCATTERING STUDIES 251

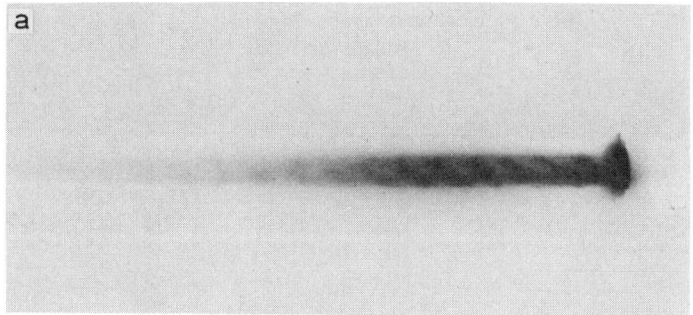

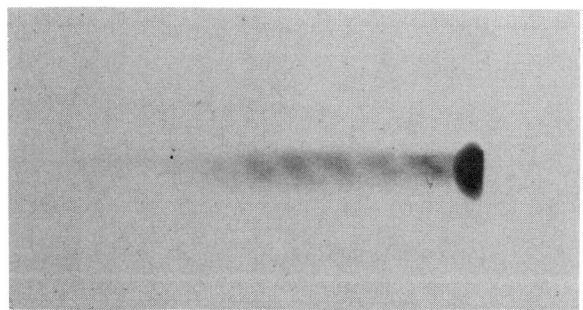

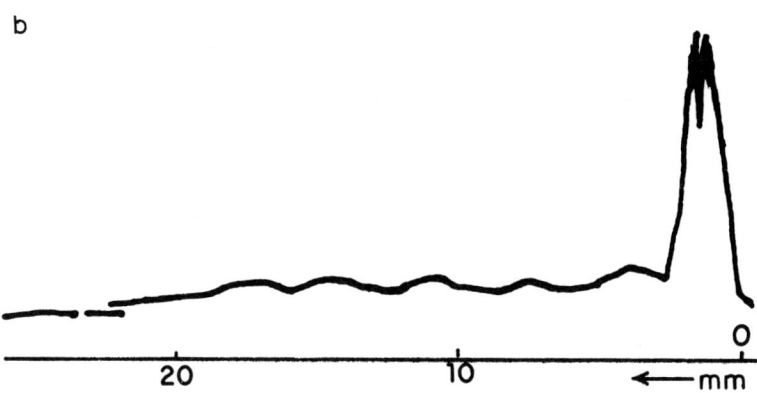

Fig. 16. (a) Two of many diffraction patterns of phase grating of Fig. 15, with 6000 lines [mm^{-1}] prepared with carbon radiation in arrangement of Fig. 8a. The curvature of the grating on the drum is equivalent to rocking a crystal through angles $0° < \theta < 12°$. Evaluation of pattern in Table 4.
(b) Microdensitometer trace of one of the patterns.

Table 4. Evaluation of patterns of Fig. 16 by Bragg equation

Order of diffraction	$r(mm)$	$2\theta° =$ arctan r/D	$n \times d$ (nm)
1	4·0	1·53	167·7
2	8·0	3·05	167·8
3	11·3	4·31	178·4
4	14·7	5·60	183·1
5	17·6	6·69	191·5
6	20·5	7·78	197·6

Sample to film distance $D = 150$ mm, theoretical $d = 166·67$ nm (6000 lines per mm).

7 CONCLUSIONS

It is hoped that the examples presented, which are limited because of space restrictions, suffice to convince that:

(a) aluminium radiation is useful as an extension of copper radiation, yielding information that overlaps and supplements that from copper radiation
(b) dual wavelength X-ray scattering is unique in its ability to provide insight into crystallographic–morphological relationships
(c) carbon radiation fills part of the gap between X-ray and optical characterisation methods with some surprising results like the highly ordered ~150 nm phases.

All three methods are relatively new and have not yet found many practitioners, for various reasons. One of the reasons is instrumentation that requires constant maintenance. Electron synchrotrons as dedicated X-ray sources might relieve the researcher of the maintenance problem and also drastically reduce exposure times.[14] The other cause for reluctance is the lack of interpretation experience. Decades of intensive practical and theoretical work have led to the insight presented in other chapters of this book. The future will hopefully bring more experiments with long-wavelength X-rays of the type of the holographic phase lattice to deprive carbon radiation scattering of its mysteries and make it a tool that contributes routinely to insight into polymer morphology.

REFERENCES

1. Cescon, L. A. and Hoehn, H. H. US Patent 3,551,331, 12/29/70 also British Patent 1,177,748, Jan. 14, 1970.
2. Wunderlich, B. (1973). *Macromolecular Physics*, Vol. I, Academic Press, New York.
3. Klug, H. P. and Alexander, L. E. (1962). *X-Ray Diffraction Procedures*, John Wiley & Sons, New York, p. 654.
4. Guinier, A. (1963). *X-Ray Diffraction in Crystals, Imperfect Crystals, and Amorphous bodies*, W. H. Freeman, San Francisco.
5. Compton, A. H. and Allison, S. K. (1960). *X-Rays in Theory and Experiment*, 2nd Edn, van Nostrand, New York.
6. von Laue, M. (1960). *Röntgenstrahlinterferenzen*, 3rd Edn, Akademische Verlagsgesellschaft, Frankfurt.
7. Herglotz, H. K. (1971). *Adv. X-Ray Analysis*, **14**, 275.
8. Herglotz, H. K. (1975). *Characterization of Materials in Research, Ceramics and Polymers*, Syracuse University Press, Syracuse, Chap. 5.
9. Herglotz, H. K. USP 3,743,841, July 3, 1973.
10. Herglotz, H. K. (1980). *J. Colloid Interface Sci.*, **75**, 105.
11. Herglotz, H. K. (1967). *Nature (London)*, **214**, 263.
12. Booth, B. L. (1975). *Applied Optics*. **14**, 593.
13. Booth, B. L. (1977). *J. Appl. Photographic Eng.*, **3**, 24.
14. Winick, H. and Doniach, S., (Eds) (1980). *Synchrotron Radiation Research*, Plenum Press, New York.

ADDENDUM

Between preparation of the manuscript of this chapter and printing of the book, some advances in the utilisation of long-wavelength X-rays have been made which should be brought to the attention of the reader. The major effort was devoted to reduction of the standard deviation of the method by:

(a) higher precision in the sample arrangement
(b) reduction of the sampling error by increased sample volume
(c) evaluation of the patterns by a dedicated computerised microdensitometer, capable of data acquisition, reduction and manipulation.

The higher precision was accomplished by a device based on the dual-drum concept of Fig. 8a. An advanced sample holder of this type on which fibre samples are wound up reproducibly is shown in Fig. 17.

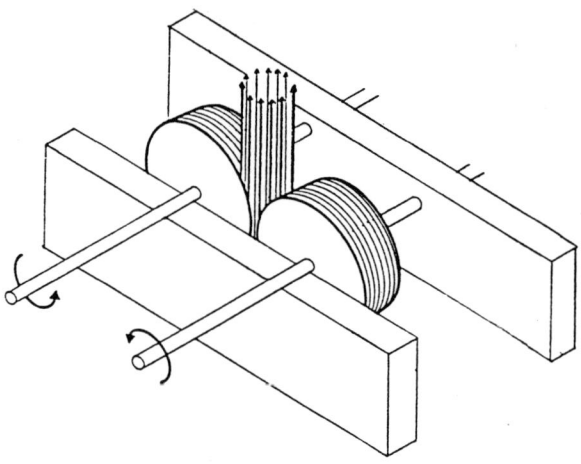

Fig. 17. Rotating double drum mechanism.

Film samples can be accommodated more easily and even polymer powders on double sticky tape have been studied in this arrangement. The low penetration depth of the long-wavelength (= soft) X-rays, particularly of carbon radiation requires precise positioning of the sample-surface in the beam. This alignment was accomplished by replacing the X-ray source by a small laser and projecting the image of the enlarged surfaces of the two drums on a screen. The adjustable double-drum mechanism is then put into a central position on the collimator so that the two surfaces are exposed to identical incident beam intensities. Then the adjusted collimator–sample mechanism is returned to the X-ray source. Quite symmetrical patterns are obtained in this way, as shown in Fig. 18. The equatorial microdensitometer track consists of a scattering function that can be described by a Guinier approximation:

$$I(\varepsilon) = I_0(\Delta\rho V)^2 \exp\left(-\frac{4\pi^2 R^2 \varepsilon^2}{3\lambda^2}\right)$$

where $I(\varepsilon)$ = intensity scattered at angle ε, I_0 = exposing intensity, $\Delta\rho$ = electron density difference between scatterer and matrix, V = total volume of scatterers, R = radius of gyration, and λ = wavelength. Superimposed is an interference maximum that can be evaluated by the Bragg equation. The data usually extracted from the maxima are

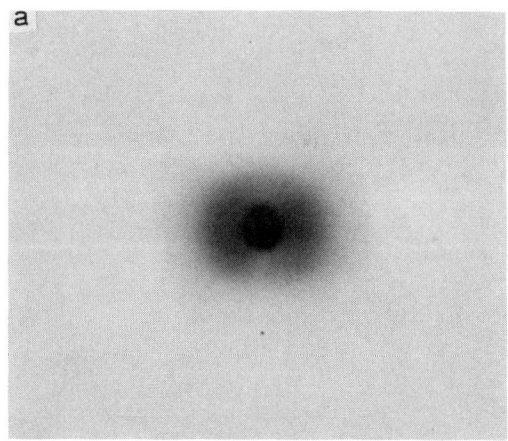

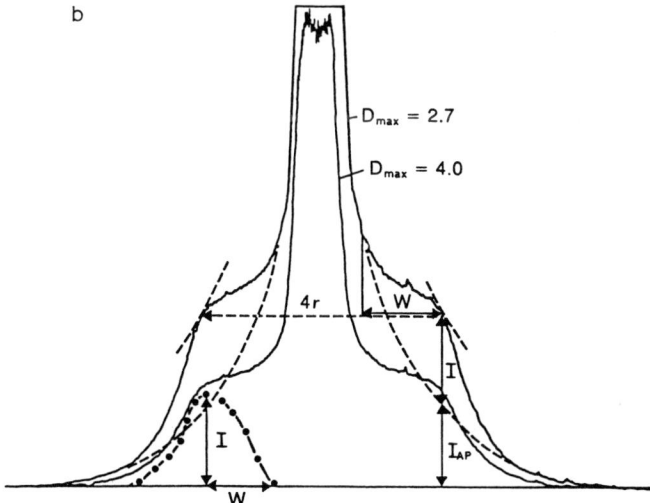

Fig. 18. (a) Pattern obtained with the arrangement of Fig. 17. Poly(*para*-phenylene terephthalamid) fibres (PPD-T) of 13 μm diameter, 7·5 cm sample to film distance, 2 cm diameter drums, carbon radiation, 8 mA, 20 kV, 2 h exposure. (b) Microdensitometer tracks with 2 densitometer sensitivities (maximum optical densities D_{max} of 2·7 and 4·0, respectively). $4r$ = measure of Bragg angle θ, I = peak height over background I_{AP}, I_{AP} = intensity at 2θ from scattering by aperiodic scatterers, W = half-width.

also indicated in Fig. 18b. It can be assumed that the interference maxima are due to repetitive electron density differences with a period d in the fibre axis direction, as suggested by the results of the experiment described in Figs 15 and 16; or by domains of size d at random locations, differing in electron density from the matrix. The repeat distances or domain sizes d in fibres measured to date, range from 23 to 58 nm. Some fibres have produced no diffraction maximum. The width W indicated in Fig. 18b is easier to measure than $FWHM$, the full width at half maximum, and serves well in the comparison of samples to indicate the fluctuation Δd of d.

Digital storage of the microdensitometer readout makes it easy to manipulate data. Figure 19 shows a curve from a fibre pattern $D = f(x)$ (D = optical density, x = equatorial linear dimension) and its logarithmic derivative $(dD/dx)(1/D) = F(x)$. The latter mode of evaluation, in a compressed x-scale has shown in some cases two orders of diffraction (Fig. 20). There is obviously a high degree of repetitiveness of the density differences in these fibres, but still far less than in the films of Figs 8c and 9a.

The comprehensive interpretation of the diffraction patterns and their fit into the logical triangle of Fig. 13 is still lacking. Therefore, additional information is desirable. An attempt in this direction is

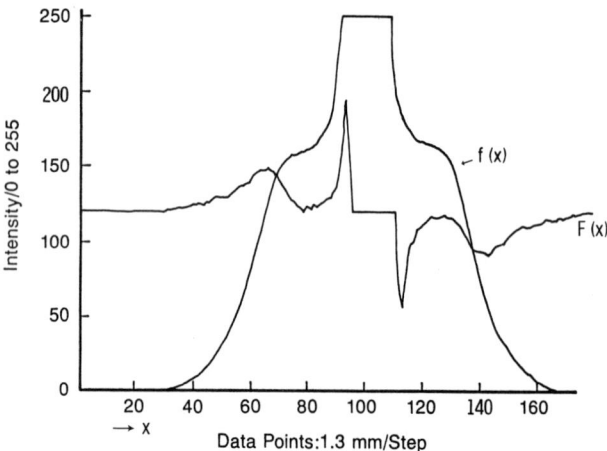

Fig. 19. Microdensitometer scan $f(x)$ and logarithmic derivative $F(x)$ (see text) of PPD-T fibre. 1 mm on recording film equals 10 mm on densitometer track. Al-radiation, 2 cm drum, $D_{max} = 2\cdot 3$.

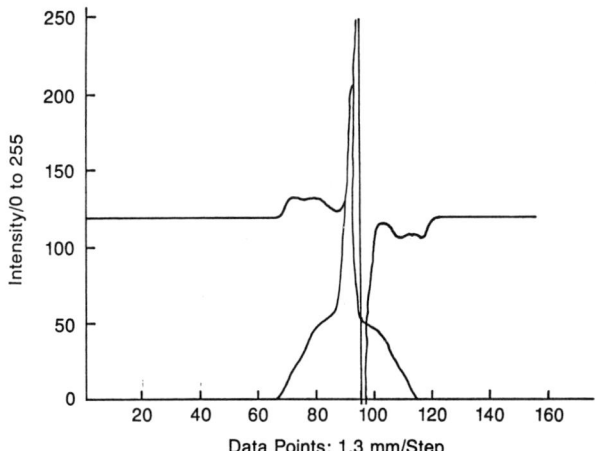

Fig. 20. As Fig. 19 but 1 mm on recording film equals 2 mm on densitometer track. C-radiation, 2 cm drum, $D_{max} = 4\cdot 0$.

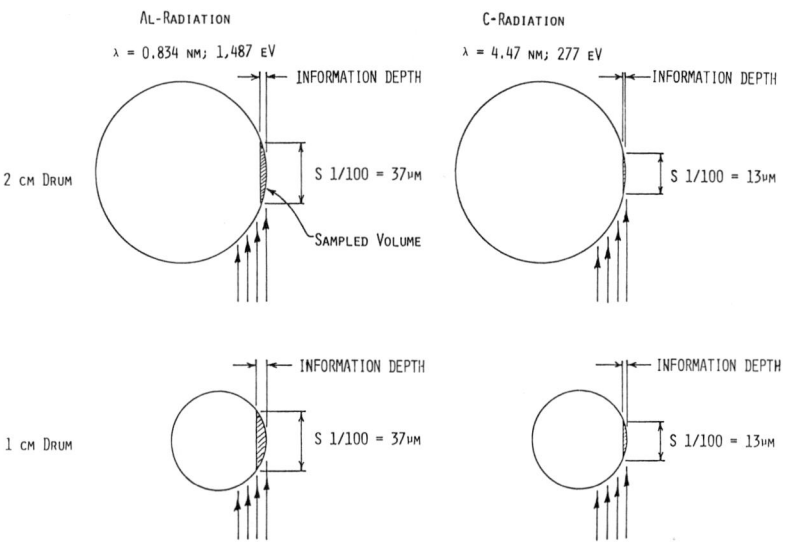

Fig. 21. Four arrangements to obtain morphological information from several levels of information depth: wavelengths $\lambda = 0\cdot 834$ nm and $\lambda = 4\cdot 47$ nm, and drum diameters of 2 cm and 1 cm (see Fig. 17), S 1/100 = path length of 1% transmission through polymer. Drum diameter and S 1/100 not to scale.

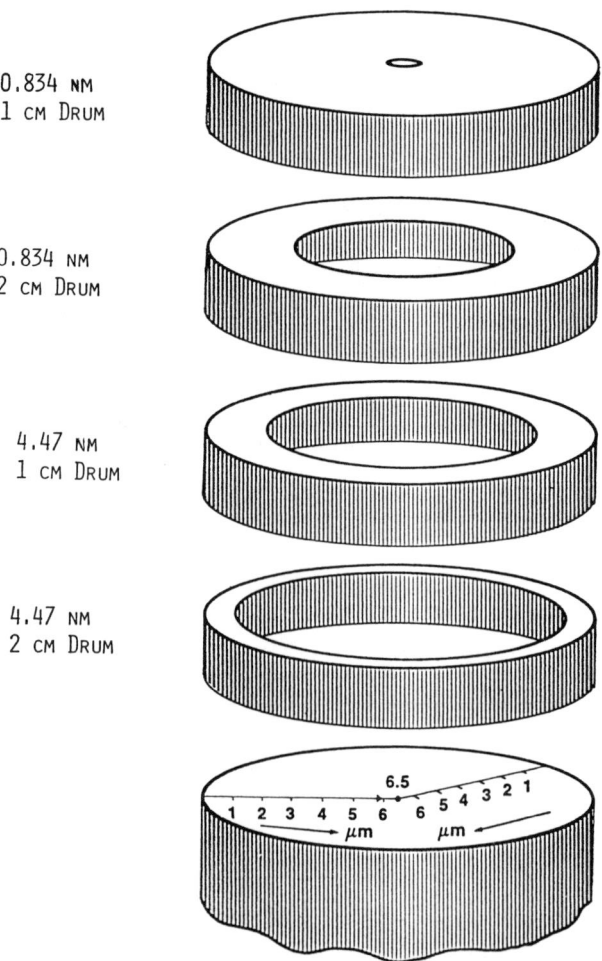

Fig. 22. Ultimate information depth obtained under the conditions of this laboratory with the four arrangements of Fig. 21. 13 µm diameter fibres of poly(ethylene terephthalate) (PET).

indicated in Fig. 21. The shallow penetration depth of both aluminium and carbon radiation allows us to observe variation of structural–morphological features from skin to core, by variation of the drum diameters of Fig. 17. The probed volume of poly(ethylene terephthalate) (PET) fibres of 13 µm diameter derived from the information depth of the four patterns in the particular arrangement defined in Fig.

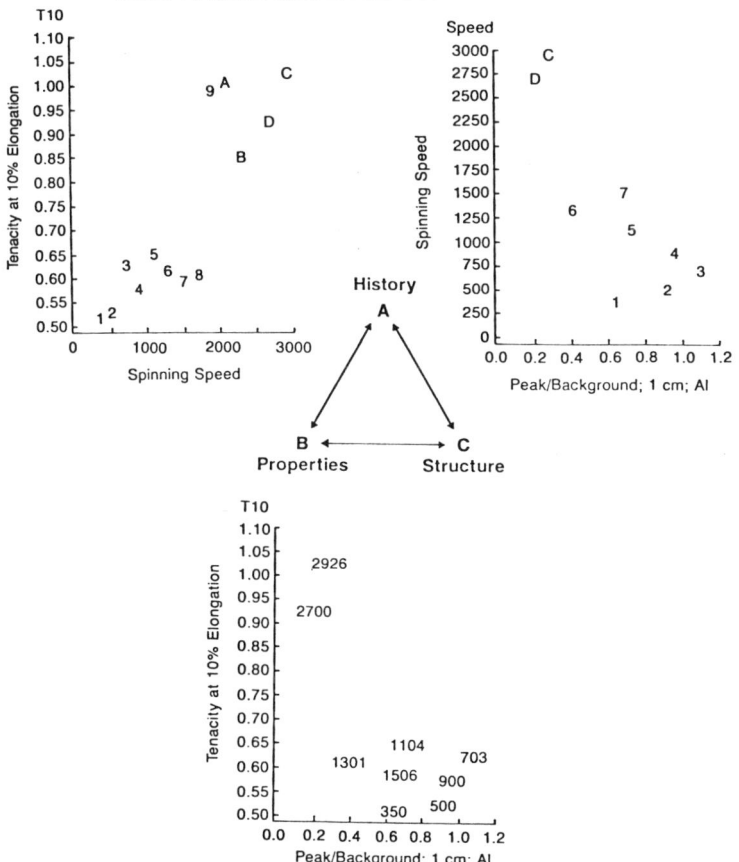

Fig. 23. Example for results with PET-fibres displayed on logical triangle of Fig. 13, demonstrating correlations of long-wavelength X-ray measurement (C) with history of the polymer (A) (= spinning speed) and property B (= tenacity). The numbers and letters 1 to D in the graphs identify the sample and its spinning speed (1 = 350 m/min; D = 3000 m/min). In the 'property' vs. 'structure' graph the sample is identified by its spinning speed. Tenacity in relative units, peak/background = I/I_{AP} of Fig. 18 (obtained with 0·834 nm radiation and 1 cm drums).

21 and with the beam divergences used in this laboratory are displayed in Fig. 22. The multiple data extracted in this way combined with the data extracted from the classical arrangement (Figs 8b, 8e, 9d and 9e), can then be searched for correlations with fibre history and properties according to Fig. 13.

In cooperation with H. L. Snyder and R. W. McClure, of E. I. du Pont de Nemours & Co. data correlations of this type led to the results of Fig. 23, which are explained in the caption of the figure. This example of poly(ethylene terephthalate) fibres spun at different spinning speeds illustrates that some features measured by long-wavelength X-rays are significant for the behaviour of solid polymers in fibres.

Chapter 7

CHAIN-FOLDING IN POLYMER CRYSTALS: EVIDENCE FROM MICROSCOPY AND CALORIMETRY OF POLY(ETHYLENE OXIDE)

C. P. BUCKLEY

Department of Mechanical Engineering, UMIST, Manchester, UK

and

A. J. KOVACS

CNRS, Centre des Recherches sur les Macromolécules, Strasbourg, France

1 INTRODUCTION

Chain-folding is the most controversial feature of the structure of semi-crystalline polymers. For years it has been the subject of a vigorous debate, in which the emphasis placed on ideal theoretical models has often outweighed careful evaluation of the experimental evidence, most of which is indirect. It is the aim of this chapter to help reverse the balance, by reviewing an extensive series of experimental observations relevant to chain-folding as it occurs during crystal growth from the melt at relatively low undercoolings.

Although chain-folding is widely believed to be a common feature of polymer crystallisation, it is often forgotten how heavily this belief relies on indirect evidence and inference. The original evidence for chain-folding, which was indeed incontrovertible, came from the combination of electron microscopy and electron diffraction applied to *solution-grown* single crystals of polyethylene.[1] By contrast, there is little evidence of such a direct nature for chain-folding in melt-crystallised polymers, and even less for the regular adjacent re-entry of the chain that is often inferred, as in the ideal model for solution-grown crystals. In fact, even in the latter case, the structure of the folds has not yet been established unambiguously, as testified by recent neutron scattering techniques.[2] The lively debate at the recent Faraday Discussion[2] shows the problem to be still unsettled. Clearly, there is no single model of chain-fold structure that is valid in all circumstances. The regularity of the folds and the nature of the fold surface rely heavily upon crystallisation conditions, structural (and stereo-) regularity of the molecule, and its flexibility and length.

Perhaps the most direct evidence for chain-folding in a *melt-crystallised* polymer was obtained by Skoulios and co-workers,[3,4] using small-angle X-ray scattering (SAXS). They observed that low molecular weight (MW) fractions (2000 < MW < 12 000) of poly(ethylene oxide) (PEO) crystallise from the melt to form lamellae, the thickness of which decreases discontinuously with increasing undercooling in such a way that, to a fair accuracy, it is always an integral fraction (1, $\frac{1}{2}$, $\frac{1}{3}$, etc.) of the chain length λ. This step-wise reduction of lamellar thickness clearly implies that the molecules are either extended or else folded once, twice or three times etc., with their chain-ends lying at the lamellar surface, as sketched in Fig. 1. Hence, for chains composed of p monomer units, the lamellar thickness L depends on two independent molecular parameters, through

$$L(n, p) = \lambda(1+n)^{-1}, \quad \text{with} \quad \lambda = pl_u \tag{1}$$

where the number n of folds per molecule is an integer and l_u is the length (0·2783 nm) of one monomer unit along the helical chain axis, parallel to the crystal **c** axis.[5] In this system, therefore, one must conclude that crystallisation from the melt does involve chain-folding, and moreover that only a small portion of the chain is involved in the folds. This unusually clear-cut, model behaviour makes low molecular

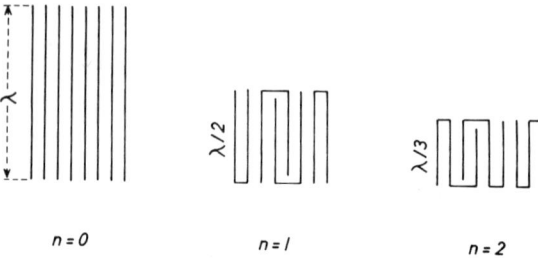

Fig. 1. Schematic diagram of possible molecular conformations in lamellar PEO single crystals. Thicknesses are integral fractions of chain length λ; n is the number of folds per chain molecule.

weight PEO fractions a powerful tool for investigating chain-folding and unfolding in detail that is not possible at present with other polymers.

In this chapter the results of applying various experimental techniques to the PEO system will be reviewed, in order to shed light on three central issues regarding the nature of chain-folding.

1. When and why is folded-chain crystallisation preferred to extended-chain crystallisation (i.e. why do chains fold)?
2. What is the structure of the folded-chain crystal surface?
3. What determines the stability of folded-chain crystals (i.e. when and why do chains unfold)?

The data analysed here were gathered chiefly by (a) optical microscopy of individual single crystals during their growth from the melt and subsequent thickening, and (b) by differential scanning calorimetry (DSC) of mature crystals. The investigations were mostly carried out at the Centre de Recherches sur les Macromolécules, Strasbourg, France, over a period of some ten years. Here, the experimental procedures will only be outlined: for details the reader should refer to the original papers. We shall, however, highlight those aspects of technique which are essential to the power of the methods employed, and those which impinge directly on the reliability of the results.

2 EXPERIMENTAL DETAILS

2.1 Materials

PEO is an unusual polymer since it is commercially available as low MW fractions in the critical range of MW (1000–20 000) where the mode of crystallisation changes from extended-chain to folded-chain (within the time-scale of laboratory experiments). Details of the samples that will be referred to are given in Table 1, which also includes a rough fraction of PEO of high MW. These samples all consist of chains terminated at each end by an OH group. Their number average MW ($\bar{M}_n$) was determined by OH analysis, and their weight average ($\bar{M}_w$) by light scattering from methanol solution. The number average chain-length λ is calculated from $\lambda = (\bar{M}_n/44)l_u$. Table 1 shows the low MW fractions used are rather sharp ($\bar{M}_w/\bar{M}_n \simeq 1.1$), although far from being 'monodisperse'.

Table 1. Characteristic molecular parameters of the PEO fractions studied.

PEO	$\bar{M}_n$	$\bar{M}_w/\bar{M}_n$	p	λ calc. (nm)	$T_m(0, p)$ (°C)
2 000	1 890	1·27	43	11·9	52·7
3 000	2 780	1·08	63	17·6	57·6
4 000	3 900	1·11	89	24·7	60·4
6 000	5 970	1·06	136	37·7	63·3
6 200	6 220	1·03	141	39·2	63·7
8 000	7 760	1·19	176	49·1	64·3
10 000	9 970	1·20	227	63·0	65·4
150 000	152 000[a]	($\simeq 2$)	—	—	—

Number average MW ($\bar{M}_n$), polydispersity ($\bar{M}_w/\bar{M}_n$), degree of polymerisation (p) and chain length (λ) calculated from eqn. (1). $T_m(0, p)$: melting temperature of extended-chain crystals ($n = 0$).
[a] viscosity average MW from solution in benzene.

2.2 Specimen Preparation for Microscopy

The prerequisite for the work reviewed here was the discovery[6–8] that lamellar single crystals, large enough for examination under the *optical* microscope, can be grown from low MW PEO melts. To accommodate optical microscopy of the crystals, they were usually grown in c. 10 μm thick molten films, sandwiched between thin glass cover slips (0·15 mm). At low undercooling, however, the intrinsic concentration of heterogeneous nuclei for initiating crystal growth in the PEO fractions is very low (c. 10^{-2} mm^{-3}). It was therefore necessary to 'self-seed' the melt[8,9] prior to isothermal crystallisation, to raise the concentration of nuclei to c. 5×10^2 mm^{-3}, yielding approximately five crystals per mm^2 of the film. Self-seeding was achieved by heating a conventionally crystallised specimen to just above (c. 0·1 K above) the melting-point of extended-chain crystals, leaving tiny extended-chain crystal fragments to act as 'seeds' in the subsequent crystallisation step. It should be pointed out, however, that even when this procedure is used, crystals grown to a lateral size of, say, 200 μm still occupy only a very small fraction of the volume of the specimen (e.g. 4×10^{-4} for 20 nm thick lamellae).

The resulting lamellar crystals usually grow with their planes parallel to that of the cover slips (their growth being restricted in the third direction). Consequently they are optically transparent with negligible birefringence within the crystal plane and thus barely visible by the usual techniques of optical microscopy (though detectable when thick enough by phase contrast). It is therefore necessary to render them visible for observation. This is achieved by using a very effective 'self-decoration' procedure.[6-8]

At a chosen moment during isothermal crystallisation the sandwich (polymer between cover slips) is rapidly quenched into a dry-ice/acetone mixture, resulting in two effects. Firstly, all growth faces of the crystals act as nuclei for the formation of spherulitic outgrowths around the crystal perimeter, as do any protrusions from the large crystal surfaces. Thus the crystal edges and protrusions become clearly delineated, and can be easily observed between crossed polars or by phase or interference contrast. Secondly, the residual molten polymer crystallises with a very high concentration ($c.\ 10^{13}\ mm^{-3}$) of presumably homogeneous nuclei,[10] thus giving rise to uncorrelated growth of very small crystals (about 30 nm in size) and hence to an optically transparent background, with negligible birefringence.

An example of a folded-chain crystal obtained in this way is shown in Fig. 2a for the 6000 PEO fraction. It shows that self-decoration not only makes possible the observation of crystal edges, and thus measurement of their size in appropriate crystallographic directions, but also reveals some characteristic features of their surface texture. The latter detail, in particular, makes it possible to distinguish folded-chain crystals from extended-chain crystals. Figure 2b shows in fact a crystal of the same fraction as Fig. 2a, but crystallised with extended-chains, at lower undercooling. The decoration of the crystal surface is clearly seen to be much less intense and uniform than for the folded-chain crystal of Fig. 2a; in particular the fine speckling of the surface of the latter is absent. The intensity of the surface decoration generally increases with molecular weight and presumably also with the polydispersity of the fraction, as can be seen by comparing the extended-chain crystals in Figs 2b and 2c; the latter, obtained with PEO 3000, displays only a few decorating units, including that of the central seed.

This unusual procedure of observing melt-grown single crystals *in situ* has numerous advantages over the classical method of observing the growth of spherulites. Growth of the various prism faces can be followed separately, as well as any changes in crystal morphology that occur during isothermal growth. Furthermore, the range of growth-rate measurements can be extended to cover more than six decades (from about $10^3\ nm\ s^{-1}$ to $10^{-3}\ nm\ s^{-1}$),[11-15] whereas with classical methods

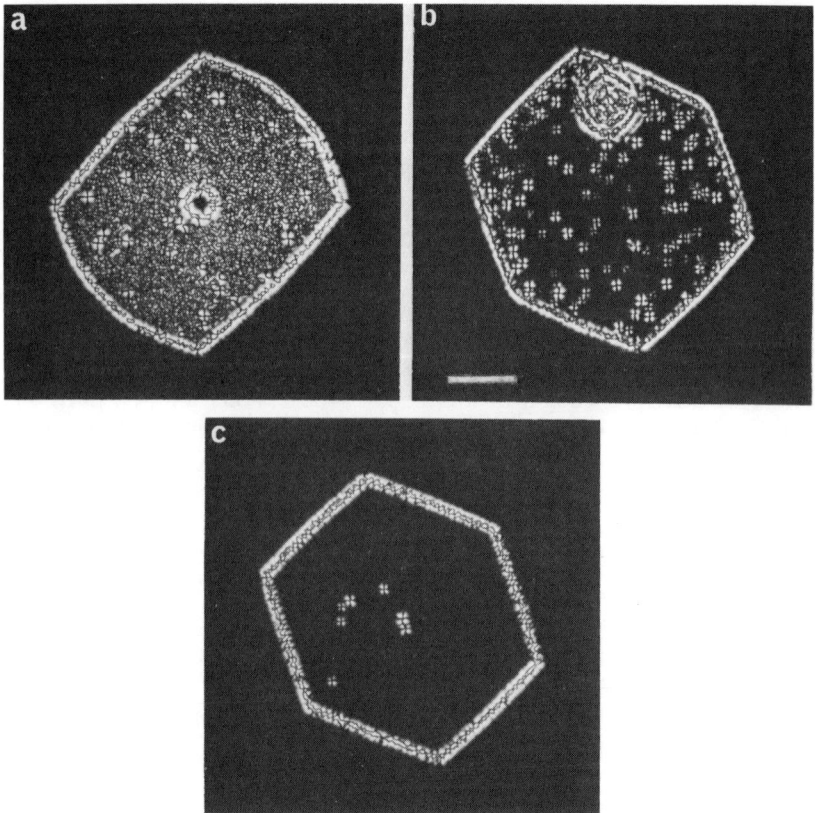

Fig. 2. Typical self-decorated PEO single crystals, as seen under the optical microscope between crossed polars. (a) PEO 6000, $T = 58.5\,°C$, $t_c = 17\,h$; (b) PEO 6000, $T = 61.56\,°C$, $t_c = 288\,h$; (c) PEO 3000, $T = 55.8\,°C$, $t_c = 21.5\,h$. Scale bar: 20 μm, common to all three crystals.

one rarely covers more than two decades of growth rate. Another notable feature of the method is that by employing a two-step thermal sequence before quenching, in which the second temperature lies above the melting-point of the crystals grown at the first temperature, it is possible to follow the course of isothermal melting of crystals. In the work described here, the melting rate was measured in this way and provided a means of determining the melting-point of crystals by the 'zero growth rate' criterion, that is by finding the temperature at which the crystals neither grow nor melt.[11]

The method does, however, require considerable care as compared to other procedures. In particular, the thermostatted baths in which self-seeding and crystal growth are carried out must be controlled to within ±0·01 K over periods which may extend to several weeks. Some critical phenomena to be discussed below are even sensitive to temperature changes of the order of ±0·003 K.[15] A corollary of the need for precise temperature control is the need to avoid water absorption in this hydrophilic polymer. Thus all the samples were stored *in vacuo* over P_2O_5 for several months prior to use and in most experiments the subsequent melting and isothermal crystallisation were also carried out *in vacuo*.[13-15]

At high undercooling where the crystal growth rate exceeds about 10^2 nm s^{-1}, the single crystal morphology gives way gradually to hedrites and spherulites.[6-8] The latter appear to be merely degenerated forms of crystal development, caused by the increased number of screw dislocations, and the subsequent interaction between spiral terraces. In addition, the concentration of active nuclei rises to c. 1 mm^{-3}, making self-seeding unnecessary. For these reasons rapid spherulitic growth was followed directly under the optical microscope using a thermostatted hot-stage in the usual way.[8]

Occasionally, where bubbles were trapped between the cover slips, it was possible to observe crystals grown from the ultra-thin PEO films (c. 100 nm) adhering to each glass surface. In such cases the melt may become completely depleted by the growing crystal, preventing self-decoration on quenching (which is not in fact necessary for delineating the outline of these crystals). Crystals obtained in this way appear similar to solution-grown crystals (Fig. 3a), and can be examined in the electron microscope, in particular by electron diffraction which shows that the PEO chains are normal to the lamellae (cf. Fig. 3b).

2.3 Crystallography

The PEO crystal is monoclinic with chains in 7/2 helical conformation. Its unit cell parameters are $|\mathbf{a}| = 0·805$ nm, $|\mathbf{b}| = 1·304$ nm ($\simeq 2 |\mathbf{a}| \sin \beta$), $|\mathbf{c}| = 1·948$ nm and $\beta = 125·4°$;[5] the projection of the unit cell onto the plane normal to **c** is represented in Fig. 3d. The principal prism faces of the overall hexagonally shaped single crystals, such as shown in Fig. 2, have been identified on the basis of electron diffraction of crystals grown from very thin molten films (cf. Fig. 3b) to be (100) and {140}, sometimes with (010) faces truncating {140}. This assignment of the growth faces has been confirmed by the observation of (100) and

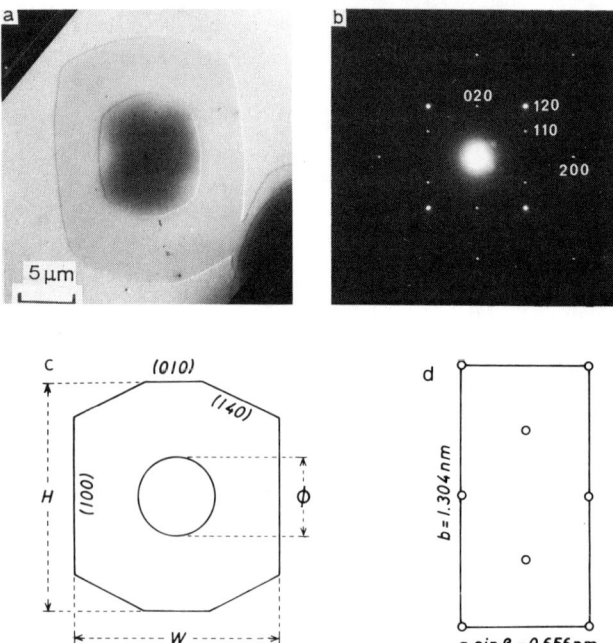

Fig. 3. (a) Bilayer PEO 10 000 single crystal grown from a thin film at 58·0 °C (electron micrograph, Pt shadowed at 15°), and (b) its indexed electron diffraction pattern. (c) The overall hexagonal habit of PEO crystals with indexed $(hk0)$ growth faces, displaying a central thickened portion of diameter ϕ, and (d) schematic projection of the PEO unit cell onto the plane normal to the chain axis.[5] (a) is after Kovacs, A. J., Gonthier, A. and Straupe, C. (1975). *J. Pol. Sci.*, **C50**, 283.

(120) twin modes,[12] giving rise to eight basic varieties of twinned crystal habit.[11] The prism faces, and the crystal outline, may however also become rounded (cf. Figs 2a and 3a).

The characteristic dimensions measured for determining the crystal growth-rate, G, were the height H and width W along the two mirror symmetry planes, which are parallel to **b** and **a*** respectively, as is clear from Fig. 3. For all the fractions investigated, the aspect ratio (H/W) of the crystals varies with crystallisation temperature in a systematic manner, ranging between 0·8 and 1·5. *During* isothermal growth, however, it remains invariant, independent of crystal size.

In a folded-chain monolayer crystal, grown from an extended-chain seed, chain-unfolding starts from the seed without any time-lag and spreads outwards in all directions at the same rate. It thus gives rise to a central circular thickened portion, of diameter ϕ (cf. Figs 2a and 3c). When unfolding results in full chain extension, the edge of the thickened portion also becomes delineated by the self-decoration described above. In this way ϕ can be easily measured during crystallisation and hence its rate of expansion G_ϕ determined.

Spiral growths generated by screw dislocations frequently occur (see Fig. 2b). Since their number increases with the size of the basal lamellae, monolayers are rarely found with a size larger than 200 μm. Though the spirals grow at the same rate as monolayers, their thickening and melting behaviour is quite different, clearly involving more complex mechanisms than those operative in monolayers, to be discussed below.

3 CALORIMETRY

Calorimetric measurements contribute to knowledge of chain-folding by providing information on the surface free energy σ_e of lamellar crystals, as deduced from measurements of their melting temperatures T_m. This deduction, however, depends critically on three intermediate steps.

1. Heating of polymer crystals to their melting region clearly constitutes a heat-treatment, which may give rise to structural modifications of the initial crystals, since folded-chain polymer crystals are always far from stable thermodynamic equilibrium. In particular, the lamellar thickness L (cf. eqn. (1)), crucial to the determination of σ_e from T_m, may (and generally does) increase during heating.[16] One must be able to assign a value of L confidently to any particular measured value of T_m.
2. One needs a reliable theory for relating σ_e to T_m and L.
3. One also needs reliable values for the thermodynamic parameters which enter into the theory.

The PEO system is again unusual, in that all these requirements can be satisfied: they will now be considered in turn.

3.1 Assignment of the Lamellar Thickness

Folded-chain low MW PEO crystals thicken in a step-wise manner, from exactly say n folds per molecule to $n-1$ folds per molecule.[8,17] It follows that, when a specimen of partially thickened crystals is melted in the DSC, two (or more) melting-peaks are found corresponding to the two (or more) discrete crystal thicknesses. When these peaks are sufficiently separated along the temperature scale to be resolved individually, chain-unfolding on the time-scale of the DSC experiment is clearly visible as a change in the ratio of the areas under the peaks as the heating-rate $dT/dt = s$ is varied (Fig. 4). When the peaks are so close that they merge together to form a single peak, thickening during heating appears as a characteristic increase of the peak temperature T_m as s decreases, as often found with high MW polymers.[16] Both of these effects are found with the PEO fractions.[18]

On the other hand, some crystals with high values of n, although believed to grow from the melt, thicken so rapidly during crystallisation and/or subsequent heating in the DSC that the melting-peak corresponding to their initial thickness is absent. (A similar situation can arise when SAXS is used,[3] since like DSC it reveals only mature crystals, already thickened during crystallisation). Conversely, some crystals, with low values of n, are found to thicken so slowly that the they appear stable on the time-scale of laboratory experiments. For these it is also possible to determine their melting-points $T_m(n, p)$ using the zero growth rate criterion[11,13] mentioned above.

Accordingly, one can classify PEO crystals into one of three categories with respect to their thermal stability under well-defined conditions. Thus some crystals appear stable (S) when their melting peak does not change for $s \geq 0.5$ K min^{-1}. Others appear unstable (U) and thicken during heating in the DSC, as in the case of once-folded chain H4000 shown in Fig. 4. Finally, even more ephemeral crystals (O) are never revealed by DSC, at least at heating-rates $s \leq 32$ K min^{-1}. The resulting classification is given in Table 2 and reveals that for any fraction the thermal stability increases with lamellar thickness (i.e. as n decreases). Furthermore, the critical degree of instability (denoted $U - O$) occurs in the range of a minimum lamellar thickness, of about 12 nm, independent of chain length. Clearly, such well-defined behaviour means that the assignment of melting-point T_m to crystal thickness L is more reliable for low MW PEO than for any other polymer system.

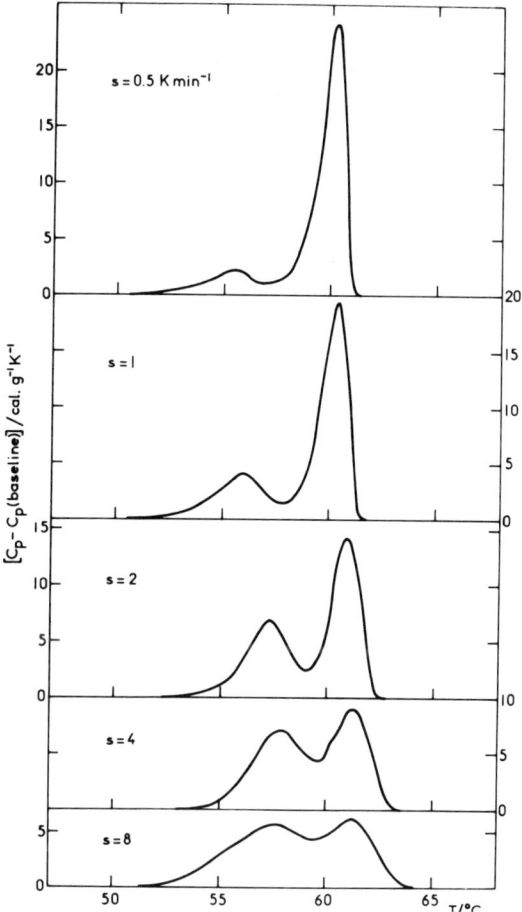

Fig. 4. Melting endotherms of PEO 4000 obtained at various heating rates s (in K min^{-1}), after full crystallisation at 42·9 °C in once-folded conformation. After Buckley, C. P. and Kovacs, A. J. (1976). *Colloid and Pol. Sci.*, **254**, 695.

3.2 Analysis of Melting Data

The appropriate theory for analysing the data listed in Table 2 was able to be thoroughly checked, because of the availability of extended-chain and n times folded-chain crystal specimens with various chain lengths. Lamellar PEO crystals have been shown to melt by sequential

Table 2. Thermal stability and DSC melting temperatures $T_m(n, p)$ (at scanning rate $s \to 0$) of folded-chain PEO crystals.[18,35]

PEO	n	L calc. (nm)	Stability	$T_m(n, p)$ (°C)
10 000	1	31·5	S	64·0
10 000	2	21·0	S	62·9
10 000	3	15·8	U	61·8
10 000	4	12·6	U–O	60·8
10 000	5	10·5	O	—
8 000	1	24·5	S	62·5
8 000	2	16·4	U	61·0
8 000	3	12·3	U–O	59·8
8 000	4	9·8	O	—
6 000	1	18·9	S	60·7
6 000	2	12·6	U–O	59·0
6 000	3	9·4	O	—
6 200	1	19·6	S	60·8
6 200	2	13·1	U	58·8
6 200	3	9·8	O	—
4 000	1	12·3	U	55·9
4 000	2	8·2	O	—
3 000	1	8·8	O	—

S, stable in DSC for $0.5 \leq s \leq 32\, \text{K min}^{-1}$; U, unstable in DSC for $0.5 \leq s \leq 32\, \text{K min}^{-1}$; U–O, unstable in DSC, only observed for $s \geq 8\, \text{K min}^{-1}$; O, not observed by DSC with $s \leq 32\, \text{K min}^{-1}$.

detachment of molecular stems from corners around the crystal perimeter. Direct microscopic evidence for this was obtained as part of the study reviewed here.[11,13] In terms of the model crystal depicted in Fig. 5, therefore, melting occurs when a molecule from a corner site such as (c) entering the melt involves no net free enthalpy change. This criterion is expressed by:[18,19]

$$(n+1)abL\Delta f - 2ab(\sigma_{e,e} + n\sigma_{e,f}) - T[S_{(mix)l} - S_{(mix)c}] = 0 \quad (2)$$

where the lengths a and b are defined in Fig. 5 (and are not to be

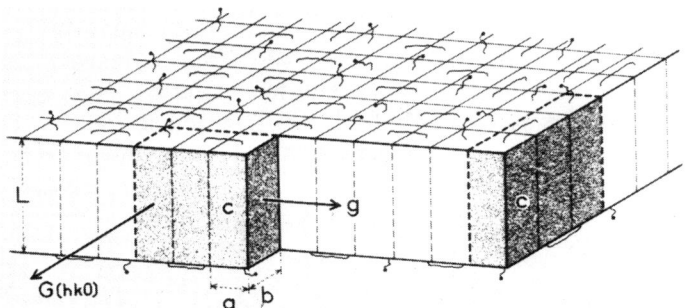

Fig. 5. Schematic model for a folded-chain lamellar crystal (with $n = 2$), showing the $(hk0)$ growth face and corner, the dimensions a, b and L of one molecular stem, and the growth rates normal (G) and parallel (g) to the $(hk0)$ face. Note that in reality L is much greater in proportion to a and b than shown above.

confused with crystal lattice parameters of Fig. 3). The first term in eqn. (2) is the bulk free enthalpy of melting per chain molecule, Δf being the free enthalpy of melting per unit volume of crystal. The second term represents the free energy contributions of the two chain-ends and n chain-folds, $\sigma_{e,e}$ and $\sigma_{e,f}$ being the surface free energies of chain-ends and chain-folds respectively. The square bracket in the third term represents the entropy difference between randomly dispersing a given monomer unit U_i of a chain (say one at the end) in the melt, and the entropy of dispersing U_i among its possible positions when the chain is incorporated into the crystal. The first of these is known to be $k \ln Cp$, in which C is the number of statistical segments per monomer unit and k is the Boltzmann constant.[20] The second, $S_{(mix)c}$, is not known and is rarely discussed. Two extreme situations can be envisaged, however:

(A) $S_{(mix)c} = S_{(mix)l}$
(B) $S_{(mix)c} = $ constant (independent of MW) $\quad$ (3)

Case (A) is that usually assumed in dealing with high MW polymers, when site (c) is occupied by only a small fraction of the chain. Case (B) would apply to low MW polymers, if the nature of the crystal surface were independent of p (end-pairing of molecules[20] would be a special case). In this case, however, it is impossible to separate $S_{(mix)c}$ from the entropic part of $\sigma_{e,e}$.

In the absence of *a priori* preference for assuming either of the above possibilities to apply, Buckley and Kovacs[19] carried out an objective test, using melting-point data for *extended-chain crystals* of the PEO fractions listed in Tables 1 and 2, together with other such data from the literature for similar low MW PEO fractions. Only Case (B) was found to result in the correct dependence of T_m on degree of polymerisation p. The resulting expression, extended to allow any number n of folds per chain, takes the form:[18]

$$T_m(n, p) = T_m^0 \left[1 - \frac{2\Sigma^{+0}(1+\alpha T_m^0)}{pl_u \Delta h^0} \right] \left[1 - \frac{2\alpha T_m^0 \Sigma^{+0}}{pl_u \Delta h^0} + \frac{kT_m^0}{abl_u \Delta h^0} \frac{\ln p}{p} \right]^{-1}$$
(4)

where Σ^+ represents the sum $\sigma_{e,e} + n\sigma_{e,f} + (kT/2ab) \ln C$, α is $-d \ln \Sigma^+/dT$ and Δh is the bulk enthalpy of fusion per unit volume of crystal. The superscript ^{0} refers to the melting-point T_m^0 of a large perfect PEO crystal, containing neither chain-ends nor chain-folds.

From the exhaustive fit of eqn. (4) to the data, and taking into account the dependence of latent heat on p (yielding Δh^0 and the enthalpic part of $\sigma_{e,e}$, for $n=0$) Buckley and Kovacs derived the following values for the thermodynamic paramaters[18,19]

$$\left. \begin{array}{l} T_m^0 = (68\cdot 9 \pm 0\cdot 4)\,°\text{C}; \quad \Delta h^0 = (2\cdot 42 \pm 0\cdot 04) \times 10^8 \text{ Jm}^{-3} \\ \alpha = (0\cdot 0129 \pm 2 \times 10^{-4})\text{ K}^{-1}; \quad \sigma_{e,f} = (2\cdot 02 \pm 0\cdot 25) \times 10^{-2} \text{ Jm}^{-2} \\ \sigma_{e,e} + (kT_m^0/2ab) \ln C = (2\cdot 30 \pm 0\cdot 17) \times 10^{-2} \text{ Jm}^{-2}, \text{ for } n=0 \\ \sigma_{e,e} + (kT_m^0/2ab) \ln C = (3\cdot 14 \pm 0\cdot 37) \times 10^{-2} \text{ Jm}^{-2}, \text{ for } n \geqslant 1 \end{array} \right\}$$
(5)

The variation of T_m with L^{-1}, for various fixed values of n, predicted by eqn. (4) with the above values of the thermodynamic parameters, is shown in Fig. 6 (solid and dashed lines). Also plotted are the experimental data listed in Tables 1 and 2, including three recently measured T_m values. The calculated and experimental values of $T_m(n, p)$ can be seen to be in excellent agreement.

Before making use of the parameters given in eqn. (5), it is right to ask whether the experimental data could be fitted equally well starting from a different set of assumptions. In fact, as the reader will have noted, the theoretical treatment given above assumes $\sigma_{e,e}$ to be independent of the chain length (p). Clearly, this assumption will not be reasonable if the nature of the extended-chain crystal surface varies with p. Booth and co-workers[21,22] relaxed this assumption when analysing similar data for PEO fractions, and started instead from a

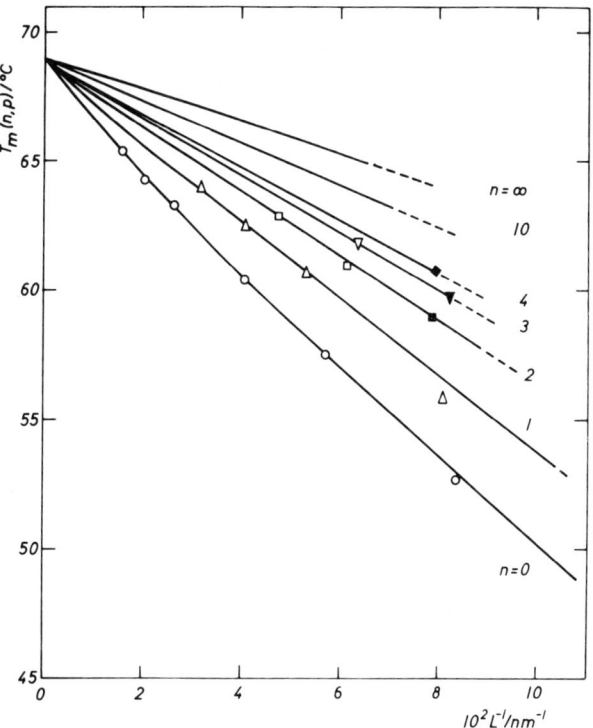

Fig. 6. Dependence of the melting temperatures $T_m(n, p)$ on L^{-1} and n as predicted by eqn. (4) (full and dashed lines), together with 16 experimental data points (Tables 1 and 2).[18,19,35] After Buckley, C. P. and Kovacs, A. J. (1976). *Colloid and Pol. Sci.*, **254,** 695, with the addition of new data[35] (filled symbols).

different viewpoint. A value for T_m^0 of 76 °C was assumed *a priori* and this led to derived values for $\sigma_{e,e}$ (for $n = 0$) which strongly increased with increasing chain length. This effect was ascribed to increasing roughness of the crystal surface, resulting from polydispersity (note that the fractions listed in Table 1 are far from being monodisperse) assuming all the chains to co-crystallise, irrespective of their length. It certainly is possible to fit the T_m data by choosing for T_m^0 a value above 69 °C and values for $\sigma_{e,e}$ which increase, essentially linearly, with MW. T_m data in isolation cannot decide which approach is correct, and one must seek other evidence to resolve the issue. Buckley and Kovacs[19] rejected the approach of Booth and co-workers on three grounds.

1. There was no indication in the data of any correlation of T_m or latent heat with polydispersity.
2. The undercoolings used for crystallisation were relatively low, where some molecular segregation by chain length is known to occur in low MW PEO fractions,[23,24] even though such segregation is not complete.[25]
3. The value of 76 °C for T_m^0 has little experimental support, whereas there is independent confirmation of a value close to 69 °C, e.g. from self-seeding studies.[9]

For these reasons, the interpretation put forward by the present authors appears to be better supported by available evidence.

The results described above show the melting behaviour of low MW PEO fractions to reveal a wealth of detail unprecedented for a polymer system. The step-wise crystal thickening makes it an unusually simple matter to identify and take account of crystal thickening in the DSC. The fact that specimens can be prepared with different combinations of n and p has enabled a thorough check to be made of the correct theory to use in analysing melting data, and also made possible a study of the variation of σ_e with the number n of folds per chain. Thus $\sigma_{e,f}$ was found to be independent of n,[18] and $\sigma_{e,e}$ was found to differ for extended-chain and folded-chain crystals (see eqn. 5). In the discussion below, the values of $\sigma_{e,f}$ and $\sigma_{e,e}$ listed in eqn. (5) will be used to obtain crucial information on the nature of the crystal fold surface, and in particular the fold contour length.

4 FORMATION OF FOLDED-CHAIN CRYSTALS

The formation (i.e. growth) of folded-chain lamellar crystals has been explained using two different theoretical approaches based on equilibrium considerations[26,27] and kinetics.[28,29] Of these two, the former is now out of favour, partly because it predicts only a shallow minimum in the free energy at the lamellar thicknesses observed (and in any case semi-crystalline polymers never reach equilibrium), but mainly because of the success enjoyed by kinetic theories in correctly predicting the temperature dependence of the crystal growth rate for high MW polymers.[30] It will be shown below, however, that the present kinetic approaches still contain basic flaws which remain unexplained. It is wiser at this stage, therefore, to consider the issue not yet settled, and to examine the validity of the premises on which the theory is based.

We thus begin this section by focusing on some experimental evidence obtained with PEO which supports the kinetic origin of chain-folding.

4.1 Evidence Supporting the Kinetic Origin of Chain-folding

Kovacs and co-workers[8,11,13–15] made extensive measurements of the crystal growth rate, G, for the fractions listed in Table 1, as a function of crystallisation temperature, T, using the methods described in Section 2.2. Their data span an unusually wide range of G (more than 6 decades). Some of these are reproduced in Fig. 7, where a remarkable feature emerges. While for the 150 000 sample the $\log G$ v. T curve displays the usual form observed with high MW polymers,[30] for the low MW fractions all the curves show at least one sharp break, where the temperature coefficient $d \log G/dT$ undergoes a discontinuity. Some of the critical temperatures, $T^*(n, p)$, which like T_m depend on both the chain length and the number of folds per molecule, are indicated by arrows in Fig. 7.

4.1.1 Variation of n with Crystallisation Temperature

When the data shown in this figure are combined with the lamellar thickness L, determined by direct observation on single crystals grown from thin films at the same temperatures,[8] it becomes clear that each 'growth branch' of the low MW fractions corresponds to a particular thickness and thus to zero or an integral number, n, of folds per chain (cf. eqn. (1)) as indicated in Fig. 8 for PEO 6000. These observations thus provide clear evidence that the 'degree' of chain-folding (i.e. n) is kinetically controlled. Figure 8 shows that, as the undercooling, $\Delta T(0, p) = T_m(0, p) - T$, is increased from zero, first extended-chain crystal growth prevails, until the growth rate of once-folded-chain crystals exceeds that of extended-chain crystals. Below this critical *growth transition temperature*, denoted by $T^*(0)$, once-folded-chain crystallisation takes over, giving way in its turn to twice-folded-chain crystallisation at $T^*(1)$, below which the growth rate of twice-folded-chain crystals becomes the greatest, and so on. In short, any given PEO fraction is found to crystallise from the undercooled melt, at any temperature, with that number of folds per molecule which maximises the rate of crystal growth, as required by the kinetic theories.

4.1.2 Variation of n with Orientation of Crystal Growth Face

Further results from the work of Kovacs et al.[11,15] provide spectacular confirmation of the link between growth kinetics and chain-folding,

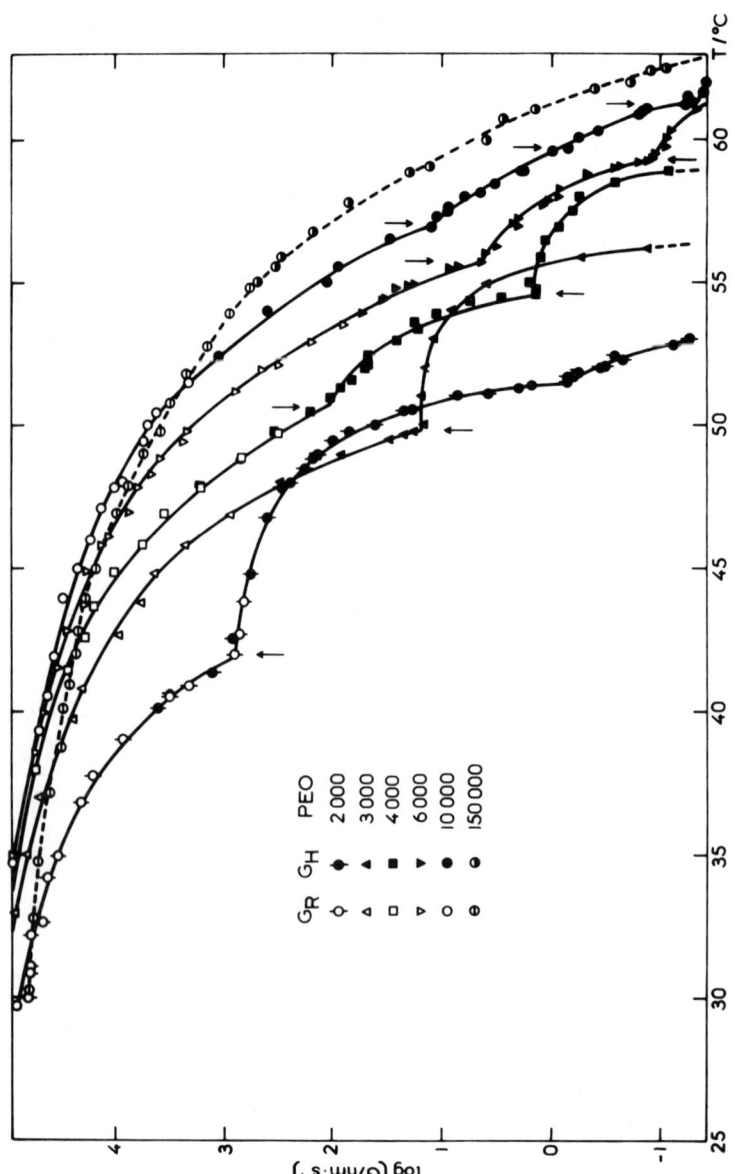

Fig. 7. Temperature and molecular weight dependence of the growth rate of PEO single crystals (G_H) and spherulites (G_R). Arrows indicate growth transition temperatures. After Kovacs, A. J., Gonthier, A. and Straupe, C. (1975). *J. Pol. Sci.*, **C50**, 283.

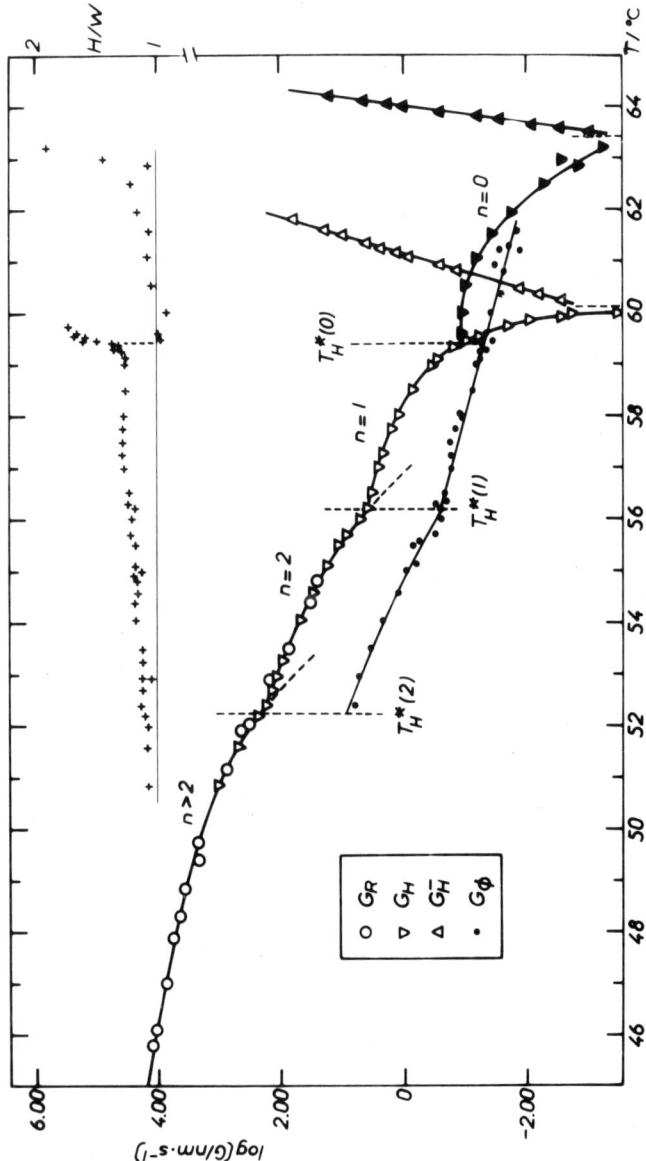

Fig. 8. Rate map of growth (G), thickening (G_ϕ), and melting (G^-) of PEO 6000 single crystals (triangles and dots) and spherulites (open circles), together with the temperature dependence of crystal aspect ratio (H/W). Growth transition temperatures $T_H^*(n)$ and melting temperatures $T_m(1)$ and $T_m(0)$ are indicated by vertical dashed lines, delimiting the growth branches. After Kovacs, A. J., Straupe, C. and Gonthier, A. (1977). *J. Pol. Sci.*, **C59**, 31.

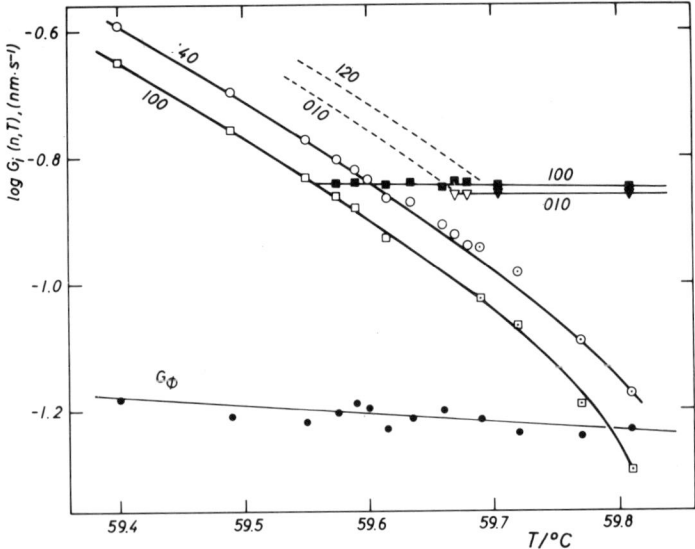

Fig. 9. Enlarged detail of temperature dependence of the rates of growth, $G_{hk0}(n, T)$, and thickening G_Φ, for the various $(hk0)$ growth faces, in the temperature range where transition from once-folded to extended-chain growth occurs in PEO 6200. Open symbols, $n = 1$; closed symbols, $n = 0$; open symbols with dot, crystals grown upon folded-chain seed crystals obtained at 59·0 °C. After Kovacs, A. J. and Straupe, C. (1979). *Faraday Disc.*, **68**, 225.

showing that the above criterion applies even under isothermal conditions, when the growth rates of different prism faces are compared. Figure 9 shows a detailed enlargement of the narrow temperature interval where the growth-transition between once-folded and extended-chain crystals occurs in the PEO 6200 fraction. When examined in such detail, made possible only by temperature control of the order of ±0·01 K during crystallisation, the different growth faces are seen to undergo their transitions at slightly different temperatures. For the extra dry PEO 6200 fraction, these were:[15]

$$\left. \begin{array}{l} T^*_{100} \simeq 59 \cdot 56\,°\text{C} \\ T^*_{140} \simeq 59 \cdot 62\,°\text{C} \\ T^*_{010} \simeq 59 \cdot 68\,°\text{C} \\ T^*_{120} \simeq 59 \cdot 70\,°\text{C} \end{array} \right\} \qquad (6)$$

where subscripts refer to the Miller indices of the respective growth faces, as indicated in Fig. 9. Once again, the degree of chain-folding on each growth face (here $n = 1$ or $n = 0$) is found to correspond to that giving the fastest growth. Figure 9 thus reveals a remarkably strict relationship between n and G_{hk0}, which excludes any loose probabilistic interpretation, and defines certainly the most precise set of transition temperatures ever found with polymers, in spite of the polydispersity of these fractions. Clearly, short PEO chains select their conformation during crystal growth in a manner which is uniquely determined by the orientation of the growth face and the temperature.

4.1.3 Variation of n with Substrate Thickness

Single crystals grown in the narrow transition interval T^*_{100}–T^*_{120} (cf. eqn. (6)) also display some spectacular morphologies ('pathological crystals'),[11] which reflect on a microscopic level the changes in the chain conformation, and are worth further analysis. In fact, according to Fig. 9 and to the kinetic criterion given above, the central extended-chain seed grows with chains either fully extended or folded once, depending on the growth direction considered and the relative magnitudes of the respective growth rates. As an example, Fig. 10 depicts two basic pathological crystal habits[11,15] obtained slightly above T^*_{100} and T^*_{140}, respectively. Within each crystal one can notice the diamond-shaped extended-chain sectors showing that, at the temperatures considered, the extended-chain growth faces retain the width of

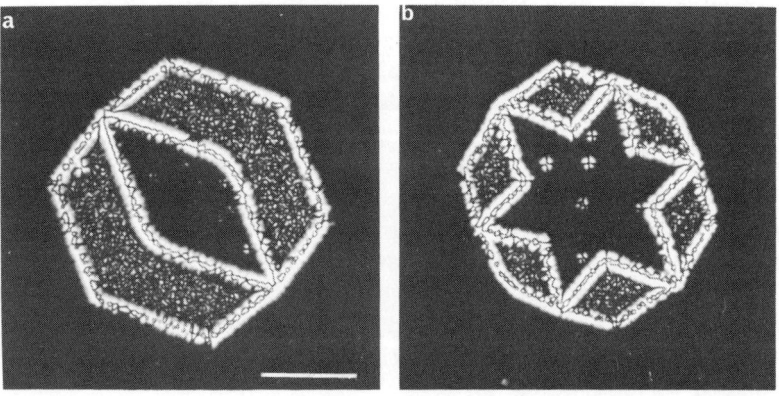

Fig. 10. Basic pathological crystal habits of PEO 6200, grown at (a) 59·59 °C and (b) 59·64 °C; $t_c = 48$ h. Optical micrographs with crossed polars; scale bar 20 μm for both crystals.

the initial extended-chain seed crystal and result in sharp tips advancing at the centre of (100) and {140} prism faces, giving rise to roof-shaped once-folded-chain prism faces. As the crystal grows, it develops, however, a triangular-shaped thickened sector behind the growth front through progressive unfolding of the adjacent folded chains, as revealed by the lack of surface decoration. This isothermal thickening, to be discussed in Section 6, thus involves a process of full chain extension progressing through the crystal at a rate G_ϕ (plotted in Figs 8 and 9), which is independent of crystallographic direction (cf. Figs 2a and 11b).[8]

Kovacs, Gonthier and Straupe[11] have also shown that when extended-chain growth at $T_1 > T^*_{hk0}$ is temporarily interrupted by cooling the sample for a short time to a temperature T_2 only slightly below T^*_{100} or T^*_{140}, the crystals do not resume their previous extended-chain growth on re-heating to T_1 again, but instead continue to grow with once-folded chains† (provided, of course, that $T_1 < T_m(n = 1)$). Although at first sight this appears to violate the kinetic criterion discussed above, in fact it provides clear support for another premise of the kinetic theory: that growth proceeds via *coherent surface nucleation*.[28-30] During the short stay at $T_2 < T^*_{hk0}$, the extended-chain prism faces are presumably covered with a few folded-chain layers which then provide too thin a substrate for subsequent deposition of extended-chains, even though the rate of deposition of the latter may be considerably greater than that of the former.

An interesting example of the shielding effect of folded-chain layers is provided by asymmetric pathological crystals, the simplest example of which is shown in Fig. 11a and sketched in Fig. 11b. Clearly, in such crystals one or several extended-chain facets (100) or {140} have been contaminated at some early stage of growth by deposition of folded-chains (presumably due to local fluctuation in T, water content or in the chain length). As a consequence, the relevant sector recovers its 'normal' habit (since it develops merely through once-folded-chain deposition) as depicted by the left-hand half of the crystal shown in Fig. 11a. The six varieties of such truncated pathological crystals have all been observed;[11,15] in fact, they can be grown intentionally above the transition temperatures T^*_{hk0}, by including a short temperature excursion slightly below the latter.

† In fact, this is the method by which once-folded chain crystals can be made to grow above the transition temperatures $T^*_{hk0}(0, p)$, to extend the growth branches above these temperatures, as shown in Figs 8 and 9 for $n = 1$.

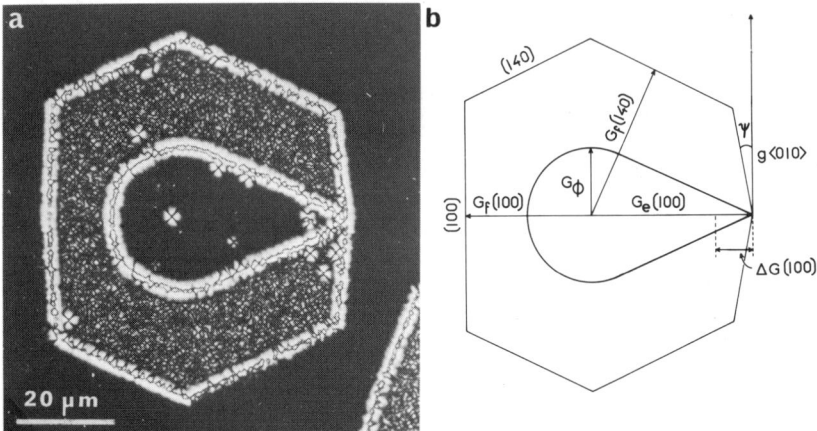

Fig. 11. (a) Asymmetrical pathological crystal of PEO 6200 grown at 59·59 °C, $t_c = 72$ h. Optical micrograph with crossed polars. (b) Schematic model for asymmetrical pathological crystals, with indication of growth rates, the angle ψ (see eqn. 9) and $\Delta G(100) = G_e(100) - G_f(100)$, reproduced from Kovacs, A. J., Gonthier, A. and Straupe, C. (1975). *J. Pol. Sci.*, **C50**, 283, with permission of the publishers, John Wiley and Sons.

4.1.4 Crystal Growth at Large Undercooling

A further feature which is in satisfactory agreement with the kinetic theory is that, for all the samples investigated, including the high MW material,[8,13] the growth branches obtained at large undercoolings and shown in Fig. 12 can all be described by the well-known and widely used relation derived from kinetic theories[28–30] viz:

$$\log G(T, p) + E/2 \cdot 303 kT = \log G_0 - KT_m(0, p)/rT \, \Delta T(0, p) \quad (7)$$

in which E is the activation energy for molecular transport in the melt, $G_0 = bkT/h \approx 3 \times 10^{12}$ nm s^{-1} is a pre-exponential factor insensitive to temperature and to chain length, h is the Planck constant, and r is a factor equal to 1 or 2, depending on whether growth occurs by mononucleation (Lauritzen's Regime I) or by multiple nucleation (Regime II) respectively.[31] The factor K is given by:[28–30]

$$K = 4b\sigma\sigma_e/2 \cdot 303k \, \Delta h \quad (8)$$

in which σ and σ_e denote the lateral and basal surface free energies, respectively, and b is the width of a growth strip (Fig. 5).

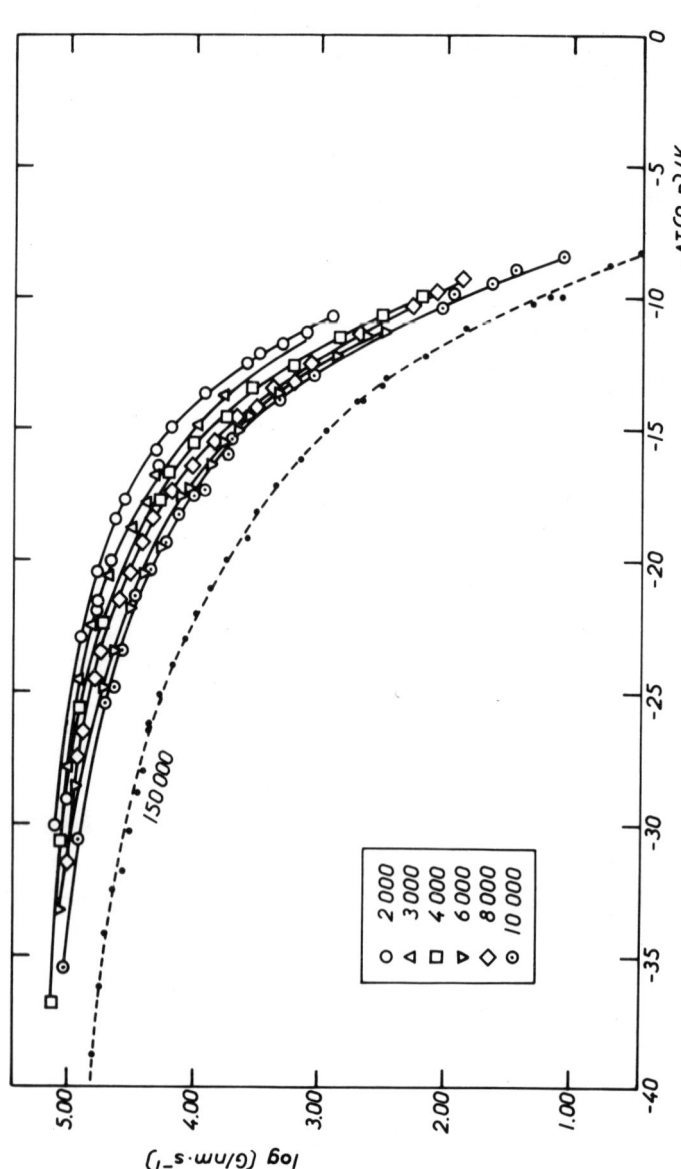

Fig. 12. Temperature dependence of the growth rate of ephemeral crystals (and spherulites) obtained at relatively high undercooling, $\Delta T(0, p) \gtrsim 10$ K. After Kovacs, A. J., Straupe, C. and Gonthier, A. (1977). *J. Pol. Sci.*, **C59**, 31.

Application of eqns (7) and (8) to the data shown in Fig. 12 yields a value for the product $\sigma\sigma_e$ which is practically independent of chain length[8,13] and is in reasonable agreement with the DSC determination of σ_e for folded chains, assuming $\sigma_e = \sigma_{e,f}$ (cf. eqn. (5)) and that growth proceeds via Regime II ($r = 2$), as predicted theoretically for large undercooling.[31]

4.2 Evidence Conflicting with Current Kinetic Theories

The observations described above support the kinetic model of chain-folding, as put forward for example by Lauritzen and Hoffman (LH)[28,30] and Frank and Tosi.[29] The intricate details of the growth of PEO single crystals from the melt, however, make possible far more searching tests of current models for chain-folding, and here the models are found to contain basic flaws. The PEO data therefore provide a challenging and at present unique experimental framework for further theoretical advances.

4.2.1 Variation of Lamellar Thickness with Crystallisation Temperature

Perhaps the most significant experimental fact to be accounted for is that the measured lamellar thickness of the growing crystals is invariant along a given branch of the $G(T)$ curve (Figs 7 and 8). In fact, it should be recalled that the kinetic theory involves a fundamental assertion: the critical length, l, of the surface nuclei determines *both* the resulting lamellar thickness L (with $L \simeq l$) and the most temperature-sensitive factor in the expression for $G(T) \propto \exp(-2\sigma bl/kT)$.[28-30] In terms of the theory, therefore, it is impossible for G to vary with T within each branch to the extent shown in Figs 7 and 8, while simultaneously L remains constant.

Two possible explanations have been tentatively put forward to reconcile this basic discrepancy with the LH model, but either one would involve some new features which need to be accommodated within the theory.

1. The most probable lengths of the growth nuclei are integer fractions of the chain length and their value increases step-wise at each growth transition temperature $T^*(n, p)$, as does the measured lamellar thickness (eqn. (1)). This model, suggested by Point and Kovacs,[32] implies that the temperature-dependence of $\log G(T)$ within each growth branch arises from terms usually

considered as being of only minor importance (as compared to the dominant one given above), resulting in a reduced temperature sensitivity, which is in fact observed.
2. The growth nuclei attach to the substrate with loose folds, their length displaying the same temperature dependence as in the LH model, and then rapidly 'zip up' to to the maximum extent possible without dragging chain-ends into the crystal lattice, as suggested by Buckley[33] and Labaig.[34]

Although the distinction between these two approaches appears crucial to the origin of the tightness of folds, they both result in similar predictions, and the present experimental data cannot discriminate between them: in fact, it has been shown that neither of the above assumptions is capable of bringing the theory into reasonable quantitative agreement with the data shown in Figs 7 and 8.[32-34] Though both treatments correctly predict increased chain-folding with increasing undercooling, they greatly over-estimate the sensitivity of $G(n, T)$ to the value of n. Thus the major discrepancy, for any given fraction, lies in the theoretical growth branches being too widely spaced along the $\log G$ axis as compared with the experimental data, which display a variation of $\log G(T)$ with n which is three times smaller. One can in fact notice in Fig. 8 that the overall envelope of the $\log G(n, T)$ curve is nearly a straight line (except in the vicinity of $T_m(1)$ and $T_m(0)$), rather than being strongly curved downwards, as predicted by both theoretical treatments.[32-34]

4.2.2 Rate of Chain Deposition Along the Growth Face

Another critical test of the kinetic theory was made by determining the rate of lateral expansion of the growth strip, $g\langle hk0\rangle$ (for a strip spreading in the $\langle hk0\rangle$ direction, see Fig. 5). Usually it is not possible to determine g experimentally, although it plays a vital role in the theory.[30,31] In the case of low MW PEO it was possible, however, from measurements of the inclination of the prism faces of pathological crystals, grown within the two lowest transition intervals (cf. eqn. (6)).

As mentioned above, and as can be seen in Fig. 10 and at the right-hand side of the crystals shown in Figs 11a and 11b, pathological crystals, grown between T^*_{100} and T^*_{010}, display a slight apex at the centre of their (100) and {140} faces, when the rate of extended-chain growth (G_e) exceeds that of folded-chain growth (G_f) by an amount ΔG (cf. Fig. 9). This is sketched in Fig. 11b for an asymmetrical crystal grown between T^*_{100} and T^*_{140}. As a consequence, a nucleation niche is created periodically on both sides of the extended-chain growth tips, and these niches initiate deposition of folded-chains, as depicted schematically in Fig. 13. They thus give rise to a roof-shaped habit through the inclination ψ of the original (100) (and possibly {140}) prism faces. Figure 11b shows that ψ is related to $g\langle 010 \rangle$ through:

$$\tan \psi = \Delta G(100)/g\langle 010 \rangle \tag{9}$$

Measured values of $\tan \psi$ and $\Delta G(100)$ yielded[11] $g\langle 010 \rangle$, a value of the same order of magnitude as the growth rate G_f (or G_e) in the transition range. In terms of the theory,[30] this means that growth proceeds through Regime II, i.e. via quasi-simultaneous, multiple surface nucleation, resulting in crenellated prism faces as depicted schematically in Fig. 13. This conclusion is consistent with the roof-shaped habit of the (100) and {140} faces of pathological crystals (Figs 10 and 11). Furthermore, it also appears to be supported by the rounded shape of

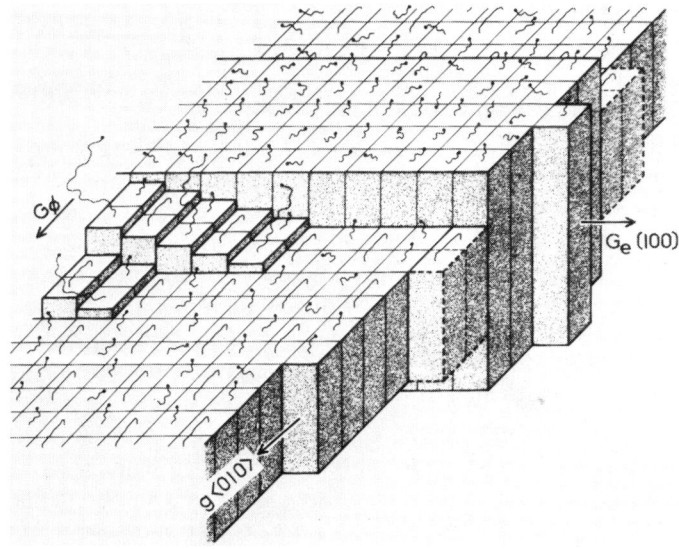

Fig. 13. Schematic model for the (100) growth face of pathological crystals growing in the lowest transition interval (eqn. (6)). Once-folded chains are depositing in the niches created by the excess growth rate $\Delta G(100)$ (see Fig. 11) and are indicated by dashed lines. The tangential growth rate g is parallel to $\langle 010 \rangle$. After Kovacs, A. J., Gonthier, A. and Straupe, C. (1975). *J. Pol. Sci.*, **C50**, 283.

extended-chain crystals grown just above the transition range, which is a common feature for all the PEO fractions investigated.[8]

Such a small value for the ratio g/G at relatively low undercooling is at variance with the theory[31] which, instead, predicts Regime I growth via mononucleation, with G being much smaller than g until the undercooling exceeds about 30 K.[33] The result given above thus reveals another conflict with the theory: relative to G, surface nuclei are much more sluggish (by a factor of about 10^5!) in spreading laterally along the growth strip, than predicted theoretically.

4.2.3 Lamellar Thickness Obtained at Large Undercooling

At undercoolings $\Delta T(0, p)$ ($= T_m(0, p) - T$) greater than c. 10 K the growth behaviour of PEO crystals becomes more ambiguous and poses other problems for the theory. The most unstable (ephemeral) crystals,[13] denoted by 'O' in Table 2, are known to form, since they can be associated with the distinct and well-defined low temperature growth branch along which $G(T)$ approaches its maximum (Figs 7 and 8). Yet these crystals have never been observed by SAXS or DSC. For example, the reported minimum lamellar thickness for PEO 6000 (crystallised between 56 °C and room temperature) corresponds to $n = 1$,[3,4] while Fig. 8 clearly shows a growth branch for $n = 2$, and at still higher undercoolings, i.e. below $T^*(2) = 52\cdot25$ °C, one can distinguish another branch, associated with $n > 2$ (though n may not be an integer in this region). Thus in a temperature range where SAXS measurements on mature crystals yield $L \simeq \lambda/2$, the magnitude of $G(T)$ varies by almost four orders of magnitude! This example further substantiates the claim made above for low undercoolings, namely that there is no close correlation between the temperature dependences of L and G, as would be theoretically expected if both were controlled directly by the critical length of the surface nucleus.

The paradox can be resolved, at least partly, by taking into account the unfolding of the chains during growth and subsequent storage or annealing. In fact, the rate of unfolding has been shown to increase appreciably as T decreases,[11,13] and more so as n increases (cf. G_ϕ in Fig. 8). This yields a minimum lamellar thickness, L_{min} (in the range of 12 nm) for folded chain crystals which appear stable on the usual laboratory time-scale. Accordingly, crystals grown at high undercoolings with more than $(\lambda/L_{min} - 1)$ folds per chain (cf. eqn. (1)) readily unfold to a thickness L_{min} and the original thickness cannot be detected by the usual techniques. Experimentally, it has recently been found to be possible to counter this effect by using thin molten specimens without self-seeding (so delaying crystal growth until the temperature reaches sufficiently low values). In fact, SAXS and DSC measurements performed on such specimens[35] showed an appreciable fraction of crystals with relatively high integral values of n, which have been assigned to crystals classified 'U–O' or even 'O' in previous work[18] (cf. Table 2). The relevant melting-points $T_m(n, p)$ are indicated by filled symbols in Fig. 6. Yet the growth branches obtained at the largest undercoolings (Fig. 12) cannot be associated unambiguously with integral values of n.[13]

The reader will note that it is the same high undercoolings which lead to the very unstable (ephemeral) crystals that were found above to produce agreement between G and eqn. (7), predicted by the kinetic theory! If this is not merely a coincidence, it suggests that growth-rate data reported in the literature for high MW polymers and obeying eqn. (7)[30] all refer to such unstable crystals. If this is the case, the measured lamellar thicknesses and melting-points do not correspond to as-grown crystals, but rather to crystals which have thickened during their growth and the subsequent thermal treatment prior to being studied by SAXS or DSC.

The possibility that crystals as measured may have thickened during crystallisation and subsequent melting has been acknowledged previously, and a simple rule is often used to take it into account. The thickness of mature crystals is assumed to be 2 to 3 times the critical length of the nucleus, independent of crystallisation temperature.[30] The data for low MW PEO show that this assumption can be grossly in error, since in this system no such simple relationship applies, even in the low temperature range where the LH theory appears to apply.[8,13]

4.3 The Mobility Objection to Regular Chain-folding

In comparing the growth of PEO crystals with current models of chain-folding, it is interesting to use the data reviewed here to test one of the objections to regular chain-folding by Flory and Yoon.[36] These authors base their argument on the inferred impossibility of the molecules rearranging their random conformation in the melt during the passage of the crystal's growth-front (especially at high undercooling) to form regular and tight chain-folds.

Consider in this respect the growth transition region for PEO 6000, occurring at $T^*(0)$ in Fig. 8. According to the discussion above concerning the magnitude of g, one can easily calculate that the time taken for the chain to attach to the growing crystal is about 3 s. On the other hand the terminal Rouse relaxation time (from the published value of the friction coefficient[37]) can be estimated as being c. 10^{-4} s. The chains, therefore, have plenty of time to change their conformation before final attachment, and they presumably do so more than 10 000 times. The mobility objection of Flory and Yoon[36] thus appears to be irrelevant in this case and even for undercoolings $\Delta T(0, p)$ up to about 10 K, i.e. in a temperature range where the data suggest that chain-folding (if any) is regular, with only a small fraction of the chain involved in the fold(s). Thus the mobility objection cannot be invoked against the occurrence of 'tight' folds during crystal growth at low undercoolings, as will be further documented below. At high undercoolings, however, where the growth rate becomes larger by a factor of up to 10^5 (cf. Fig. 8), the objection may apply, though again chain reptation,[38] which seems to be the relevant motion in the growth process, considerably enhances the chain mobility and may still account for such an increase in G.

In this respect, it is worth emphasising two significant features which emerge from the PEO data. Consider the maximum value of the crystal growth rate G_{max} occurring at temperatures below which G becomes predominantly controlled by the rapidly decreasing molecular mobility. (Recall that the latter is generally associated with the transport term[30] in eqn. (7), though at low temperatures it would involve a WLF type of temperature dependence.[39]) Figure 12 clearly shows that for the PEO system G_{max} is independent of chain length and is only slightly smaller for the 150 000 sample. These results strongly suggest that the particular molecular mobility involved in chain transport at the crystal–liquid interface is much less sensitive to chain length than the mobility involved in shear flow or stress relaxation, displaying the well-known 3·4 power law for molecular weights larger than the critical value M_c ($\simeq 7000$ for PEO).[40] The results shown in Fig. 12, therefore, provide a challenging basis for attempts to account theoretically for the effects of chain mobility on crystal growth. They show there is incomplete understanding of the type of motion and mobility required by chain molecules to attach to the crystal. The Flory–Yoon objection based upon presumably irrelevant assumptions about this mobility, must therefore be considered with caution at present. In any case for PEO crystals at undercoolings of 10 K or less, where $G <$ 100 nm s^{-1}, the rate of molecular attachment to the crystal is so slow that the objection clearly does not apply.

5 STRUCTURE OF THE LAMELLAR SURFACE

5.1 Surface Decoration Effects

The difference between extended-chain and folded-chain PEO crystal surfaces is so pronounced that it is visible even under the optical microscope when self-decorated single crystals are examined (cf. Section 2.2). In fact, these crystals display two types of decorating units,[8,15] as follows.

5.1.1 Large Decorating Units
The largest (2–5 μm) are randomly distributed hemispherical units, displaying between crossed polars positive birefringence and sharp isoclinic extinction lines parallel to the polars (cf. Figs 2 and 14). These units clearly consist of artichoke-like half-spherulites grown around relatively large and crystallographically coherent protrusions emerging from the lamella. It is likely that such protrusions are associated with

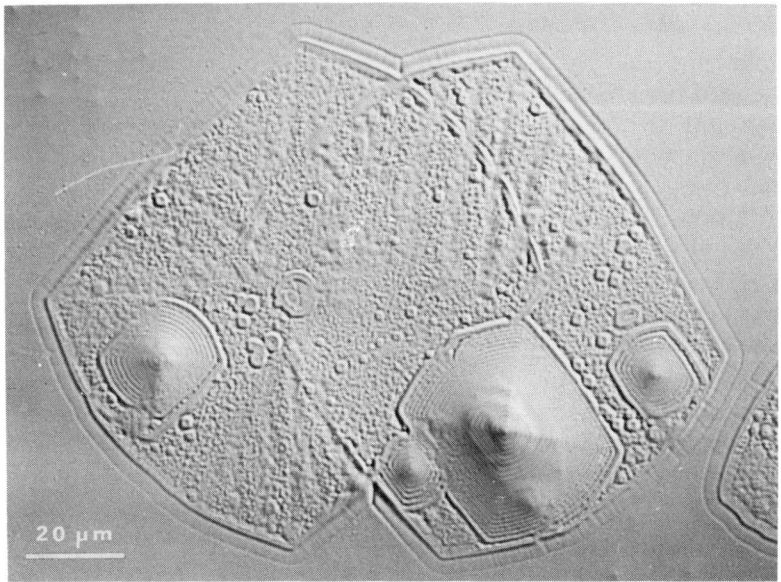

Fig. 14. Self-decorated single crystals of PEO 6000 obtained before depletion of the melt, at $T = 58 \cdot 3\,°C$, $t_c = 15$ h. Note the two different types of decorating feature, and the undecorated, growth spirals emerging from the basal lamellae. Optical micrograph with interference contrast.

the high MW tail of the chain length distribution. The longest chains, when locally concentrated, may form oriented bundles (see below) which on cooling rapidly crystallise along the chain direction and nucleate the half-spherulitic overgrowth in a manner similar to the edge decoration (Fig. 14). These large decorating units are common to both extended and folded-chain lamella and, as already mentioned, their surface concentration increases with MW and polydispersity (Fig. 2) consistently with the above interpretation of their origin. Furthermore, for a given PEO fraction their concentration increases as n decreases and as T increases, especially in the temperature range where extended-chain growth ($n = 0$) prevails. This is again consistent with the dominant role of the longest chains, the crystallisation of which becomes increasingly favoured with decreasing undercooling, owing to severe fractionation.[8]

5.1.2 Small Decorating Units

On the other hand, the surfaces of folded-chain crystals are, in addition, uniformly speckled with smaller decorating units ($\sim 1\ \mu$m), the structure of which cannot be reliably resolved under the optical microscope (see Figs 2a and 14). This type of decoration is absent in extended-chain lamellae, without exception, and thus enables one to distinguish between extended- and folded-chain conformation in an unambiguous manner. Closer inspection further reveals that these units display both positive and negative birefringence, similar to that shown by tiny three-dimensional spherulites grown from crystal fragments (seeds) in which the chain direction is normal to the light beam (i.e. parallel to the film).[8] This suggests that the specific decoration of folded-chain crystals originates from nuclei in which the chain orientation is parallel to the lamellar surface. The growth mechanism is presumably similar to that of decorating crystal halves obtained on quenching solution-grown crystals[41] as explained in previous work.[42] These nuclei must originate from bundles of chain segments lying parallel to the lamella. In bulk crystallised samples these may be hooked into the latter by the folds, and more so as n increases. If such coupling exists between folds and trapped chain segments, the quasi-homogeneous nucleation induced by quenching would preferentially start at such sites, giving rise to the rather uniform surface decoration of folded-chain crystals ($n \geq 1$) that is observed (see Figs 2a and 14). This interpretation also accounts for the absence of such surface decoration on extended-chain crystals ($n = 0$), since the conformation of the chains in the surrounding melt cannot in this case be perturbed by the entangling effect of folds. (This, of course, will be most effective when the folds are tight.)

5.2 Disordered Surface Layer of Extended-Chain Crystals

Further information about the surfaces of lamellar crystals can be deduced from the values of the surface free energies and the relevant thermodynamic parameters listed in eqn. (5). In fact, from the enthalpic parts of $\sigma_{e,e}$ and $\sigma_{e,f}$, one can estimate the numbers of monomer units contributing to dangling chain-ends (cilia) and to chain-folds, forming the disordered surface layers of the crystals.

As an example, it has been shown[18,19] that the excess enthalpy of one cilium in the disordered surface layer with respect to its value within the crystal amounts to 23·8 kJ mole^{-1}. If the excess enthalpy of one monomer unit in a cilium is assumed to be the same as in the melt (i.e. 8·7 kJ mole^{-1} [19]), one can estimate the mean cilium length to be 2·8 monomer units. Accordingly, the thickness of both surface layers of extended-chain crystals consisting entirely of chain-ends should be approximately $2 \times 2·8 \times l_u \simeq 1·54$ nm. The implied roughness of these surfaces partly reflects the polydispersity of the PEO fractions used. Such roughness would be absent in strictly monodisperse fractions.[25]

5.3 The Tightness of Chain-folds

A similar argument can be used to estimate the contour length of the chain-folds, though with a larger uncertainty, since there is no independent means for splitting $\sigma_{e,f}$ into enthalpic and entropic contributions. In this situation, Buckley and Kovacs[18] made the simplifying assumption that the temperature dependence of $\sigma_{e,f}$ is identical to that of $\sigma_{e,e}$ (which has been estimated through the value of α; cf. eqns (4) and (5)). The resulting excess enthalpy of one fold is then found to be 30·6 kJ mole^{-1}. Assuming the excess enthalpy of one monomer unit in the fold to be *at least* that which it would have in the melt, one can deduce that the fold consists of *at most* $30·6/8·7 \simeq 3·5$ monomer units. Thus its length $(3·5 \times l_u \simeq 1·0$ nm$)$ is smaller than that of two cilia. These values therefore imply rather tight folds (consisting of about ten main chain bonds), with mostly adjacent re-entry.

In fact, the maximum length of one PEO monomer unit, fully stretched into a planar zig-zag, is 0·373 nm, and hence the upper limit of fold-length amounts to $3·5 \times 0·373 = 1·32$ nm. This length should be compared with the X_{hk0} distances separating adjacent stems in various

($hk0$) growth faces. From the PEO crystal lattice (cf. Fig. 3d) one obtains:[5]

$$X_{120} \simeq 0\cdot46 \text{ nm} \qquad X_{100} \simeq 0\cdot65 \text{ nm}$$
$$X_{140} \simeq 1\cdot47 \text{ nm} \qquad X_{010} \simeq 0\cdot66 \text{ nm} \qquad (10)$$

It is immediately clear that chain-folding parallel to the {140} growth faces is inconsistent with the value of the excess fold enthalpy derived above and therefore must be ruled out. On the other hand, folding can occur in (100) or (010) planes, but must involve adjacent re-entry.

A further possibility, given the crenellation that the growth-faces are believed to have at the molecular level (cf. eqn. (9) and the relevant discussion), is that the fold-planes are in fact mainly {120}, and that the growth faces actually consist of serrated {120} sectors,[12] which cannot be resolved by microscopic observation (optical or electronic). Such micro-faceting of the prism faces, as suggested by Kovacs, Straupe and Gonthier (see Fig. 2 in Reference 13), is consistent with the {120} fold-planes of solution-grown crystals,[43] and also with the preferred fracture planes of multi-layer hedrites grown from the melt.[6,7] Comparison of the fold length derived above with X_{120} (eqn. 10) shows that folding along the {120} planes may extend to the next-nearest neighbour stem rather than being confined to adjacent re-entry.

6 CHAIN-UNFOLDING

Since folded-chain polymer crystals are generally not in thermodynamic equilibrium, a full description should include *their stability against chain-unfolding*, the mechanism of which is still not clear. This is another important aspect of chain-folding which has been elucidated by optical microscopy of low MW PEO fractions.

6.1 Morphological Modifications

Folded-chain PEO crystals are metastable with respect to extended-chain crystals. For a given fraction, the number (n) of folds per chain decreases in a stepwise manner until full chain extension ($n = 0$) is reached, as either the crystallisation temperature (T) or time (t_c) increases.[8,11] One would therefore expect the thickened portion of a lamella to be delineated by a sharp step, giving rise on quenching to the same type of self-decoration as obtained around crystal edges. This is in fact observed, but only when the thickened portion corresponds to fully extended chains. In that case, the remarkable optical contrast between the surface decoration of folded and extended-chain crystals discussed previously (Fig. 2) allows one to distinguish readily the thickened and unthickened portions of the crystal. Unfortunately, and presumably due to rapid chain-unfolding when the thickened region also consists of folded chains, it is not distinguishable in this way under the optical microscope. (Nevertheless, it can be detected in the electron microscope, on shadowed surface replicas of mature, undecorated crystals.[7,44]) For this reason, the description which follows is restricted to the process of full chain-extension in once- or twice-folded-chain crystals.

The typical morphological modifications resulting from isothermal thickening during growth of PEO crystals (monolayers) are depicted in Fig. 15, from the work of Kovacs, Gonthier and Straupe.[8,14] This figure shows, for PEO 6000, the succession of crystal habits observed at increasing t_c, at 59·35 °C (slightly below $T^*_{100}(0)$, see Fig. 8), where growth proceeds via once-folded chain deposition $(n = 1)$. Figure 16 shows the concomitant variation of the characteristic dimensions of such crystals, along the two mirror symmetry planes, the relevant symbols being defined in Figs 3c and 15.

Clearly, at all times t_c, the central undecorated portion in each crystal depicted in Fig. 15 consists of fully extended chains (see Fig. 17a), whereas the intensely decorated annulus, appearing around this region at $t_c > 48$ h, originates from extended-chain islands randomly emerging from the basal, folded-chain lamella (see Fig. 17b). The diameter of these circular islands increases linearly with t_c, involving the same rate, G_ϕ, as that characterising the radial expansion of the circular step around the central extended-chain seed, during the initial stage of growth (cf. Fig. 16 at $t_c < 48$ h and Fig. 17a).

Furthermore, the width of the peripheral, once-folded-chain annulus, given by $H - H'$ (and $W - W'$), and the width of the intensely decorated intermediate zone, $H' - H''$ (and $W' - W''$), reach constant values after about 40 and 50 hours, respectively, as shown by the H (and W) isotherms of Fig. 16. Therefore, as t_c further increases, the outlines of these annuli take up the overall hexagonal shape of the crystal (Fig. 15). Both H' and W' extrapolate to zero at the same time τ', which defines the *incubation period* necessary for stable extended-chain nuclei to emerge from the basal folded-chain lamella (see below). A similar situation prevails for H'' and W'', which extrapolate to zero at τ'', defining the time-lag after which the expanding thickened islands (Fig. 17b) eventually merge.

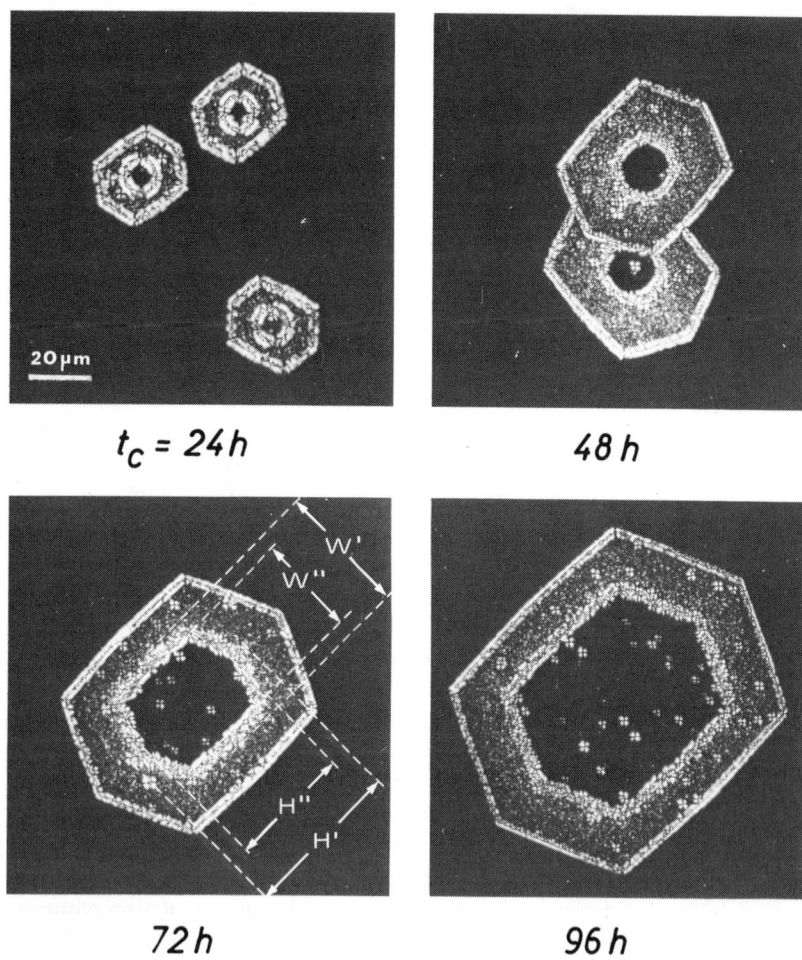

Fig. 15. Isothermal thickening. Modifications in the morphology of self-decorated single crystals of PEO 6000 during isothermal growth at 59·35 °C, at increasing crystallisation times (t_c) as shown. Optical micrographs with crossed polars; 20 μm scale bar applies to all four micrographs. After Kovacs, A. J. and Straupe, C. (1980). *J. Crystal Growth*, **48,** 210.

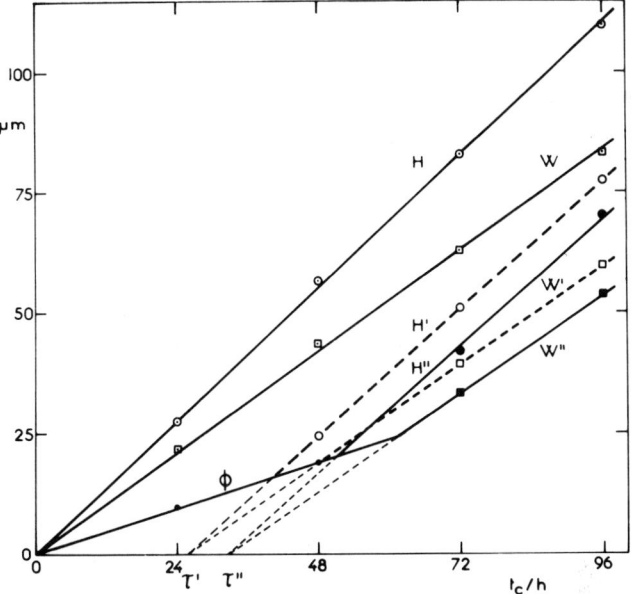

Fig. 16. Isothermal changes in the average dimensions of PEO 6000 single crystals measured along their two mirror symmetry planes, during growth at 59·35 °C. For symbols see Figs 3 and 15. After Kovacs, A. J. and Straupe, C. (1980). *J. Crystal Growth*, **48**, 210.

6.2 Mechanisms of Unfolding

These features reveal unequivocally that lamellar thickening results from migration of chain-ends through the crystal and proceeds by longitudinal diffusion of chain segments within the crystal lattice. It implies significant molecular mobility along the direction of the chain axis. Furthermore, since the mass per unit area of the lamella increases (in the present example by a factor of 2), molecules from the melt must be entering the lattice to fill the tubular vacancies left by migrating chain-ends. Otherwise this process would result in a prohibitive increase in the free energy, amounting to $2(a+b)l\sigma$ per vacancy (with $0 < l < \lambda/2$), which would resist migration of the chain-ends and hence prevent the unfolding.

The situation is quite different, however, if another chain-end is dragged in from the melt (Fig. 18), since after the energy barrier for entering the lattice vacancy is overcome, simultaneous migration of

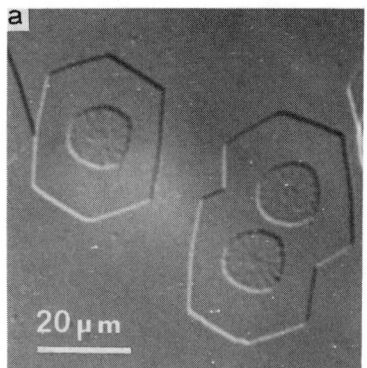 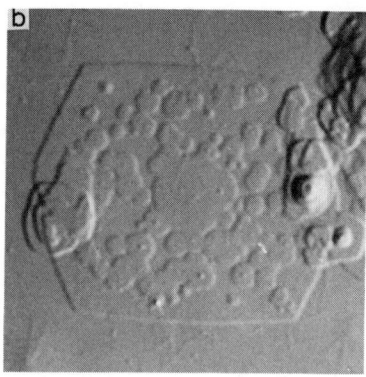

Fig. 17. Undecorated once-folded-chain single crystals of PEO 6000 grown from thin films. (a) Crystals grown at 58·8 °C, displaying a central fully extended-chain portion. (b) Crystal grown at 59·0 °C displaying, in addition to the central fully extended-chain portion, circular thickened islands grown around extended-chain nuclei (some of which are still visible) emerging randomly from the basal once-folded-chain lamella. The uneven distribution of these islands, compared to those implied by Fig. 15 (at $t > 48$ h) can be attributed to uneven depletion of the melt. Optical micrographs with interference contrast; scale bar applies to both micrographs. Reproduced from Kovacs, A. J. and Gonthier, A. (1972). *Kolloid Z.u.Z. Polymere*, **250,** 530, with permission of the publisher, Dr Dietrich Steinkopff Verlag.

both chain-ends would not increase the free energy of the system further. In this case, two situations may be envisaged.

1. The unfolding chain segment deposits coherently upon an adjacent, already organised extended-chain prism face, or corner, as in Fig. 18. This would result in a decrease of the free energy. Clearly, this mechanism is consistent with the observed initial expansion of the thickened circular step around the extended-chain seed (cf. Figs 15–17a). In fact, when the seed consists of a folded-chain crystal fragment, this first stage of unfolding is absent, and apparently no unfolding occurs until $t_c \geqslant \tau'$. This interpretation suggests that the initial thickening mechanism, like crystal growth, is controlled by coherent surface nucleation. Moreover, the relevant rate, $G_\phi = (\tfrac{1}{2}) \, d\phi/dt_c$, like $G(T)$, decreases as T increases (in the transition range, near $T^*(0)$, the ratio G_e/G_ϕ is approximately equal to 2 (see Fig. 9)), while its temperature coefficient is of comparable magnitude to that of G, and

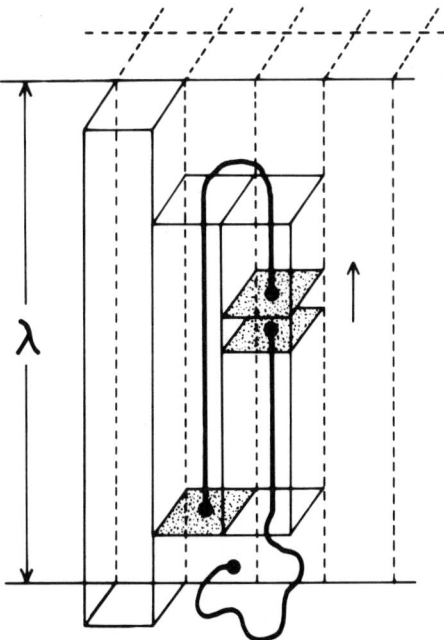

Fig. 18. Schematic model for full chain extension of once-folded chains neighbouring an extended-chain substrate and surrounded by the melt. The arrow indicates the direction of the driving force dragging the molten chain into the crystal lattice. After Point, J. J. and Kovacs, A. J. (1980). *Macromolecules*, **13,** 399.

even displays a discontinuity at $T^*(1)$, similar to that displayed by the growth rate (Fig. 8).[13] One can thus conclude that this chain-unfolding mechanism involves an energy barrier similar to that for crystal growth, and vice versa. This similarity further suggests that growth proceeds through the reeling in from the melt, via reptation, of suitably oriented chains into the growth-face.[38]

2. The second stage of unfolding which involves an incubation period τ' (Fig. 16) reveals another mechanism. This may be imagined as a primary nucleation process, depicted schematically in Fig. 19. Here, the first line represents a growth layer consisting of once-folded chains, deposited at $t_c = a$. By longitudinal diffusion of chain ends as described above, the folded chains may

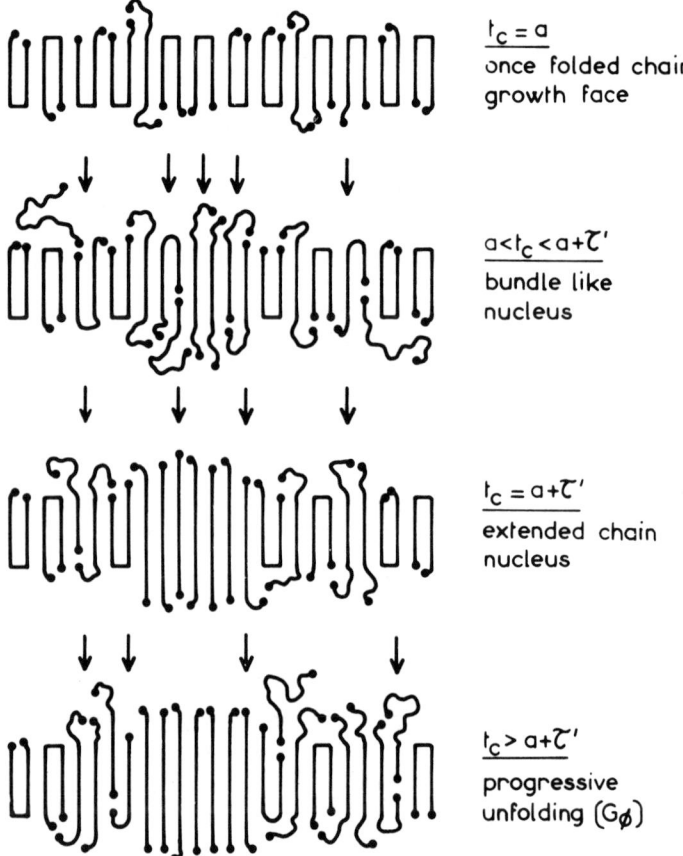

Fig. 19. Schematic model of a once-folded-chain growth layer deposited at $t_c = a$, and its progressive unfolding at increasing times after deposition, showing the nucleation process for full chain extension and subsequent growth of the thickened region. Arrows indicate sites of changes in the chain conformation. After Kovacs, A. J., Gonthier, A. and Straupe, C. (1975). *J. Pol. Sci.*, **C50**, 283.

unfold, and give rise to half crystallised chains with dangling cilia (cf. 2nd line in Fig. 19). Local accumulation of a sufficient number (say *c*. 10) of such chains (the probability of which is proportional to $1/\tau'$) then generates a bundle-like nucleus. At $t_c = a + \tau'$ the emerging cilia suddenly crystallise along the chain axis to form a stable extended-chain nucleus, similar to the initial extended chain seed. Such nuclei would then initiate growth of the extended-chain regions in just the same manner as unfolding around the extended-chain seed (Figs 15–17), involving the same mechanism and rate, G_{cb}, as described above (Fig. 18).

Clearly, both mechanisms imply reeling in of molecules from the surrounding melt to fill the lattice vacancies. In fact, when the melt is depleted both mechanisms are interrupted, and the lamellae no longer thicken (which is the case in Figs 17a and 17b). For the same reason, in folded-chain multilayer spirals (and in other closely stacked lamellae), the thickening mechanisms discussed above are operative only until the adjacent layer prevents contact with the melt. It appears that appreciable chain-unfolding can only occur in the presence of molten chains (or long cilia). If this is so, premelting must be a prerequisite for lamellar thickening in bulk samples.

Both parameters, G_{cb} and $1/\tau'$, characterising the rate of chain-unfolding, have negative temperature coefficients, which are of the same order of magnitude. This implies, consistently with the above models (Figs 18 and 19), that the rate-controlling process is nucleation rather than molecular transport. Accordingly, one must conclude that PEO crystals with integral numbers of folds per molecule are *metastable* rather than unstable, since they can only reach their thermodynamic equilibrium, extended-chain, conformation through a nucleation step.

7 IS PEO A SPECIAL CASE?

The importance of the PEO results for semi-crystalline polymers in general hinges upon whether this polymer is typical in its chain-folding behaviour, or is a special case for some specific reason (e.g. hydrogen bonding of the OH chain-ends). There is, however, no evidence so far that PEO should be treated as a special polymer. The only other polymer to have been studied comprehensively in the molecular weight range where chain-folding takes over from extended-chain crystal growth, is polyethylene (PE). Yet, according to presently available data,[34,45] PE does not exhibit quantised lamellar thicknesses in the manner of PEO (cf. eqn. (1)). In low MW PE crystals, therefore, either the chain-ends enter the lattice, or they form long, dangling cilia (with length of the order of L), while the folds may also form loops of comparable contour length, in order to accommodate L values which are non-integral fractions of the chain length. These three possibilities, which may occur in various combinations, should be compared with the situation which prevails in PEO crystals grown at low undercooling, where chain-ends lie at the surfaces of the lamellae, as depicted schematically in Fig. 1. Which of PE and PEO is the more typical for semi-crystalline polymers remains an open question, until a wider variety of crystallisable polymers becomes available in the appropriate range of MW. Meanwhile, the folding behaviours of PE and PEO probably represent two extreme situations.

A similar comment can also be applied to the observation of distinct growth branches, $G(n, T)$, for PEO crystals, uniquely related to the stepwise changes in L. The fact that these have only rarely been reported for other systems† does not prove that the behaviour of PEO is exceptional. It may, instead, simply indicate that the relevant data have not yet been obtained with sufficient precision in the appropriate range of T or MW. In fact, the occurrence of discrete growth branches (Figs 7 and 8) was completely missed even in early investigations of PEO, either because the range covered by growth-rate measurements was insufficient,[47–49] or because the scatter of the data points obtained was excessive.[6,50] Furthermore, the pathological crystals (Fig. 10) growing within a temperature interval of 0·1 K would certainly have been missed in the work reviewed here, if there had not been a deliberate exploration of the critical temperature range where transition from once-folded to extended-chain growth occurs.

If PEO does turn out to be special in its crystallisation behaviour, this will presumably be associated in some way with its chain-ends. For moderate and high MW PEO crystals, where the density of chain-ends is less, the temperature-dependence of the growth-rate (cf. (Fig. 7, dotted lines), and the lamellar thickness, *both* display the usual behaviour observed with other polymers.[30] On the other hand, the conditions for obtaining integral n values (and thus discrete L values and growth branches) as found with low MW PEO, are: firstly, that the chain-ends must be rejected from the crystal lattice onto the surface layers of the lamellae, and secondly, that the folds must connect neighbouring (though not necessarily adjacent) molecular stems, and be

---

† Takayanagi and Yamashita[46] have observed similar breaks in the $G(T)$ curves for (OH-terminated) poly(ethylene adipate). One of these presumably originates from a discontinuous change in the lamellar thickness, although this interpretation was not envisaged by the authors.

tight. The first condition clearly implies that the free energy of a pair of chain-ends incorporated within the crystal lattice (as in Fig. 18) is larger than that associated with two adjacent chain-ends at the crystal surface. Such a situation will certainly be favoured by polar and/or bulky end-groups, which appreciably perturb short range molecular interactions within the lattice.

The extent to which the chain-folding behaviour of PEO relies on the presence of OH end-groups has been tested by chemical modification of the chain-ends. Thus stepwise changes in L and discrete growth branches in the $G(T)$ curve were observed with PEO chains in which one or both OH end-groups had been substituted by C_2H_5O-,[51] or by $CO_2H-C_2H_4-CO_2-$,[35] or by a diphenyl radical at one end.[14] Although there are some differences in detail between the folding behaviour of these materials and of the hydroxyl-terminated chains discussed above, these differences are minor, and the main features appear to be common to them all.[35,51] This implies that hydrogen bonding between the OH end-groups (if any) at the lamellar surface does not contribute significantly to chain-folding in PEO crystals.

Whatever the case, the PEO system certainly yields information relevant to polymers such as PE, which show only gradual variations in lamellar thickness. It seems highly implausible that the molecular mechanisms involved in chain-folding and unfolding are totally different in the two cases. Rather, the differences must lie in the way these mechanisms manifest themselves. In PEO they involve effects which are discontinuous, dramatic and experimentally detectable, for some reason not yet proven, whereas in PE they do not. We therefore contend that the conclusions arrived at with PEO, concerning the molecular mechanisms of chain-folding and unfolding, are likely to be applicable, with appropriate adaptation, to other crystallisable polymers.

8 CONCLUSION

The previous sections relate a remarkable story. Low MW fractions of PEO have produced a wealth of detailed information on how crystallisation takes place in this polymer from the melt, with and without chain-folding, sometimes even simultaneously within the same single crystal.

With this system it has been possible to confirm some of the fundamental premises of kinetic theories upon which current understanding of chain-folding is based. The transition from extended-chain to folded-chain crystallisation, and the increasing number of folds per molecule with increasing undercooling (thus the actual chain conformation) have been shown beyond any doubt to result from crystal growth kinetics. Furthermore, the fact that the number of folds in depositing chains critically depends upon the substrate thickness (at least for $n = 1$) implies that coherent surface nucleation is a prerequisite for polymer crystal growth and thickening. Present theoretical treatments of these phenomena cannot be accommodated to the PEO data in a quantitative manner, however.

In fact, when the detailed predictions of the current kinetic theories and their recent modifications, are compared with experiment, some irreducible discrepancies appear. The tangential (lateral) growth-rate (g) of the surface nuclei has been determined to be smaller by a factor as large as 10^5 than that predicted by theory. Furthermore, although the theory correctly predicts increasing chain-folding with decreasing crystallisation temperature, it greatly exaggerates the growth-rate ratios (again by up to $10^5 \times$) for different integral numbers of folds per molecule. Clearly, there is a need for a new kinetic approach based on coherent surface nucleation, to produce more comprehensive and possibly quantitative agreement with experiment over the whole range of temperature and molecular weight investigated.

In the past development of the kinetic theory, the apparent correlation between crystal thickness and the predicted length of surface nuclei has often been invoked in its favour. When new data appeared to break this rule, the theory was even modified to reduce the length of the secondary nuclei to re-establish agreement with the measured crystal thickness. The PEO data, however, clearly show that the thicknesses of growth nuclei and of mature crystals may not be related in a simple way. As an example, there is no necessity for the crystals to thicken during isothermal growth by a constant factor. In fact, for low MW PEO crystals thickening occurs in a sepwise manner, and involves nucleation and growth of thickened regions. Only thicknesses corresponding to an integral fraction of the chain length are sufficiently stable to be observed, regardless of the thickness of growth nuclei.

Experimentally, the stepwise thickening of these crystals gives rise to rather unambiguous melting behaviour during heating in the DSC. This has enabled the determination of values for the surface free energy for various integral numbers of folds per chain, with an unusual accuracy (as compared with other polymers). In this way, the free energy of one chain-fold has been found to be comparable with, though smaller than, that of two chain-end cilia. From the free energy of a chain-fold its average length can be estimated to be no more than a few monomer units. This suggests that the folds are long enough to stretch only to nearest neighbours or next-nearest neighbour stems.

Finally, as already pointed out by Keller,[2] the remarkable determinism which governs chain-folding and unfolding in the PEO system, producing visible morphological modifications in the crystal habit and surface texture, suggests corresponding regularity and perfection on a molecular level, and hence in the folding mechanism and fold structure. Moreover, the unprecedented precision (<0.01 K, cf. Fig. 9) in the definition of the growth transition temperatures, $T^*_{hk0}(n, p)$, at which the number of folds per chain abruptly changes by one in a given ($hk0$) prism face, reveals that polymer chains select their conformation in a remarkably precise manner. 'There is no room for statistical groping, the chains seem to know exactly what to do and when!'[2] In this respect, low MW PEO is certainly not unique. Its peculiarity is that variations in chain conformation produce large, experimentally detectable effects, which have made possible a study of chain-folding in detail that has not yet been achieved for any other polymer system.

ACKNOWLEDGEMENTS

The authors are indebted to Mme C. Straupe and Miss L. Callaghan for their help in preparation of the manuscript, and to Dr A. Schierer for the photographic work.

REFERENCES

1. Keller, A. (1957). *Phil. Mag.*, **2**, 1171.
2. Keller, A. (1979). *Faraday Disc.*, **68**, 145.
3. Arlie, J. P., Spegt, P. and Skoulios, A. (1967). *Makromol. Chem.*, **104**, 212.
4. Spegt, P. (1970). *Makromol. Chem.*, **140**, 167.
5. Takahashi, Y. and Tadokoro, H. (1973). *Macromolecules*, **6**, 672.
6. Gonthier, A., Vidotto, G. and Kovacs, A. J. (1970). *IUPAC Symposium, Leiden 1970, Book of Abstracts*, Vol. 2, Inter Scientias Inc., The Hague, p. 797.

7. Wickjord, A. G., Gonthier, A. and Kovacs, A. J. (1970). *IUPAC Symposium, Leiden 1970*, Book of Abstracts, Vol. 2, Inter Scientias Inc., The Hague, p. 801.
8. Kovacs, A. J. and Gonthier, A. (1972). *Kolloid Z.u.Z. Polymere*, **250**, 530.
9. Vidotto, G., Levy, D. and Kovacs, A. J. (1969). *Kolloid Z.u.Z. Polymere*, **230**, 289.
10. Price, F. P. (1959). *IUPAC Symposium, Wiesbaden 1959, Communication 1 B2*, Verlag Chemie GmbH., Weinheim.
11. Kovacs, A. J., Gonthier, A. and Straupe, C. (1975). *J. Pol. Sci.*, **C50**, 283.
12. Kovacs, A. J., Lotz, B. and Keller, A. (1969). *J. Macromol. Sci. Phys.*, **B3**, 385.
13. Kovacs, A. J., Straupe, C. and Gonthier, A. (1977). *J. Pol. Sci.*, **C59**, 31.
14. Kovacs, A. J. and Straupe, C. (1980). *J. Crystal Growth*, **48**, 210.
15. Kovacs, A. J. and Straupe, C. (1979). *Faraday Disc.*, **68**, 225.
16. Wunderlich, B. (1980). *Macromolecular Physics*, Vol. 3, Academic Press, New York.
17. Spegt, P. (1970). *Makromol. Chem.*, **139**, 139.
18. Buckley, C. P. and Kovacs, A. J. (1976). *Colloid and Pol. Sci.*, **254**, 695.
19. Buckley, C. P. and Kovacs, A. J. (1975). *Progr. Colloid and Pol. Sci.*, **58**, 44.
20. Flory, P. J. and Vrij, A. (1963). *J. Am. Chem. Soc.*, **85**, 3548.
21. Beech, D. R., Booth, C., Pickles, C. J., Sharpe, R. R. and Waring, J. R. S. (1972). *Polymer*, **13**, 246.
22. Ashman, P. C. and Booth, C. (1972). *Polymer*, **13**, 459.
23. Arlie, J. P., Spegt, P. A. and Skoulios, A. (1966). *Makromol. Chem.*, **99**, 160.
24. Gilg, B. and Skoulios, A. (1971). *Makromol. Chem.*, **140**, 149.
25. Marshall, A., Domszy, R. C., Teo, H. H., Mobbs, R. H. and Booth, C. (1981). *Euro. Polymer. J.*, **17**, 885.
26. Peterlin, A., Fischer, E. W. and Reinhold, C. (1962). *J. Chem. Phys.*, **37**, 1403.
27. Petersen, J. M. and Lindenmeyer, P. H. (1968). *Macromol. Chem.*, **118**, 343.
28. Lauritzen, J. I. and Hoffman, J. D. (1960). *J. Res. N.B.S.*, **A64**, 73.
29. Frank, F. C. and Tosi, M. (1961). *Proc. Roy. Soc.*, **A263**, 323.
30. Hoffman, J. D., Davis, G. T. and Lauritzen Jr, J. I. (1976). In *Treatise in Solid State Chemistry*, Vol. 3, (N. B. Hannay (Ed.)), Plenum Press, New York.
31. Lauritzen, J. I. (1973). *J. Appl. Phys.*, **44**, 4353.
32. Point, J. J. and Kovacs, A. J. (1980). *Macromolecules*, **13**, 399.
33. Buckley, C. P. (1980). *Polymer*, **21**, 444.
34. Labaig, J. J. Thesis, Univ. Louis Pasteur, Strasbourg, 1978.
35. Gonthier, A. and Kovacs, A. J. Unpublished.
36. Flory, P. J. and Yoon, D. Y. (1978). *Nature*, **272**, 226.
37. Yin, T. P., Lovell, S. E. and Ferry, J. D. (1961). *J. Phys. Chem.*, **65**, 534.
38. Hoffman, J. D. (1982). *Polymer*, **23**, 656.
39. Ferry, J. D. (1980). *Viscoelastic Properties of Polymers*, 3rd Edn, Wiley, New York.

40. Pierson, F. Thesis, Fac. Sci., Strasbourg, 1968.
41. Bassett, D. C. and Keller, A. (1962). *Phil. Mag.*, **7,** 1553.
42. Lotz, B., Kovacs, A. J. and Wittmann, J. C. (1975). *J. Pol. Sci. (Phys. Edn)*, **13,** 909.
43. Lotz, B. and Kovacs, A. J. (1966). *Kolloid Z.u.Z. Polymere*, **209,** 97.
44. Barnes, W. J. and Price, F. P. (1964). *Polymer*, **5,** 283.
45. Hoffman, J. D., Frolen, L. J., Ross, G. S. and Lauritzen Jr, J. I. (1975). *J. Res. N.B.S.*, **A79,** 671.
46. Takayanagi, M. and Yamashita, Y. (1956). *J. Pol. Sci.*, **22,** 552.
47. Barnes, W. J., Luetzel, W. G. and Price, F. P. (1961). *J. Phys. Chem.*, **65,** 1742.
48. Jain, J. N. and Swinton, F. L. (1967). *Euro. Polymer J.*, **3,** 371.
49. Hay, J. N., Sabir, M. and Steven, R. L. T. (1969). *Polymer*, **10,** 187.
50. Beech, D. R., Booth, C., Hillier, I. H. and Pickles, C. J. (1972). *Euro. Polymer J.*, **8,** 799.
51. Fraser, M. J., Marshall, A. and Booth, C. (1977). *Polymer*, **18,** 93.

INDEX

Aluminium radiation, 229–30, 233–7, 242–6
Amplitude contrast, 89, 100–1
Anisotropic scattering, 143–6
 high molecular anisotropy, 145–6
 moderate anisotropy, 144–5
Arc correction factors, 49
Aromatic polyesters
 chain conformation of, 68–71
 isolation of allomorphs, 57–9
 orientation of molecules within unit cell, 72–5
 structure determination method, 40–57
 structures of, 57–76
 tilting of molecular chain axes from fibre axis, 75–6
 unit cells, 59–65
 wide-angle X-ray diffraction patterns of, 39–78
Asherov distribution, 219
Asymmetric distribution function, 197–8
 both phases, on, 217
 one phase, on, 219
Atomic scattering factors, 6

Background scatter correction, 211
Benzene ring, 69, 72, 73, 74

Bombyx mori silk fibroin
 intensity transform of, 23
 X-ray diffraction pattern of, 19
Bond
 angles, 50–1
 lengths, 50–1
 parameters, 52, 54
Bragg angle, 233, 238
Bragg condition, 106, 115
Bragg equation, 248, 250
Bremsstrahlung-maximum wavelength, 237
Burger's vector, 99, 100

Calorimetric measurements, 269–76
Carbon radiation, 233, 234, 236, 237, 242, 243, 246–50
Chain
 conformation of aromatic polyesters, 68–71
 deposition rate, 286–8
Chain-folding, 160–1, 170, 229, 230, 261–307
 concept of, 261
 direct evidence for, 262
 kinetic origin of, 277–85
 kinetic theories, 285–9
 theoretical prediction, 304

Chain-folding—*contd.*
  tightness of, 293–4
  see also Folded-chain crystals
Chain-unfolding, 294–301, 305
  mechanisms of, 297–301
  stability against, 294
Coherent surface nucleation, 282
Collimation
  broadening function, 203, 204
  system, 201, 212
Computer
  analysis of X-ray diffraction patterns, 1–37
  mapping
    film space, in, 15–17
    reciprocal space, in, 17
Computer-aided data collection, 15
Conformation comparisons, 163–4
Conformational energy calculation, 55
Conventional transmission electron microscopy (CTEM), 79–124
  contrast mechanisms, 89–120
  image
    intensification, 88
    interpretation, 89–120
  imaging technique, 81–3
  microscopy technique, 80–9
  special concerns for polymer specimens, 83–9
Convergent beam microdiffraction, 107, 108
Copper radiation, 229, 230, 232–5, 244
Correlation function, 135, 142, 151, 190–2, 223
Crenellated structure, 170
Crystal
  defects, 109–17
  growth rate, 267
Crystalline particles
  extraction of information from intensity maps, 20–6
  intensity transform, 2–8
  fibrous assembly of, 9–12
  simulation of fibre diffraction patterns, 32–4
Crystalline polymers
  neutron scattering by, 125–80
  unit cell determination, 40
Crystallinity, 130, 141
Crystallisation
  behaviour, 302
  low supercoolings, of, 166
  mechanisms, 128
  PE from solution, of, 167
Crystallographic direction, 129

D11 diffractometer, 146–7
Debye equation, 161
Debye function, 173
Deformation, 173–5
Differential scanning calorimetry, 263
Diffraction contrast, 105–7
Dimethyl terephthalate, 56
Disorder effects, 5–7, 170, 293
Distribution function, 188, 196–8, 224
  application to models, 198–9

Electron
  density, 181, 191, 199, 256
  diffraction, 47, 80, 107
  microscopy, 79–124
Etchants, 105
Ethylene glycol diacetate, 69
Extended-chain
  crystals, 264, 274, 276, 277, 280, 291, 293, 294
  growth, 282
  nucleus, 300
  seed, 301

Fibre
  diffraction patterns, simulation of, 32–4
  morphology, 117
  transmission characterisation, 240
  ultrastructure, 117–20
Film space, computer mapping in, 15–17
Fold
  effects. *See* Chain-folding; Chain-unfolding
  surface structure, 131

Folded-chain crystals, 265, 269, 276, 282, 288, 291, 294, 298, 304
Fourier transform, 2, 4, 91, 135, 142, 191, 195, 198
Freezing-in effect, 164

Gaussian coil, 164
Gaussian distribution, 196, 224–6
Glycol terephthalates, 57
Growth transition temperature, 277–9
Guinier approximation, 235, 246
Guinier plot, 242, 247
Guinier range of q, 139–40

Helical particles, 12–14
  extraction of information from intensity maps, 26–32
Helical patterns, simulation of fibre diffraction patterns, 34
High Flux Reactor, 146

Incubation period, 295
Intensity
  data collection, 15–19
  distribution, 186, 191, 200, 201, 207–8, 213–20, 222, 224
  measurement, 185, 215
  monitor, 239
  transform, 2, 17–19
    Bombyx mori silk fibroin, of, 23
    crystalline particles, of, 2–8, 20–6
    cylindrically averaged, 7–8, 14
    definition, 4
    disorder effects, 5–7
    fibrous assembly of crystalline particles, of, 9–12
    helical particles, of, 12, 26–32
    simulation, 32–4
    stationary particle, of, 2–5
Interference function, 4, 5, 194
Intermediate range of q, 140–2

Jacobian evaluation, 19

LALS computer program, 56
Lamellar crystals, 129
Lamellar surface structure, 291–4
Lamellar thickness, 129, 270, 285–6, 288, 297, 302
Lattice
  factor, 235
  images, 96–7
Least-squares refinement procedure, 48
Long-wavelength X-ray scattering, 229–60
  advances in, 253–60
  applications, 244–50
  difficulties with, 231–5
  dual wavelength scattering, 244
  instrumentation, 236–9
  pattern recording, 242–4
  reasons for, 231–5
  sample arrangement, 242–4

Mass
  absorption coefficients, 234, 242
  thickness contrast, 101–3
    enhancement methods, 103–5
Mechanical degradation, 94
Melt
  crystallisation, 158–9, 261
  grown crystals, 152–4
Melting data analysis, 271
Melting-point determination, 266
Metal
  decoration, 105
  shadowing, 104
Methylene bonds, 54
Micro-area electron diffraction, 107–9
Microbeam diffraction, 107, 108
Miller indices, 280
Moiré fringe patterns, 97–100
Molecular conformation, 125–80, 166
Molecular deformation, 173–5
Molecular scattering factor, 139–42
Molecular weight, 94
Monte Carlo method, 159, 160, 163
Morphological features, 229–60
Morphological models, 193–200, 249

Morphological modifications, 294–5, 305
Mosaic blocks, 109–11
Mutual chain exclusion, 172

Neutron scattering, 262
 anisotropic, 143–6
 crystalline polymers, by, 125–80
 deformation, 173–5
 experimental conditions and results, 151–8
 interpretation of results, 158–72
 measurements, 146–50
  calibration of intensities, 149–50
  data reduction, 146–7
  degree of uncertainty of, 150
  instruments, 146
  subtractions, 147–9
 mixed systems, from, 136–8
 process, 132–4
 structure and, 134–6
 theory, 131–46
Non-random mixing, 142–3
Nylon, 244

Oligomethylene glycol derivatives, 50, 69
Optical density, 15, 20, 21
Orientation
 density function, 9, 20–2
 effects, 208–11, 213, 277–81

Paracrystalline effects, 210
Paracrystalline models, 194–5
Patterson function, 135, 191
Phase contrast, 89
 imaging, 90–6
Phase-grating, 250, 251
Point defects, 109
Polybutadiene, 153
Polybutylene terephthalate, 109
Poly(caproamide), 221
Polyethylene, 99, 114, 115, 116, 129, 130, 151–5, 159, 163, 167, 171, 173, 174, 176, 187, 213, 214, 215, 217, 230, 234

Poly(ethylene adipate), 71
Poly(ethylene oxide) (PEO), 158, 261–307
Poly(ethylene terephthalate) (PET), 39, 61, 94, 234, 241, 247, 249, 250, 258, 259, 260
Polyoxymethylene (POM), 110, 111, 213, 217
Poly(*para*-phenylene benzobisthiazole) (PBT), 96, 119
Poly(*para*-phenylene terephthalamide) (PPTA; PPD-T), 96, 98, 118, 234, 241, 243, 255
Poly(*para*-xylene), 96, 97
Polypropylene, 153, 155–6, 173, 232, 245
Polystyrene, 94, 103, 105, 156–8
Poly(tetramethylene terephthalate), 214
Porod's Law, 152
Projected mass thickness, 89
Projected phase change, 89

Q-space, 12
Quench crystallisation, 159
Quenching effects, 153

Radiation damage, 84, 85
Reciprocal lattice
 concept, 2
 nets, 43, 46
 points, 4
Reciprocal space
 computer mapping in, 17
 coordinates, 17, 18, 42
 position vector, 3
Reflection profile, 11
Reinhold distribution, 218, 224–6
Relaxation rates, 164
Replication technique, 105
Reverse osmosis, 246
Rotating drum microdensitometer, 15
Rouse relaxation time, 290

Scanning transmission electron
microscopy (STEM), 79–124
  basic components of, 82
  contrast mechanisms, 89–120
  image
    intensification, 88
    interpretation, 89–120
  imaging technique, 81–3
  microscopy technique, 80–9
  special concerns for polymer
    specimens, 83–9
Scherrer equation, 221
Screw dislocations, 114–17, 267, 269
Segregation effects, 151–2
Selected area diffraction (SAD),
  107–8
Signal-to-noise ratio, 87
Small-angle X-ray scattering, 131,
  181–228, 288, 289
  assessment of reported
    experiments, 211–13
  broadening
    due to beam collimation, 201–4
    due to detector, 205–6
    due to recording system, 206–7
    due to specimen size, 205
    due to spectral distribution of
      X-rays, 204–5
  comparison of theoretical and
    experimental intensity
    distributions, 213–20
  correction
    background scatter, for, 211
    specimen orientation, for,
      208–11
  direct methods, 183–90
  experimental procedures, 200–13
  indirect methods, 190–3
  morphological models, 193–200
  resolution loss, 200–7
  three-dimensional morphology,
    220–3
Solution grown crystals, 154, 157, 261
Stacking faults, 111–14
Staining methods, 104
String of pearls structure, 94

Structure factor, 2, 4, 5, 12, 32, 106
  data
    current practice, 20–2
    profile fitting, 22–6
Structure refinement, 56–7
Substrate thickness effects, 281–3
Supercooling effects, 172
Surface
  decoration effects, 291–3
  free energy, 269
  layer disorder, 293
  roughness effect, 171
Symmetric distribution function, 196
  both phases, on, 214–17
  one phase, on, 219

Taylor's theorem, 139
Terephthalic acid, 74
Thermal stability, 272
p-Toluene sulphonate (PTS), 113
Transition zones, 220
Transmission electron microscopy,
  79–124
Tri-X film, 238, 239
Trial models, 51–6

Undercooling effect, 283–5, 288–9
Unit cell, 59–65, 229, 230
  determination, 40–8

Wide-angle X-ray diffraction patterns
  of aromatic polyesters, 39–78

X-ray
  diffraction pattern
    Bombyx mori silk fibroin, of, 19
    computer analysis of, 1–37
  reflection intensity measurement,
    48–50

Zimm plot, 143